Ethics of Science and

Volume 47

The series *Ethics of Science and Technology Assessment* focuses on the impact that scientific and technological advances have on individuals, their social lives, and on the natural environment. Its goal is to cover the field of Science and Technologies Studies (STS), without being limited to it. The series welcomes scientific and philosophical reviews on questions, consequences and challenges entailed by the nature and practices of science and technology, as well as original essays on the impact and role of scientific advances, technological research and research ethics. Volumes published in the series include monographs and edited books based on the results of interdisciplinary research projects. Books that are devoted to supporting education at the graduate and post-graduate levels are especially welcome.

More information about this series at http://www.springer.com/series/4094

Carl Friedrich Gethmann · Georg Kamp ·
Michèle Knodt · Wolfgang Kröger ·
Christian Streffer · Hans von Storch ·
Thomas Ziesemer

Global Energy Supply and Emissions

An Interdisciplinary View on Effects, Restrictions, Requirements and Options

Carl Friedrich Gethmann
Universität Siegen
Forschungskolleg Siegen
Siegen, Germany

Michèle Knodt
Institut für Politikwissenschaft
Technische Universität Darmstadt
Darmstadt, Germany

Christian Streffer
University Clinics Essen
Essen, Germany

Thomas Ziesemer
Maastrich University
Maastrich, The Netherlands

Georg Kamp
Würselen, Nordrhein-Westfalen
Germany

Wolfgang Kröger
ETH Zürich
Zürich, Switzerland

Hans von Storch
Helmholtz Zentrum Geesthacht
Institute of Coastal Research
Geesthacht, Germany

ISSN 1860-4803 ISSN 1860-4811 (electronic)
Ethics of Science and Technology Assessment
ISBN 978-3-030-55357-9 ISBN 978-3-030-55355-5 (eBook)
https://doi.org/10.1007/978-3-030-55355-5

This Springer imprint is published by the registered company Springer Nature Switzerland AG
The registered company address is: Gewerbestrasse 11, 6330 Cham, Switzerland

Preface

In 2012, the Institute for Advanced Study at the University of Siegen, Germany, began planning and started preliminary studies for the establishment of an interdisciplinary working group on questions of global energy supply. The intention was to broaden the energy policy perspective from a purely national German perspective to European and global view. As a interdisciplinary project group, "Societal challenges posed by the interactions between energy system transformation, raw material demand and climate change in a global perspective (GlobEn)" was funded.

The project group was composed of:

- Carl Friedrich Gethmann (University of Siegen; ethics of science) (chair)
- Georg Kamp (Research Centre Jülich; philosophy)
- Michèle Knodt (Technical University Darmstadt; political science)
- Wolfgang Kröger (ETH Zurich; mechanical engineering sciences)
- Hans von Storch (Institute for Coastal Research of the Helmholtz Centre Geesthacht; climate research)
- Christian Streffer (University Clinics Essen, medical faculty; medical radiobiology)
- Thomas Ziesemer (Maastricht University; environmental economics).

Furthermore, Harry van der Laan (University of Utrecht, astrophysics; climate research) and Karl Josef Koch (University of Siegen; economics) have participated in early stages of the discussion of the project group. At times, the project was supported by the research assistants Jan Mehlich, Jochen Sattler and Hendrik Kempt active.

The group held a total of 19 plenary meetings in Siegen, Cologne and Zurich in the years from 2013 to early 2018. The aim was to produce a monograph based on the concept of interdisciplinary cooperation with transdisciplinary objectives developed within the framework of the European Academy Bad Neuenahr-Ahrweiler.[1] The authors take responsibility for the text in collective

[1] C.F. Gethmann, M. Carrier, G. Hanekamp, M. Kaiser, G. Kamp, S. Lingner, M. Quante, F. Thiele, *Interdisciplinary Research and Trans-disciplinary Validity Claims,* Heidelberg u.a. 2015.

authorship. Some of the members of the GlobEn working group had previously presented a study on nuclear disposal, among other things, within the framework of the Europäische Akademie Bad Neuenahr Ahrweiler.[2]

The authors would like to thank the Institute for Advanced Study of the University of Siegen for the financial support of the study and Springer-Publishers for their cooperation in preparing the publication of the study.

Inevitably, there has been a time gap between the end of the scientific work and the time of publication. At the same time, the subject area is developing at an accelerated rate. The authors have made every effort to keep the data stock up to date. However, not in every area could the latest developments be taken into account.

On behalf of the authors:
Carl Friedrich Gethmann

Siegen, Germany
February 2020

[2]C. Streffer, C.F. Gethmann, G. Kamp, W. Kröger, E. Rehbinder, O. Renn, K.-J. Röhlig, *Radioactive Waste. Technical and Normative Aspects of its Disposal,* Berlin: Springer 2011. Further preliminary work: U. Steger, W. Achterberg, K. Blok, H. Bode, W. Frenz, C. Gather, G. Hanekamp, D. Imboden, M. Jahnke, M. Kost, R. Kurz, H.G. Nutzinger, Th. Ziesemer, *Nachhaltige Entwicklung und Innovation im Energiebereich,* Berlin 2002 (*Sustainable Development and Innovation in the Energy Sector,* Berlin 2005); C. Streffer, C.F. Gethmann, K. Heinloth, K. Rumpff, A. Witt, *Ethische Probleme einer langfristigen Energieversorgung,* Berlin 2005; B. Droste-Franke, H. Berg, A. Kötter, J. Krüger, K. Mause, J.-C. Pielow, I. Romey, T. Ziesemer, *Brennstoffzellen und Virtuelle Kraftwerke. Energie-, umwelt- und technologiepolitische Aspekte einer effizienten Hausenergieversorgung*, Berlin 2009; B. Droste-Franke, P. Paal, C. Rehtanz, D. U. Sauer, J.-P. Schneider, M. Schreurs, T. Ziesemer, *Balancing renewable electricity. Energy Storage, Demand Side Management and Network Extension from an Interdisciplinary Perspective*, Berlin 2012; C.F. Gethmann, G. Kamp, "Globale Energiegerechtigkeit. Ethische Fragen", in: J. Nida-Rümelin, D. von Daniels, N. Wloka (Hgg), *Internationale Gerechtigkeit und institutionelle Verantwortung,* Berlin 2019, 311–340.

Contents

About the Authors

Carl Friedrich Gethmann, studied philosophy at Bonn, Innsbruck and Bochum and obtained a lic. phil. in 1968 from Institutum Philosophicum Oenipontanum—Universität Innsbruck. He obtained his Dr. phil. from the Ruhr-Universität Bochum in 1971. In 1978, he completed the habilitation in philosophy at the University of Konstanz. In 2003, he received the honorary degree of doctor of philosophy (Dr. phil. h.c.) from the Humboldt-Universität Berlin. In 2009, he was appointed as an Honorary Professor at the University of Cologne. During his career, he has served as a scientific assistant (1968), Professor of Philosophy at the University of Essen (1972), a private lecturer at the University of Konstanz (1978) and as Professor for philosophy at the University of Essen (1979). He also held numerous lectures at the universities of Düsseldorf and Göttingen. Invited to join the Board of Directors at the Akademie für Technikfolgenabschätzung Baden-Württemberg and to receive a full professorship of Philosophy in 1991, he refused. He refused full professorship offers from other universities. In 1991, he accepted the offer of Full Professorship at the University of Essen. Since March 2013, Gethmann has served as Professor at the Institute for Advanced Study at the University of Siegen. Between 1996 and 2012, he was Director of the *Europäische Akademie zur Erforschung von Folgen wissenschaftlich-technischer Entwicklungen GmbH*, Bad Neuenahr-Ahrweiler, Germany. He has also been Member of the Academia Europaea (London), of the Berlin-Brandenburgische Akademie der Wissenschaften and of the German National Academy of Sciences Leopoldina (Deutsche Akademie der Naturforscher). Between 2000 and 2013, Professor Gethmann has served as a member of the Bioethics Commission of Rhineland-Palatinate, Germany. From 2006 to 2008, he was President of the German Association for Philosophy "Deutsche Gesellschaft für Philosophie e.V.". Since 2008, he has been a member of the German Academy of Science and Engineering (German: Deutsche Akademie der Technikwissenschaften) "Acatech." He is currently a member of the German Ethics Council (2013–2021). His main fields of research include: linguistic philosophy and philosophy of logic; phenomenology and practical philosophy, ethics of medicine, ethics of environment and technology assessment.

Georg Kamp, carried out an apprenticeship and worked as a retail salesman between 1979 and 1984. During 1987–1993, he was studying philosophy, German literature and linguistics in Bochum, Duisburg and Essen. From 1993 to 1998, he served as a scientific assistant at the Institut für Philosophie at the Universität Duisburg-Essen. In 1998, he completed his Ph.D. studies at the University of Essen with a thesis about logics in normative contexts. From 1999 to 2002, he served as a member of the scientific staff of the *Europäische Akademie zur Erforschung von Folgen wissenschaftlich-technischer Entwicklungen GmbH*, in Bad Neuenahr-Ahrweiler, Germany. During 2002–2005, he worked as a freelance consultant, lecturer and editor. During 2005–2006, he attended a cooperative education program on "Master of Mediation" at the Fernuniversität Hagen. During 2007–2012, he worked as an assistant lecturer in philosophy at the University of Duisburg-Essen. Since 2005, he has been active as a scientific coordinator, participant and manager of numerous interdisciplinary projects on sustainability and energy topics. Moreover, he was a staff member of the Niederrhein University of Applied Sciences (2005–2006) and of the Europäische Akademie in Bad Neuenahr-Ahrweiler (2006–2015). Since 2015, he has been a staff member of the Jülich Research Centre.

Michèle Knodt studied political sciences at Technische Universität Darmstadt (TU Darmstadt) and completed there her Master in political science in 1992. She received her Dr. phil. from the University of Mannheim in 1997 and worked as a research assistant at the Mannheim Center for European Social Research (MZES) (1997–2000). She completed her habilitation in political sciences in 2005 at the University of Mannheim. In 2000, she was appointed Assistant Professor at the University of Mannheim. In 2005, she obtained Full Professorship from TU Darmstadt, where she has been serving as Professor for Comparative Politics and European Integration to date. She was a guest professor at University of Massachusetts Boston, USA (1998), at the Institut d'Etudes Politiques de Lille, University of Lille, France (2003), at the Institute for Advanced Studies (IHS), University of Vienna, Austria (2007) and at the University of Pondicherry, India (2010). In 2011, Prof. Knodt was awarded with a Jean Monnet Chair *ad personam* from the European Commission. She is Director of the Jean Monnet Centre of Excellence "EU in Global Dialogue" (CEDI), Vice-Director of the Energy Center (German: Profilbereich Energiesysteme der Zukunft) at TU Darmstadt, Co-Leader of the Loewe Centre of Excellence "emergenCITY" at TU Darmstadt as well as Co-Leader of the DFG Research Training Group (GRK) KRITIS. She is currently leading the Cost Action ENTER "EU Foreign Policy Facing new Realities" (17,119). She is also President of the German European Community Studies Association (ECSA-Germany) and has been leading several international and interdisciplinary projects supported by DFG, VW foundation, EU Commission, BMBF, BMWI, among others. She published around 50 peer-reviewed journal articles, co-authored 8 monographs, co-edited 16 books and wrote more than 100 chapters in edited volumes. Her main fields of research include energy governance, EU multi-level governance, energy transition and participation.

Wolfgang Kröger studied mechanical engineering, specializing on nuclear technology, at the RWTH Aachen. He completed his diploma degree in 1972, his doctoral degree in 1974 and the habilitation in 1986. From 1974 to 1989, he worked at the Institute for Nuclear Safety Research at the German Research Center Jülich, leading research projects aimed at the development of advanced reactor concepts, and on related methods for comprehensive safety assessment. In 1990, he became Full professor of safety technology at the ETH Zurich. Simultaneously, he was appointed director of the Research Department on Nuclear Energy and Safety at the Swiss National Paul Scherrer Institut (PSI). At ETH, as director of the Laboratory of Safety Analysis, he contributed to the development of risk and vulnerability analysis methods of complex cyber-physical systems, including energy supply infrastructure. He has been approaching risk assessment and interdependence issues in a multidisciplinary, trans-sectorial way, and shaped the concept of sustainability and resilience. On his initiative, the International Risk Governance Council was established in Geneva in 2003, and he became Founding Rector. After his retirement in 2011, he was nominated Executive Director of the ETH Risk Center.

Professor Kröger is currently working as a senior scientific advisor and member of distinguished national and international committees, such as the Swiss Academy of Engineering Sciences (SATW), the International Review Group of the Japanese Nuclear Safety Institute (JANSI), and the project of "energy systems of the future" (ESYS). Senior Fellow of the Institute for Advanced Sustainability Studies (IASS) Potsdam, Distinguished Affiliated Professor at TU Munich, he is also author and co-author of numerous publications, including books.

Christian Streffer studied chemistry and biochemistry at the universities of Bonn, Tübingen, Munich, Hamburg and Freiburg. He received his Ph.D. in biochemistry in 1963. He was a postdoctoral fellow at the Department of Biochemistry, University of Oxford. In 1971, he was appointed Professor for Radiobiology at the University of Freiburg, Germany. From 1974 to 1999, he served as a full Professor for medical radiobiology at the University of Essen. During 1988–1992, he was Vice-Chancellor of the same university and received the title of Emeritus in 1999. He was Guest Professor at the University of Rochester, N.Y., USA, in 1985, and at the University of Kyoto, Japan, in 2000. As an Honorary Member of several scientific societies, he received an Honorary Doctor from the University of Kyoto in 1995. Professor Streffer is a member of the Institute for Science and Ethics of the University of Bonn and an Emeritus member of the International Commission on Radiological Protection (ICRP). He received several scientific awards, such as: the Roentgen Plakette (English: Roentgen Medal) by the City of Remscheid (1985), a prize awarded to people who have made great contributions to the progress and usage of X-ray in science and practice; the Bacq–Alexander Award of the European Society for Radiation Biology in 1996; the Sievert Award of the International Radiation Protection Association (IRPA) in 2008; and the Distinguished Service Award of the Radiation Research Society, USA, in 2009.

Professor Streffer's main research interests include: radiation risk, especially during the prenatal development of mammals; genomic instability after radiation exposure; combined effects of radiation and chemical substances; and experimental radiotherapy of tumors and especially individualization of cancer therapy by radiation.

Hans von Storch is Director Emeritus of the Institute of Coastal Research of the Helmholtz Zentrum Geesthacht (HZG), Professor at the University of Hamburg and Guest Professor at the Ocean University of China (Qingdao). From 1987 to 1995, he was Senior Scientist and leader of the *Statistical Analysis and Modelling Group* at the Max Planck Institute for Meteorology. He also served as Director of the Institute of Coastal Research. His research interests included climate diagnostics and statistical climatology, and regional climate change and its transdisciplinary context. He published twenty books, including "Statistical Analysis in Climate Research," co-authored with Francis Zwiers and "Die Klimafalle" (English: The Climate Trap), co-authored with the ethnologist Werner Krauss. He also authored numerous articles. Editor-in-chief of the Oxford Research Encyclopedia of Climate Science, Oxford University Press, he is also a member of a number of editorial and advisory boards. He was the lead author of Working Group I of the Third Assessment Report and of Working Group II of the Fifth Assessment Report of the Intergovernmental Panel on Climate Change (IPCC) and chaired the efforts for a climate change assessment for the Baltic Sea Catchment (BACC II). Professor Hans von Storch is also a foreign member of the Polish Academy of Sciences and holds an honorary doctorate from the University of Göteborg. The significance of his work was also recognized with the Order of Merit of the Federal Republic of Germany in 2019.

Thomas Ziesemer has been serving as Associate Professor of Economics at Maastricht University, the Netherlands, since December 1996. After studying economics at the universities of Kiel (1974–1975) and Regensburg (1975–1978), in Germany, he was employed at the University of Regensburg (1982–1989), where he completed his doctoral dissertation on the topic of "Economic Theory of Underdevelopment" in 1985. Starting in December 1989, he has been successively appointed Assistant Professor of International Economics, Associate Professor of Microeconomics and Associate Professor of Economics, from the School of Business and Economics at Maastricht University. In November 1996, he completed his "Habilitation" at the Freie Universitaet Berlin. He also serves as Senior Researcher at United Nations University—Maastricht Economic and Social Research Institute on Innovation and Technology UNU-MERIT. His fields of interest include development, international and environmental economics, growth and technical change.

List of Figures

List of Tables

Chapter 1
Introduction

1.1 Initial Situation

A secure and at the same time cost-effective, environmentally friendly and resource-saving energy supply is an important prerequisite both for the further development of the countries of Asia, South America and Africa and for maintaining living conditions in the industrialized countries. This entails a wide range of national and regional design tasks, but also those which, in view of the competition for resources on a globalized market and in view of the consequences of the release of emissions during the production and consumption of useful energy, can only be adequately considered in the light of global developments. At the same time, there is a need to broaden the view of more complex interrelationships that go far beyond energy supply issues. For example, the provision of energy in dry zones close to the coast allows the extraction of fresh water, which can be used for agricultural purposes and to develop settlement areas. This is expected to have an impact on social, economic, political, and demographic developments, which may have a direct impact on living conditions in industrialized countries in the form of a reduction in migration movements and an increase in trade activities. At the same time, the type and extent of energy production and use have an impact on climatic developments, which, according to the current state of knowledge, will in turn influence the expansion and distribution of dry zones, among other things. This means that national energy policy decisions and measures, if they are to be taken not just for the sake of short-term effects but prudently and responsibly, must be based on a foundation that goes far beyond the technical interrelationships and the respective requirements of regional markets and incorporates ideas of longer-term global development.

If these challenges are to be met, then both the supranational steering possibilities and the specific local conditions, the disparate goals and the diversity of options, the unequal distribution of potential (e.g., technical, financial, social and cognitive resources), and the unequal distribution of opportunities and risks must be considered. For this purpose, scientific input on a broad interdisciplinary basis

C. F. Gethmann et al., *Global Energy Supply and Emissions*,
Ethics of Science and Technology Assessment 47,
https://doi.org/10.1007/978-3-030-55355-5_1

is indispensable, which should be carried out from a scientific-technical, political, social science, economic and philosophical perspective on the basis of representative selected regions (China, India, Brazil, and Europe), especially in the fields of science, technology, politics, economics, and philosophy. For this, the following outline will be followed: (i) critically reconstructing the target systems and the technical, economic, ecological, and social conditions for achieving the targets, (ii) developing and refining criteria, benchmarks and methods for responsible energy policy decisions and their effective implementation, and (iii) developing cross-disciplinary, coordinated, sustainable, and promising recommendations for action for a prudent and long-term energy policy from the perspective of all disciplines involved.

The development of viable strategies for a sustainable energy supply raises not only questions of technical feasibility and economic viability but also manifold questions of ethical justifiability and political responsibility, which extend far beyond national borders and the present day and can often only be adequately answered on a global scale and in an intergenerational long-term perspective.

1.2 Energy Policy and Climate Targets

The International Energy Agency (IEA) puts the total volume of energy-related CO_2 emissions for 2018 at 33.1 Gt and gives a clear indication of the relevance of energy policy and energy management decisions for climate change, which can only be adequately considered a global phenomenon.[1]

This represents an increase of more than 40% over the 23.2 Gt reported for the year 2000, and since 2005 emissions have risen by more than 22% despite the economic downturn. The IEA has calculated an increase of 1.8% for 2018 alone. According to the UNEP "Temperature Briefing" (2010) "there is a medium likelihood to stay within the 2-degree limit if the following conditions are met:

- Global emissions peak sometime between 2015 and 2021.
- Global emissions in 2020 are approximately 40.0–48.3 Gt CO_2 eq/yr.
- By 2050 global emissions decrease by 48–72% relative to 2000".

According to the calculations of the Intergovernmental Panel on Climate Change (IPCC), in order to meet the 1.5 °C target set by the Paris Agreement, which came into force in 2016, there would even have to be negative emissions.[2] The resulting

[1] "Global energy-related CO_2 emissions grew 1.7% in 2018 to reach a historic high of 33.1 Gt CO_2" https://www.iea.org/reports/global-energy-co2-status-report-2019/emissions#abstract (accessed 13-Dec-2019).

[2] "All pathways that limit global warming to 1.5 °C with limited or no overshoot project the use of carbon dioxide removal (CDR) on the order of 100–1000 $GtCO_2$ over the twenty-first century. CDR would be used to compensate for residual emissions and, in most cases, achieve net negative emissions to return global warming to 1.5 °C following a peak (high confidence). CDR deployment of several hundreds of $GtCO_2$ is subject to multiple feasibility and sustainability constraints (high confidence). (IPCC 2018).

problems become even more apparent when comparing the developments in OECD countries and countries with accelerated catchup developments such as China, Brazil, or India: For the year 2019, the OECD countries' share of energy-related emissions is 35%.[3] However, they contribute only 25% to the rate of increase, with increases being recorded above all in non-OECD countries. Although their inhabitants produce only a fraction of the per capita emissions for which OECD citizens are responsible (10 t/a compared with 5.8 t/a in China, 1.9 t/a in Brazil or 1.6 t/a in India), in view of the rapidly growing populations and the accelerated mechanization of these countries, compliance with the projected targets will not be possible, or not only through abatement strategies within the OECD countries. Rather, energy policy measures should also be geared towards the development, testing, and refinement of options that offer competitive and attractive offers in the developing countries to achieve their prosperity goals while at the same time reducing climatic and other risks.

1.3 Energy Management and Energy Technologies

The choice of technologies for energy production and use has a central influence on climate development. At the same time, this raises questions of environmental protection and air pollution control, questions of resource availability and fair distribution, and elementary questions of generating and maintaining prosperity and development. Questions of safe and efficient energy supply, as they arise for modern civilizations, are particularly determined by over-complex decision situations. Even if the first warning cries raised in the 1970s turned out to be too premature and dramatic, there is no denying that in the long term, there will be a gradual shortage of essential resources, be it oil or rare earths required for the development of highly efficient turbines. At the same time, the world population has grown from about 1.6 billion people (around 1900) to 7.6 billion in little more than a hundred years—not least because of the progress made in many areas of life. The projections of the UN (Department of Economic and Social Affairs, 2017 revision) fluctuate between a shift of about 9.6 billion by the end of the century and a further increase in the world population to 13.2 by 2100 and a further increase beyond that.

The course of development will also depend to a large extent on the availability of energy: A secure and cheap availability of energy is necessary to turn the expected billions of people into producers who can provide for themselves and their families with what they produce and buy. Countries such as China, India or Brazil have in some cases made breathtaking developments here in recent decades and have caught up with the Western industrial nations, but have also increased the pressure on the demand for energy sources and contributed to the further scarcity of resources, to the increased volume of emissions and thus to an intensification of environmental problems.

[3]http://www.oecd.org/environment/environment-at-a-glance/Climate-Change-Archive-December-2019.pdfmber-2019.pdf (accessed 13-Dec-2019).

In the Sustainable Development Goals, which came into force in 2016 and are addressed globally to all countries, the UN has consequently defined more than one of its goals (Goal 7): "Ensure access to affordable, reliable, sustainable, and modern energy for all."[4]

Coal, hydrocarbons, nuclear energy, and renewable energies are the main sources of useful energy. Since an efficient use of these sources is only possible in complex processes and within the framework of large-scale installations, all these options are subject to specific technical, economic, ecological, and social risks. The risks vary considerably in size and type, and their assessment depends on numerous assumptions and prerequisites, not least of all on methodological decisions about what kind of consequences of action are included in the risk assessment and how they are evaluated: from landscape consumption and health impacts during the extraction of resources, to the dangers to man and the environment during normal operation and in the event of accidents, to the long-term effects of emissions and radiating waste. This raises the question of whether mortality rates, reductions in life expectancy, impairment of quality of life, or others should be used as a basis for the assessment. In the latter case, it must be decided whether one should be guided by the subjective perception of impairments, an operationalized willingness to pay, or by criteria that can be shown intersubjectively. In the same way, methodological decisions are used to determine the opportunities that counteract the risks—for example, market opportunities through technological developments, the improvement of living standards, future market opportunities through the low-cost availability of energy in developing countries or the increase in technical options for future generations must be weighted and weighed against each other. Additionally, various parameters are in tension with each other: for example, it must be clarified which risks must be accepted in order to benefit from useful energy. It is also necessary to weigh up how—from a short-term and long-term perspective—security of supply and equity of supply are related. It is possible that resources that can be used for energy purposes will have to be set aside in favor of future material use options. For example, one could demand that current users accept radiation risks in order to relieve members of future generations of the risks of polar ice melting and the increase in extreme weather events. Risk research shows that the perception and assessment of risks and opportunities depend on a wide range of social and cultural factors and often vary considerably from one country and region to another. It can also be assumed that the willingness to make provisions for the future and to consider future generations is at least also influenced by the standard of living achieved in each case.

1.4 Regional Specifics

By comparing leading industrialized regions such as the USA and Europe and key countries with accelerated catch-up development, each with very different conditions,

[4]Cf. http://www.un.org/sustainabledevelopment/energy/ (accessed 13-Dec-2019).

the global challenges for energy policy and the energy industry are to be identified. India and Brazil, in particular, ought to be examined more closely: Both countries have high population growth rates but belong to different economic areas and have very different levels of their own resources. Both countries are pushing ahead with their development, but they differ considerably in their status quo: While India meets its energy needs predominantly from fossil fuels (about 92% have been calculated for 2018 (BP 2019)), Brazil has already been able to build up a high proportion of renewable energies (30% hydroelectricity, 8% wind, solar, bio (BP 2019)). Both countries rely heavily on research and development and therefore deserve special attention both in terms of their rapidly growing energy needs and their increasingly important position in the international community. At the same time, due to cultural backgrounds, the social conditions for implementing energy policy strategies differ both among themselves and from those in the developed regions of North America and Europe.

In the USA and Europe, the development and implementation of strategies for energy supply, especially in connection with the planning of the necessary technical installations, is now largely dependent on the involvement of stakeholders and other parts of the population. However, if, in addition to those aspects whose relevance is locally and directly perceptible, the complex global and only very indirectly retroactive effects are to be incorporated in an appropriate manner into energy policy decisions, there is a danger that citizens willing to participate will be considerably overburdened. The same applies concerning the long-term effects of the decisions to be taken—with the difference that retroactive effects are not to be expected here from the outset, but that the decisions are to be made with appropriate assessment and consideration of the requirements of future generations. This requires both the scientific support of society with regard to the responsible design of participation procedures and the support of citizens and their representative decision-makers concerning understanding the complex prerequisites for responsible energy policy decisions.

1.5 Collective Action Problems of Global Scope

The fact that there are problems of humanity that cannot be solved locally, but also not regionally (nationwide—European), but only globally, has been commonplace for many years. It is based on the undeniable experience that, for example, the spread of pollutants in water and air or gaseous emissions in the atmosphere does not take into account national borders or other social settlements. Perceptible steps to solve such problems can only be achieved through coordinated action by all or at least many actors. They, therefore, present themselves as collective problems of action.[5] As a rule, although not necessarily, it is required to agree on the relevant cause-effect relationships. The action-oriented knowledge required to cope with problems of global scope, ranging from proven experience in the real world to scientific knowledge of

[5]Cf. Gethmann (2017) on the term "collective problems of global scope".

considerable complexity, does not result solely from the individual or small group perspective, and it would be an excessive demand on the individual actors to expect the acquisition of such knowledge and the production of uniformly coordinated action strategies based on it as the basis for their coordination of action.

Finally, by far not all controversies have been settled in the scientific disciplines. Even if there is often broad agreement on the baselines, dissent about the quality of the data, their appropriate interpretation, the methods to be applied, and the reliability of the results are among the basic constituents of science. Even if a glance at the history of science justifies the optimism that many of these dissenting views will sooner or later be overcome, the need for action on problems of global scope is usually so urgent that one must start working on them immediately and despite the existing ambiguities. Decisions often have to be made under uncertainty and with the acceptance of considerable risks. Accordingly, the discussion on political strategies must always strive for the greatest possible reversibility of decisions.[6]

However, there is usually disagreement not only about the means to be taken to tackle problems of global scope but also about the objectives—particularly the urgency of the need for action. On the one hand, the prerequisites for action by the actors are very different; on the other hand, they will be affected by the effects of the measures to be taken collectively in very different ways and will sometimes be found more on the side of the beneficiaries, sometimes more on the side of the cost bearers. And both in terms of expert assessment and in terms of the different interests involved, it is likely to be futile to make a generally accepted prioritization between problems of action with a global reach—from prevention against natural or civilizational threats to larger population groups to the supply and disposal problems of entire continents.[7]

Even where there is widespread agreement on the need for action, the very nature of these problem constellations often means that it cannot be assumed that individual actors will react appropriately: For example, the climate impacts achieved by a single individual who has decided to "make his or her contribution to climate protection" and reduce his or her CO_2 emissions will remain far below even the measurable level in view of the total global volume—while the subjectively experienced effects on his or her lifestyle will generally be considerable. For each individual, provided that he is guided by rational considerations, the motivation for individual action is

---

[6]Cf. the position paper of the German Scientific Counsel (Wissenschaftsrat 2015) "Zum wissenschaftspolitischen Diskurs über große gesellschaftliche Herausforderungen" (On the science policy discourse on major societal challenges) that deals primarily with climate change. See also World Economic Forum: Global Challenges (URL: https://www.weforum.org/agenda/2015/01/10-global-challenges-10-expert-views-from-davos/ (accessed 13-Dec-2019); the "New High-Tech Strategy—Innovation for Germany" (BMBF 2014) of the German Federal Government names as "global challenges": Climate change, demographic development, spread of widespread diseases, securing the world's food supply, and the finite nature of fossil raw material and energy sources.

[7]Cf. for example the debate on the efforts of the so-called "Copenhagen Consensus Center" to prioritize the urgency of the need for action—oriented on the expected return of investment of the tools to be used. A prominent leading question is "If you had $ 75 billion for worthwhile causes, where should you start? "(cf. http://www.copenhagenconsensus.com/copenhagen-consensus-iii, accessed 13-Dec-2019, on the approach and the discussion of the results https://en.wikipedia.org/wiki/Copenhagen_Consensus, accessed 13-Dec-2019).

correspondingly low due to this problem structure, so that the necessary collective reaction to the problems usually does not come about even when the actors are convinced of its necessity.[8] Those who—because they feel obliged to do so on the basis of moral convictions they have gained, because the recognition structure in their small group offers them a sufficient incentive, or for other reasons—nevertheless decide to "make their contribution" individually, will then, however, in view of the complex material flows and interdependencies in production, trade, and consumption, hardly be in a position to make the optimal decisions in terms of their goals.[9] In many cases, the cost of obtaining reliable information on the origin of the substances, the energy balance for production and transport per unit of use, or the "ecological footprint" would be so considerable that the cost of obtaining this information would be disproportionate to the benefit of consumption. And where mutually incompatible standards such as infrastructure damage, land consumption, and biodiversity loss, short-term losses in prosperity and securing the basis of life for future generations, preservation of habitats and forms of life, mortality and morbidity risks, etc. are all at the same time decisive for a consumer decision or the like, but their application and their super-ordination and sub-ordination is controversial, the individual must be overburdened.

In the context of collective action problems of global scope, the division of labor and the training of collective actors is, therefore, indispensable for both the decision-making and the action level. Within the framework of such a division of labor, material flows, for example, can be observed by specialists and reliably assessed based on standards that have been uniformly decided by expert advice from elected representatives. Based on the information thus provided, appropriate measures can then be efficiently designed by the responsible agencies, or the information can be made available to consumers in an appropriately easy-to-understand form. Last but not least, such collective actors can change the framework conditions for individual action in such a way that individuals can work efficiently towards their goals, either individually or in manageable groups, without having to fear excessive information demands or exploitation by free riders. This can be done, for example, by institutional safeguarding of mutually given agreements, by establishing systems of sanctions or by changing economic incentive systems. Central actors and "interactors" are often state, supranational and non-governmental organizations. However, this never implies a uniform will, which the states and their institutions only need to bundle and implement—conversely, they must first and foremost create a balance of interests in a variety of ways, contribute internally to the decision-making process and co-ordinate external approaches with other collective actors. The countries of the world have different prerequisites for responding to climate change, and they would also

---

[8]See, for example, the commentary by Vaze (2009, p. 8) on a representative survey conducted in England in 2007, A large majority of people (70%) believe that if there is no change the world will soon experience a major environmental disaster. But this widespread belief that disaster looms around the corner and that we are responsible through our actions does not translate into changes in behavior. When asked about what they had done over the past year to reduce climate change, most people said 'nothing'.

[9]Cf. the comments in Grunwald 2012 as an example.

be affected by the predicted development to very different degrees. In some regions, in the face of urgent need, provision for the future is a luxury that one simply cannot afford, while others invest considerable effort to make the acquired prosperity secure for the future in an intergenerational perspective, and some base their prosperity precisely on the extraction and sale of fossil fuels, from which others want to become independent.

1.6 Problems of International Distributive Justice

Analogous to the interaction problems between individuals or small groups, the unavoidable international cooperation between states can solve a difficult free-rider dilemma. Although there are larger and smaller, more important and less important "players" in principle, the efforts of one nation N to reduce its emissions of greenhouse gases have no relevant effect unless a sufficient number of other nations reduce their emissions to a relevant extent. However, if the others reduce their emissions, then there will be no significant difference if N continues to emit as before. Since the avoidance of greenhouse gas emissions is linked to investments, and the funds used for this purpose are not available to the collective, which has to raise these investments through taxes or higher energy prices, for anything else, not for the development of prosperity, not for the construction or maintenance of infrastructure, not for social concerns, there is a strong incentive to shift the burden onto the other actors. If, for example, nations and their respective economies are in competition with each other in other respects, and if the welfare of the populations and the approval rates of the representatives leading the negotiations depend on their respective shares in the global markets, then this incentive will be all the greater. Accordingly, effective measures depend—at the level of individual actors as well as at the level of collective actors—not only on sound and enlightened knowledge of the factual relationships. Rather, the consultations must also be organized in such a way that the resolutions can then be effective. This requires open and fair processes involving all available options, results that take into account the needs of the stakeholders and are perceived as fair, and decisions that are sufficiently binding to effectively break the dilemmatic incentive structure. Here the international community has created a considerable, albeit fragile, basis with the Paris Agreement of 2015—despite all justified criticism of details.

Chapter 2
Executive Summary

The current global situation in the energy sector is characterized by the dominance of fossil fuels, particularly in the area of electricity generation. Per capita, final energy consumption varies greatly depending on a country's level of development. Almost 1.3 billion people remain without access to modern forms of energy supply, especially electricity. In general, a slight increase in primary energy demand and a stronger increase in electricity demand are expected for the coming decades—with regional differences; the development of electromobility is one of the uncertainty factors. Future energy technologies should (i) be largely free of CO_2 and other climate-impacting pollutant emissions and should depend as little as possible on limited/rare raw materials, (ii) replace fossil power plants and be able to meet growing demand, and (iii) be accessible and affordable to large parts of the world, including countries that have been underserved to date.

2.1 Power Engineering

With regard to energy sources, a distinction is made between

- renewable energy sources (RES), either constantly and quasi simultaneously growing with the consumption such as wind, sun and practically also water, or with a delay, such as biomass;
- spent fuels that are available in large or even practically unlimited quantities and can be recovered through technical processes, such as nuclear energy (uranium/thorium fuel cycle, burn and breed reactors);
- Consumption of fuels available in large but finite quantities, such as oil and natural gas or coal, whereby reserves can be stretched by means of technical processes (gas fracking);
- Others, like geothermal energy.

C. F. Gethmann et al., *Global Energy Supply and Emissions*,
Ethics of Science and Technology Assessment 47,
https://doi.org/10.1007/978-3-030-55355-5_2

Wind and solar energy could potentially meet the world's energy needs alone. However, it is pointed out, for example, by the OECD-IEA in its World Energy Outlook 2016, that this is practically impossible: the technical exploitation of this potential is hampered by location conditions, their intermittent nature, and technical, ecological and economic barriers. Hydropower has great potential but is already largely exhausted; tidal power plants play only a modest role globally. Geothermal energy also has great potential, albeit with strong regional variations, which could often only be exploited to a greater extent with quite large risks/costs.

In terms of specific CO_2 emissions, each based on the results of life cycle analyses, today's coal-fired power plants rank at the end of the negative scale at about one kilogram per kWh and gas-fired power plants slightly below; PV (roof-mounted) and biogas plants are more than an order of magnitude below this, while wind (offshore) and nuclear power plants have CO_2 emissions almost 100 times lower than coal-fired power plants; hydropower plants have even lower values. Accordingly, the reduction of CO_2 emissions when using renewable energy sources, especially PV and biogas, is one of the goals of technical development that PV systems are considered achievable. Coal-fired and gas-fired power plants can only meet acceptable limits for CO_2 emissions when equipped with CO_2 capture and safe storage (CCS) facilities, bioenergy with CCS even allows negative emissions. In principle, the technologies required for this are available, and their economic use within the framework of an overall concept can be expected, if at all, after 2030 at the earliest.

As already explained, research and development work can be identified in the field of renewable energy sources (RES), which, in addition to the "search for something completely new," aims at the further development of today's technologies—towards (i) efficiency increases (PV) and/or higher working temperatures (solar thermal) while at the same time reducing material dependencies, consideration of environmental aspects and cost reductions (PV cells), (ii) efficiency improvements and more environmentally friendly conversions (wind, especially biogas), and (iii) further evolutionary steps towards commercialization and reliability. We assume that most renewable energy sources will be usable on a large scale during the period under review and will be able to generate electricity competitively in more widespread siting regions, which is already the case for wind power in favorable locations. However, a similarly positive, secure picture does not yet emerge for the required infrastructures, above all for the storage and transport/distribution systems, whose development has been ongoing for a long time but is still waiting for decisive breakthroughs.

Some nuclear power plants—technologically and safety-related "evolutionary" or "revolutionary" further developments—are already under construction today or will be available by 2030 at the latest, when they will be sufficiently mature. In addition to social acceptance, the financing and lack of flexibility of large plants is a problem that small, modular units promise to help overcome. Such concepts are also available in the medium term, especially if they are based on the technology of today's light-water reactors. More "exotic" concepts which promise a further increase in safety, i.e., the total exclusion of serious accidents due to inherent physical mechanisms, are under development which can also prevent or eliminate extremely long-lived waste products and breed fissile material; however, they can at most play a decisive role in an energy mix in the long term, i.e., beyond 2030.

A number of energy sources and the corresponding technology are therefore already available, or their development is reasonably certain. They can be deployed according to the circumstances of the countries that need and use them, whereby each region must and may find its optimum for the deployment strategy via suitable processes, as long as overriding goals such as global climate and environmental protection are not jeopardized.

2.2 Environment

People depend on intact ecosystems to satisfy their needs for food, clean air, and drinking water. Energy systems must therefore not endanger the natural foundations of environmental life in the longer term.

Those parts of the biosphere that are functionally necessary for the survival of the human species deserve special attention and protection:

- the atmosphere (clean air),
- clean water,
- agricultural land,
- stable climate and vegetation zones (as a prerequisite for the preservation of the environment in its basic substance, biodiversity, and the worldwide cultivation of cultivated plants).

A key role in keeping the air clean, in the production of drinking water and in the stability of climate zones, is played by the preservation of large ecosystems, especially large forest areas and seas. The preservation of the greatest possible biodiversity is also an indicator of the environmental compatibility of energy systems. Although natural biodiversity does not appear to be directly necessary for human survival at the existing scale, it is important for the stability of ecosystems and as a potential genetic and pharmaceutical resource. When intervening in landscapes, the effect on landscape scenery must be taken into account as they are also valued as recreational areas and for aesthetic reasons.

Energy systems can also pose direct health risks to people in the form of pollutant emissions from normal operation and major technical accidents. And while major accidents do not always damage people and the environment to a large extent and over large regions, health risks to humans must generally also be treated from the perspective of global environmental compatibility.

When pollutants are released, especially into the atmosphere, the global effects must always be taken into account. There is no doubt that the different economic and social circumstances of industrial, emerging and developing countries must be taken into account in a differentiated manner. These issues concern fossil fuels (hard coal, lignite, oil, natural gas), renewable energies (various solar energy options, wind power, hydropower, biomass, tidal and ocean wave energy, geothermal energy), nuclear energy and nuclear fusion.

2.3 Energy and Climate

In the current phase, climate change is dominated by human influences, namely

- on a global scale through greenhouse gases, in particular carbon dioxide,
- on a regional scale also by aerosols,
- on a local scale through land use changes, in particular urbanization.

These statements are described as practically certain in the constantly up-dated assessments of the UN Climate Council. The first step is to determine the extent to which the changes under consideration can be explained within the framework of global natural climate fluctuations. With regard to the changes that cannot be explained in this way, one speaks of "detection" and assumes an external cause. In order to determine these, it is examined to what extent the effects of the various possible "drives"—mainly anticipated in climate models—are relevant to the changes under consideration. The most plausible mix of reasons is then selected as the best current explanation and assigned to the observed climate change (attribution). This approach is used in particular with regard to global and regional changes, but hardly ever with regard to local changes.

Climate models play an essential role in determining the causes of climate change. The exact quantification of the "sensitivity" (i.e., the answer to the question of how much the global air temperature will rise if the carbon dioxide concentration is doubled, for example) as well as the effect of the interaction between radiation and clouds or the role of oceanic eddies is uncertain and remains the subject of current research. Changes in the assessment in detail are still possible. The statement "The science is settled" is therefore misleading. In fact, the scientific community does agree on the essential core statements, such as those made by the IPCC. However, there are many details, such as the development of tropical cyclones, regional distributions of altered precipitation regimes or sea-level development, which are the subject of current research. It is therefore correct to say that "some science is settled, some other is not yet."

Human-induced climate change can basically be controlled by changing the input of greenhouse gases, aerosol-forming substances, and changes in land use. Economic developments have led to significant increases in the release of greenhouse gases, but political measures in recent decades have also reduced the growth in releases. With regard to aerosols, significant reductions have been achieved in Europe and North America, while China, for example, has seen significant increases. In urban planning, the limitation of this "city effect" has only placed a role in pilot projects so far, for example in limiting urban heat island.

In addition, other reduction measures are conceivable, in particular those that lead to the removal of carbon dioxide from the atmosphere. However, large-scale implementation of these options is unrealistic for the foreseeable future. Other measures to limit global or regional climate change through geoengineering (e.g., capture and storage measures or measures to modify the radiation budget) are currently in experimental test phases and are subject to considerable public reservations.

The controllability of global and regional climate change is limited. The impact variable for global climate change is essentially the total emissions accumulated over many decades; past emissions have already been administered to the system and represent unchanging facts for the foreseeable decades. The possibility of control, therefore, only refers to future emissions. The economic system behind global emissions—industry, transport, energy, and microclimates (heating, cooling)—is sluggish and can only be changed on time scales of many (carbon dioxide) and few (aerosols) decades.

2.4 Energy and Economy

In poor countries, the lack of access to electricity because of the local absence of infrastructure excluding 70% of the people in many countries, constitutes a considerable energy problem. In contrast, the wealth of the rich countries leads to high energy consumption and CO_2 emissions.

We establish the following empirical results: (i) A one percent increase of world GDP leads to a one percent increase in CO_2 emissions; (ii) temporary deviations from this global one-to-one impact have not endured; (iii) world GDP growth is predicted to be 2.7% in the long run implying the same growth rates for emissions without policy changes or other structural breaks; (iv) poorer countries have lower emissions; (v) falling emissions in the EU, especially France and Germany; (vi) a higher international-trade/GDP ratio increases emissions; (vii) the impact of GDP on emissions can be reduced—almost one-to-one—through a higher share of renewables and nuclear (% of total electricity production) and through a higher energy productivity.

However, the share of renewables has increased less strongly than the shares of waterpower and nuclear have fallen, and, by implication, the shares of fossil-based electricity and energy have increased in spite of large investments in renewables. This trend is not in line with the carbon budget remaining under a 2-degree goal or a lower one.

CO_2 taxes or permit prices should be determined in line with consideration of social costs, uncertainties and the 2-degree goal, and the corresponding carbon budget. They should be implemented in line with principles of sound economic policies in order to avoid unnecessary costs of the 2-degree goal.

Although it is clear that there is huge uncertainty regarding the adequate CO_2 price or tax, it is clear that increasing shares of fossil-fuel-based energy indicate that, so far, the price or tax is too low to achieve the 2-degree goal. This also holds for the limited reductions in CO_2 emissions of the EU. Large parts of the world economy are not under any tax or permit system.

Environmental policies come at some costs, which are partly avoidable:

(a) National fear for a loss of international competitiveness has led to policies avoiding emission pricing. We provide evidence showing that international differences in European electricity prices lead to lower net inflows of foreign direct investment. These costs could be avoided through international coordination.
(b) If developing countries could reduce energy poverty through more growth, they would generate more emissions. Some part of the carbon budget has to be reserved for the growth of poor countries and the related emissions.
(c) The investments in new technologies may boost demand for relevant natural resource inputs and increase their prices. If this can be avoided, at least temporarily, through opening of new mines as it happened to occur recently with rare earths used in wind turbines, it may damage the natural environment near the mines. In short, policies to solve or mitigate one environmental problem may reinforce a related environmental problem.
(d) Recent policies subsidizing the clean alternatives fail to punish or reward differences in emissions per KWh electricity when comparing gas and coal. The same holds for differences in KWh production per Euro of investment among the renewables. This system obviously is unnecessarily expensive in regard to emission reductions. Transition to a more efficient system could also reduce uncertainty about future policies in the capital markets and the business decisions.
(e) A higher share of renewables leads to stronger fluctuations. Dampening them is mostly discussed in terms of gas-fired power stations and storage capacity. However, cross-border electricity cables may lead to more international trade in electric current and will thereby reduce the costs incurred for the dampening of the fluctuations.

We show that the share of renewable electricity has an impact on the worldwide bilateral trade flows of electric current. This indicates that fluctuations are dampened by international trade already at the currently low share of renewable electricity. The international trade mechanism could be used more strongly.

2.5 International Relations

The EU and the BIC countries do not devote sufficient commitment to the energy dialogues. As the EU does not seem as important for BICs as the Member States, commitment is less intensive. Thus, dialogues vary in intensity, frequency of meetings and outcomes.

Additionally, bilateral dialogues are still characterized by mutual mistrust, minor interest of the Emerging Powers in the dialogue and divergent perceptions of the underlying norms and goals.

Actors' motivations to enter the dialogue mostly do not converge, nor do the high expectations they maintain. While the EU in all the dialogues is motivated mostly by combating climate change and less by technology transfer, the BICs countries focus mostly on technology transfer and on the promotion of private-sector cooperation or else market entry. Thus, it might be important for EU actors to ensure how these interests might actually be met. Possibly, the interest in technology transfer exhibited by some of the Emerging Powers is not always echoed by heightened EU interests in market entry in these countries. Consequently, in the long run, the BICs might prefer to engage in technological cooperation with other countries or Member States, instead of the EU. The EU might, therefore, prefer to focus more on specific arenas dealing with issues of technological transfer or private actors' cooperation. Currently, the dialogues are driven by competing ownership: each partner wants to be the lead in terms of designing policies, setting the agenda, or suggesting concrete outcomes.

Another aspect is the mode of cooperation. Especially the EU has shown poor performance within these dialogues. It seems that the EU is very good at 'talking at' instead of 'talking with' external partners. The EU engages in top-down, one-way communication of projecting interests, norms, and values rather than developing a horizontal dialogue-led two-way communication process between equal partners.

The current development of, on the one side, strengthening the climate change-energy transformation nexus and, on the other side, the withdrawal of the US from multilateral agreements on climate change and a reorientation away from renewables towards fossil fuels, the need for a new multilateral international forum on energy transition is needed. If it does not seem possible to create one energy organization at the global level, then at least it seems to be necessary for some of the existing organizations to expand their membership towards the BIC countries. Bilateral cooperation could help to pave the way in some of the organizations.

2.6 International Distributive Justice

Questions of distributive justice are usually posed as questions of communication and interaction between individuals, groups or institutions within a social corporation (family, company, state, etc.). The long-standing discussion on questions of distributive justice in the environmental sector, for example, is mainly conducted in a domestic context. Until a few years ago, there had been grosso modo one dimension of questions that transcended this perspective, namely the problem of distributive justice between generations with regard to long-term responsibility. However, this question has a clear diachronic perspective despite a number of questions implying that the problem of distributive justice between states and state-like corporations exists in a synchronous and multilateral perspective. Here, the well-known questions of distributive justice, as they exist between individual and institutional actors, are transferred to states or state-like actors.

Human rights primarily focus on the protection of the individual and allow at best marginal conclusions to be drawn for questions of distributive justice. Furthermore, the reference to human rights is linked to the idea that economic globalization could, in a sense, prove to be a vehicle of normative universalism. On the whole, however, the discussion of recent decades shows that eco-nomic globalization by no means goes hand in hand with the globalization of universal norms, such as human rights.

The energy issue exacerbates the problems of global distributive justice mentioned above to an extent that is well known in the field of small-scale individual and institutional actors. At the same time, the energy issue, apart from individual aspects, is not yet the subject of an appropriate discussion on a global scale.

The benefits and burdens of a technical energy supply are unequally distributed, and in an increasingly networked world, conflicts are emerging increasingly. That gives rise to questions of fair distribution on a global scale and to the consideration of ethically legitimized strategies that allow people to meet their respective requirements. Justice as is discussed in ethics does not aim at an equal distribution of all goods or services or the equality of all living conditions, but the equal treatment and consideration of all raised claims. Justice in this ethical sense does not mean "equality," but "rationally justified inequality." To be rationally justified a distribution has to be convincing for everyone concerned under all given circumstances (invariantly with regard to addressees and situations). In the thematic framework of a global energy supply with its complexity of the interrelations on the one hand, and the diversity and disparity of the forms of life on the other, efforts to justify distributions rationally are faced with over-complexity. In such contexts, any attempt to demonstrate an ethically appropriate procedure based on supreme principles to those concerned is doomed to fail. Where the general ability to consent is not to be bought by losing oneself in the abstract general and thus not contributing to an operable solution of the problems, the effort to achieve acceptability must be sensitive to the specific preconditions of those affected and their respective way of life. In order to develop appropriately differentiated approaches to solutions that do justice to the manifold disparate requirements and prerequisites, one will rather think about organizational forms and procedures through which an appropriate clarification can be brought about by the participants themselves. Accordingly, there is a need for a public debate in which goals and expectations are compared, motivation and reasonableness are weighed against each other and converted into formal entitlements and obligations.

Philosophical ethics has a pivotal task in preparing and supporting these debates with proven and tested arguments. There is a historic set of material distribution principles that are, though there are no general rules at hand as to which of these principles has to be applied in certain situations, of importance for finding fair solutions. Most important though, are the formal principles of justice that require an equal role and vote for all in the debate that require consistent treatment of cases similar in the relevant aspects. The ethical requirement for universalizability, therefore, is to be understood rather as a formal principle than directed towards distribution outcomes, but towards the distribution procedure, as a requirement for its design.

Resources on the one hand, and risks and opportunities on the other, are the relevant subjects for distribution. In this context, resources are to be understood not in a materialistic sense but as means that are necessary for the achievement of set goals. A fair distribution of risks should be oriented to certain rules, as the rule of risk taking, the rule of opportunity sharing, the rule of risk allocation, and the risk provision rule.

Chapter 3
Recommendations

3.1 Preliminary Remarks

1.1 The politically determined goal of decarbonization, primarily for reasons of avoiding anthropogenic greenhouse gas and other pollutant releases, is the primary normative basis for the assessment of future energy technology. Other key points of reference for the following recommendations are climate protection, resource conservation, protection of the environment and human health, the security of supply, and economic viability.
1.2 The fulfillment of these requirements is a prerequisite for the acceptance of a global energy supply that takes cultural diversity into account.
1.3 The recommendations are committed to a universalistic approach: Policies should make it possible to connect all people of the world, notably also in poorer countries, to electricity supply networks and other modern forms of energy supply. In principle, the technical and organizational prerequisites are in place to achieve the goals set, but they require constant further development.
1.4 The available knowledge forms a sufficiently sound basis for necessary political decisions but is incomplete and prone to error. There is, therefore, a general requirement for research and development: the greater the impositions associated with political action or non-action, the more secure the knowledge base should be.
1.5 An international agreement on emissions should (i) integrate transport into emission reduction policies and (ii) ensure that this happens globally.

3.2 Energy Technology and Environment

2.1 Energy technologies must be assessed holistically, i.e., taking into account the entire life cycle and necessary infrastructures. For assessment purposes, patterns

C. F. Gethmann et al., *Global Energy Supply and Emissions*,
Ethics of Science and Technology Assessment 47,
https://doi.org/10.1007/978-3-030-55355-5_3

and methods are available, for example, in the form of sustainability criteria and their quantification, which are to be further developed/adapted and applied.

2.2 The results of previous applications show that none of the technologies known today, which can—in principle—be used, are superior to the others in all areas. With the aim of robust supply security and reduction of uncertainties should be based on diversity regarding technologies used with improved security. It should be based on a mix of different energy sources. However, coal, especially lignite, should be avoided whenever possible because of its high emissions.

2.3 Consideration processes must be organized, taking into account a variety of factors, including clearly defined decision-making preferences. Health risks for people caused by energy systems and possible interventions in biodiversity and the landscape must be in reasonable proportion to the benefits of energy supply. Possible impairments ought to be weighed against the benefits of the energy supply in each case.

2.4 The development of medium-term (2030) energy supply strategies should be based on low carbon/or even carbon-free technologies that are clearly ready for the market through large-scale experiments and demonstration plants. In the long term (2050), preference should be given to technologies whose feasibility can be justifiably assumed. Necessary development times, usually about 20 years from invention to market maturity, must be observed. Accordingly, nuclear fusion and some "exotic variants" of nuclear fission are not a source of hope.

3.3 Energy and Climate

3.1 In order to limit anthropogenic climate change, the focus should be exclusively on reducing greenhouse gas emissions, as no promising geoengineering technologies are currently discernible.

3.2 In emission reduction measures, imitation can only be achieved through economic benefits, not through moral or political claims.

3.3 Any planning to deal with expected climate change should take into account the possibility that the 1.5 °C target—as well as the 2 °C target—will not be met. Even if these are achieved, technological and other measures should be developed to adapt to changing climatic conditions and to counteract the vulnerability of social and natural living conditions.

3.4 Energy and Economy

4.1 CO_2 certificate prices or equally high taxes should be increased to reasonable levels in line with the social cost of carbon. Policies of pricing CO_2 should be extended to other greenhouse gases. Poor people in rich countries should be compensated for higher electricity bills.

4.2 Subsidies for the use of fossil fuels should be reduced.
4.3 If policies ensure that all fossil fuel using technologies are carrying their private and social costs, in particular for CO_2 emissions, nuclear power stations should pay insurance costs in line with the US system. Costs for nuclear waste and dismantling of permanently shut-down power stations have to be included in the cost-pricing.
4.4 As the electricity systems shift to more volatile supply sources, the cost-minimizing way of limiting fluctuations is to have a strong system of cross-border cables, to decrease the necessity of gas-fired power stations.
4.5 In the poorer countries, production capacity and networks should be increased in order to ensure access to electricity for all people and businesses. Some parts of the public carbon budget related to the 2 °C target have to be reserved for the growth of developing countries and the related emissions.

3.5 Politics/International (Bilateral and Multilateral) Relations

5.1 Decarbonization strategies should be designed within a new encompassing multilateral arena as an international platform to coordinate national and regional energy policies, including stakeholder dialogues.
5.2 It is not consistent with justice requirements, if national energy policy is geared to the actual acceptance of citizens but justified with "constraints" in international energy policy negotiations, such as scarcity of resources or climate change. Just solutions are universally "acceptable", i.e., rationally justifiable.
5.3 As participants in international negotiations, the nations of the world are basically to be classified as equals. Differences determined by differences in power and wealth or according to other differentiation criteria play an important role with regard to distribution issues but do not affect eligibility for participation. No state is a priori entitled to have its ideas serve the world as a model.
5.4 In addition, the full potential of existing bilateral energy dialogues should be tapped. Dialogue partners should devote more commitment and develop an ambitious long-term vision. This implies the need to build up mutual trust between partners, multi-directional learning, equal partners, and joint ownership.
5.5 The possibility to enlarge the bilateral dialogues on a trilateral or even plurilateral basis should be considered to include industrial, emerging, and developing countries.
5.6 Efforts to build EU positions on a broad consensus among the Member States should be strengthened. Soft governance instruments with harder elements, as proposed by the Commission, should be enforced. Other instruments, such as structural funds, to support decarbonization processes should be used to a greater extent.

3.6 Global Energy Justice

6.1 The indefinite commitment to the life interests of future generations does not mean that provision for the future takes precedence over spatial distance commitment. The prevention of certain expectable short-term life risks deserves priority over long-term prosperity risks expected with a certain probability of occurrence, irrespective of the geographical distance.

6.2 A fair balance must be struck between the risks that one is prepared to accept for oneself and expect others to accept by doing or not doing. Decisions aiming for a permanent intersubjective agreement ("compliance") should be based on a rational understanding of risk (enlightened evaluation of possible states x justified estimation of the respective probability of occurrence).

6.3 A procedural consensus on justice between states can lead to considerable injustices within states. Therefore, in the global justice discourse, it must be procedurally ensured that states participating in such discourse guarantee domestic justice procedures.

6.4 In order to clarify questions of historical justice such as the consideration of acquired economic prosperity due to a constant degree of state organization or scientific-technical successes, it is appropriate to go back no further than to a point of time of great historical upheavals such as the state order after the end of the Second World War and the end of colonization (approx. 1950).

6.5 Regarding the consideration of different economic and scientific-technical stages of development, different raw material deposits as well as the path dependencies of developments, and finally environmental policy merits or failures, a crediting procedure must be developed which guarantees proportionately lower profits for the better-off than the worse-off.

Chapter 4
Strategic Energy Requirements—Technological Developments

4.1 Baseline Situation

It is recognized that energy contributes significantly to the well-being of our society. Energy consumption began to rise moderately at the beginning of the nineteenth century, but almost exponentially from the middle of the last century onwards, strongly linked to the onset of population growth (see Fig. 4.1). Energy expenditure in Europe amounts to up to 10% of GNP, a reduction seems possible.[1]

Fossil fuels continue to dominate energy supply (around 80% worldwide in 2016). Almost 40% of global primary energy is used to generate electricity and heat, which also contributes almost 40% to global CO_2 emissions, followed by the industrial and transport sectors. For final energy consumption, oil products and gas dominate with about 55% (see Fig. 4.2), while the share of electricity is about 17% and will further increase.

A total of just under 49 Gt of anthropogenic greenhouse gases (CO_2 equivalent) were emitted in 2010, 65% due to the use of fossil fuels and industrial processes, with an increase of just over a third between 2000 and 2010 (see Fig. 4.3), followed by an increase of 7% from 2010 to 2017 to about 52 Gt.

The average per capita final energy consumption in 2015 was 2715 kWh per year and fluctuates depending on the level of development of the countries.[2] The latter applies in particular to the consumption of electricity, for which in 2010 almost 1.3 billion people had no access; scenario analyses suggest, depending on assumptions, a decline to approx. 319–530 million people by 2050 (see Fig. 4.4). It is

[1]The ratio of energy use to GNP has been continuously declining in many developed countries over the last decades, in the U.S. over the period of 1950–98 at an annual rate of 1.4% on average.

[2]The approximate total *consumption* (*primary energy consumption*) is broken down as follows: Asia ≈45% (of which China ≈25%), Europe ≈20%, North America ≈20% Africa ≈5%, Middle East ≈5%, Latin America ≈5% (Global Energy Statistical Yearbook 2017). Per capita electricity consumption in 2015 was 315 kWh per year on world average, 2618 in Norway, 1360 in the USA, 1116 in Australia, 815 in Switzerland, 762 in Germany, 461 in China, 79 in India and 2 W in Chad.

C. F. Gethmann et al., *Global Energy Supply and Emissions*,
Ethics of Science and Technology Assessment 47,
https://doi.org/10.1007/978-3-030-55355-5_4

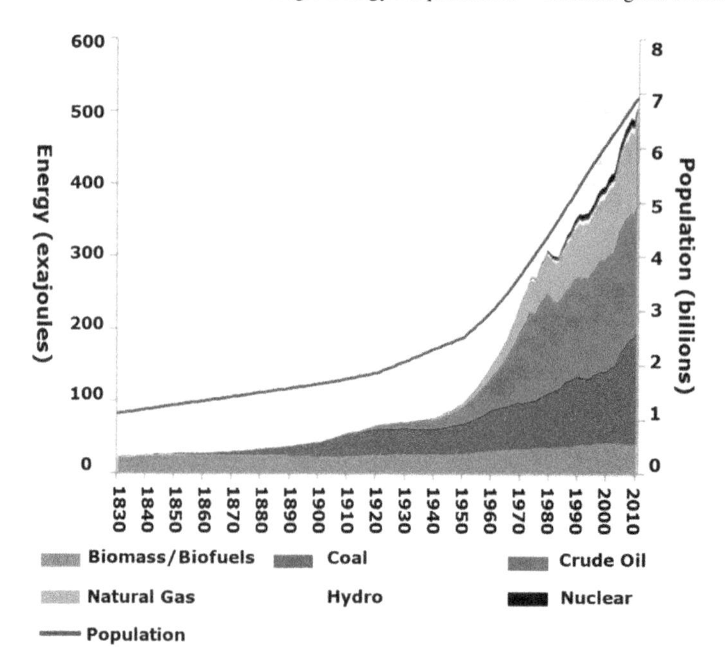

Fig. 4.1 Global Energy and Population (brown line): history of energy transitions for almost 200 years (Mearns 2014)

generally assumed that final energy consumption will rise globally and the share of electricity will increase, hopefully because more people in developing countries will have access to modern forms of energy, and not least because of new areas of application (keyword: electromobility).

The energy sector, and in particular the electricity sector as a forerunner (in the following the focus), is facing far-reaching changes in order to meet the increasing demand and ambitious targets, such as the need for decarbonization for climate protection reasons, i.e. limiting the temperature increase to an average of 2 °C by 2100. In Europe, for example, so-called energy roadmaps are being drawn up which provide for a reduction of greenhouse gas (GHG) emissions by at least 20% in 2020, an increase in the share of renewable energy sources (RES) by about 20% (compared with 1990) (European Commission 2011) and even a reduction of GHG emissions by 80–95% in 2050 (European Commission 2012).

As the driving forces behind the necessary transformations towards a far greater use of renewable energies, above all solar and wind power, which still play a very modest role today, technological (further) developments and strategic guidelines (political will, public support, etc.) must interact and be supported regionally, with varying effects on the various energy sources (see Fig. 4.5).

In the meantime, various organizations (IPCC, IEA, EU, Greenpeace, etc.) have carried out analyses which show how a limitation of the temperature increase by at most 2 °C or the CO_2 equivalent concentration to 450 ppm could be achieved,

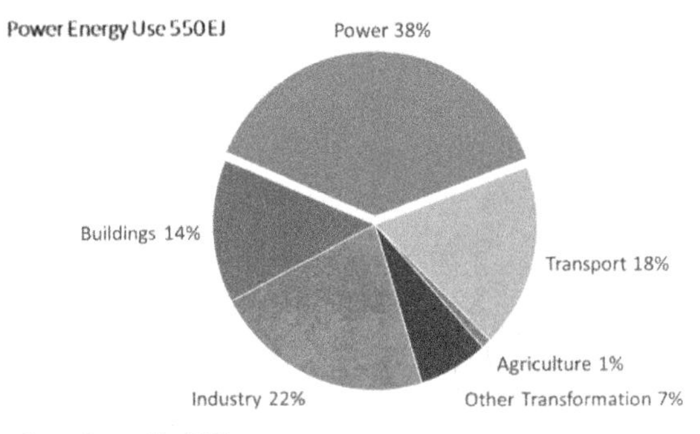

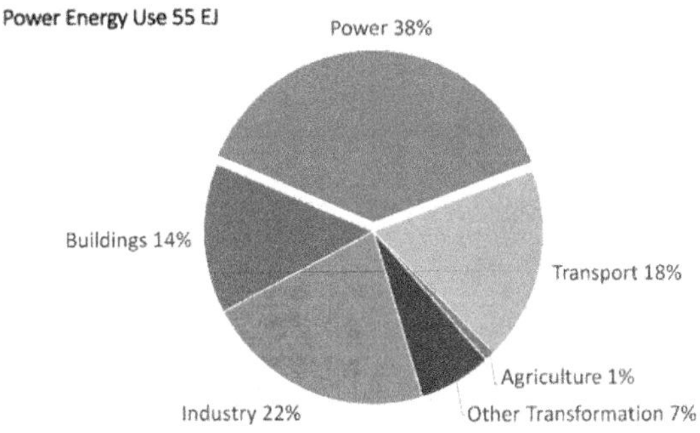

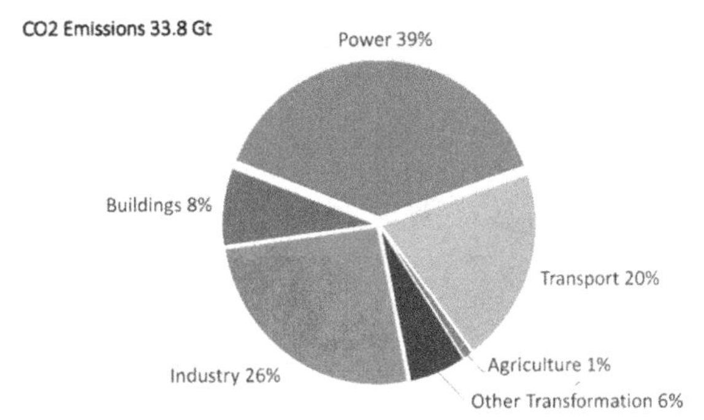

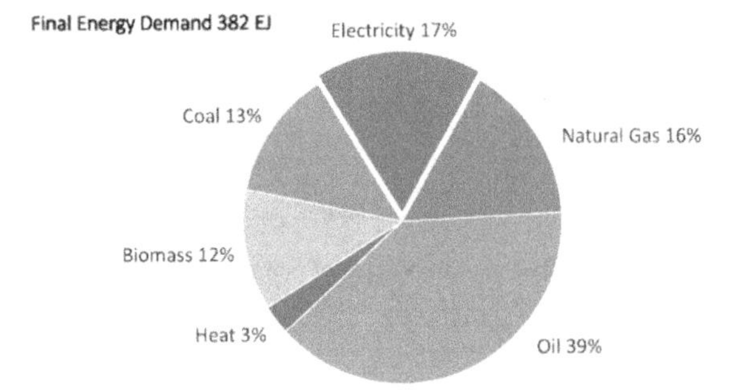

Fig. 4.2 Primary energy use by sector, CO_2 emissions by sector, and final energy by fuel in 2011 (IEA 2014)

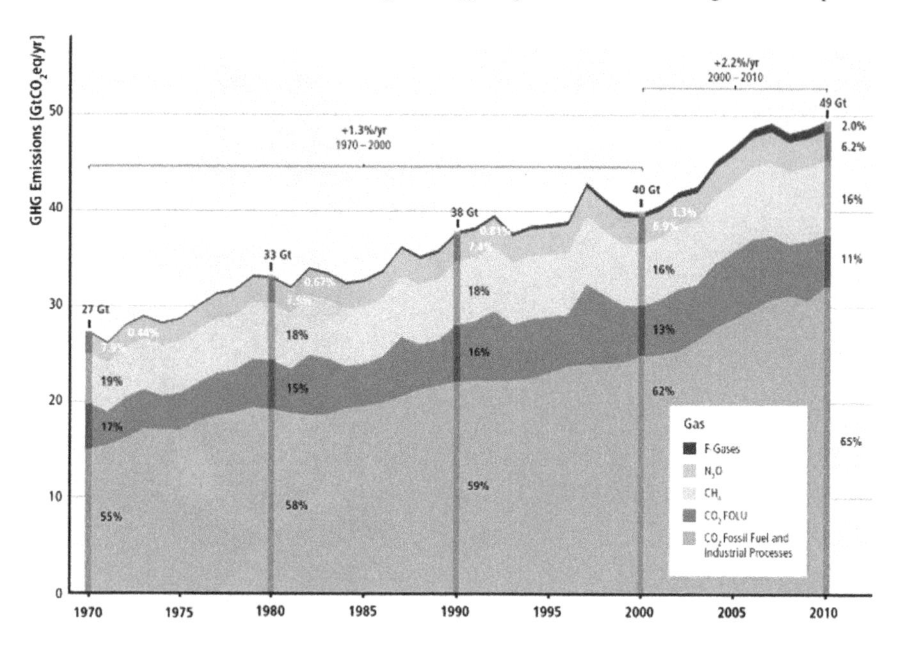

Fig. 4.3 Total annual anthropogenic greenhouse gas (GHG) emissions by groups of gases 1970–2010 (IPCC 2014b)

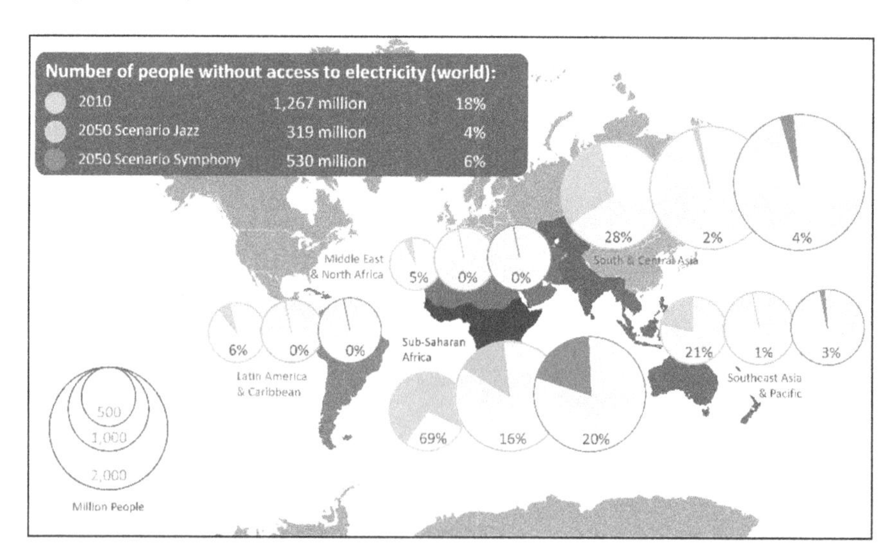

Fig. 4.4 Number and proportion of people without access to electricity in 2010 and 2050 in both scenarios; the size of the circles is proportional to the population (PSI 2013)

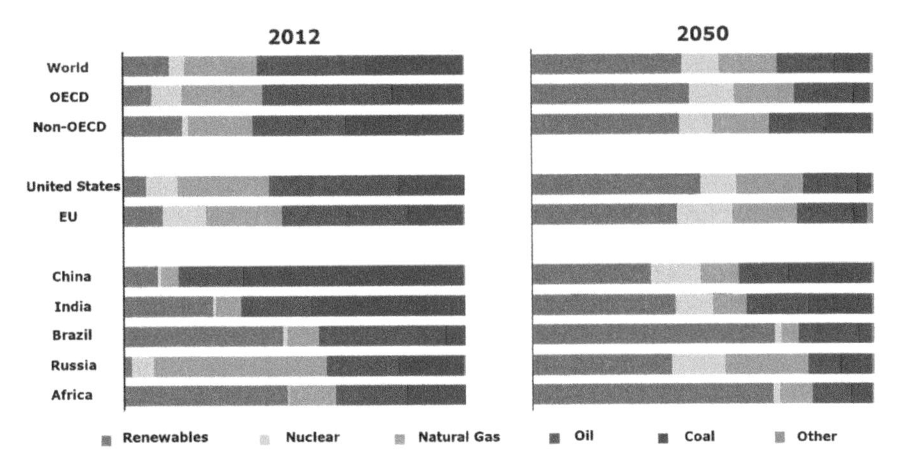

Fig. 4.5 Regional primary energy demand profiles in the 2 °C scenario (IEA 2015a)

mostly with proposals for a reduction of CO_2 emissions via an appropriate mix of different energy sources. The focus is on the massive use of renewable energies (solar, wind, water, geothermal, biomass); the use of nuclear energy is also promoted, unless excluded from the outset for ideological reasons. The transition from coal-fired to gas-fired power plants is propagated as the fastest effective mitigation step, fossil fuels ultimately only permitted in conjunction with *Carbon Capture & Sequestration/Utilization* (CCS/CCU) Technologies. Included technology recommendations often with status checks and *policy recommendations*, Table 4.1 in Sec. 4.4.1 may serve as an example. Figure 4.6 illustrates how massive the changes within the EU are or must be.

4.2 Strategic Goals, Evaluation Patterns

In industrialized countries, including the EU (European Commission 2011), the availability of safe, secure, sustainable and affordable energy is the strategic goal, while in developing and emerging countries access to modern energy is usually the main priority. As part of this work, we join these strategic objectives. "Safe, secure" means that energy is available to consumers everywhere at any time and in the desired quality and that dependence on foreign sources remains limited ("security of supply").

"Sustainable" is often still a vague term, but at least in Europe a concretization and operationalization via the three dimensions "economy", "environment" and "society" and associated indicators for the evaluation of energy sources and associated technologies, in each case including upstream and downstream processes ("life cycle analysis"), has largely prevailed. The EU project NEEDS (Hirschberg and Burgherr 2015) represents the current state-of-the-art of such an evaluation framework with a

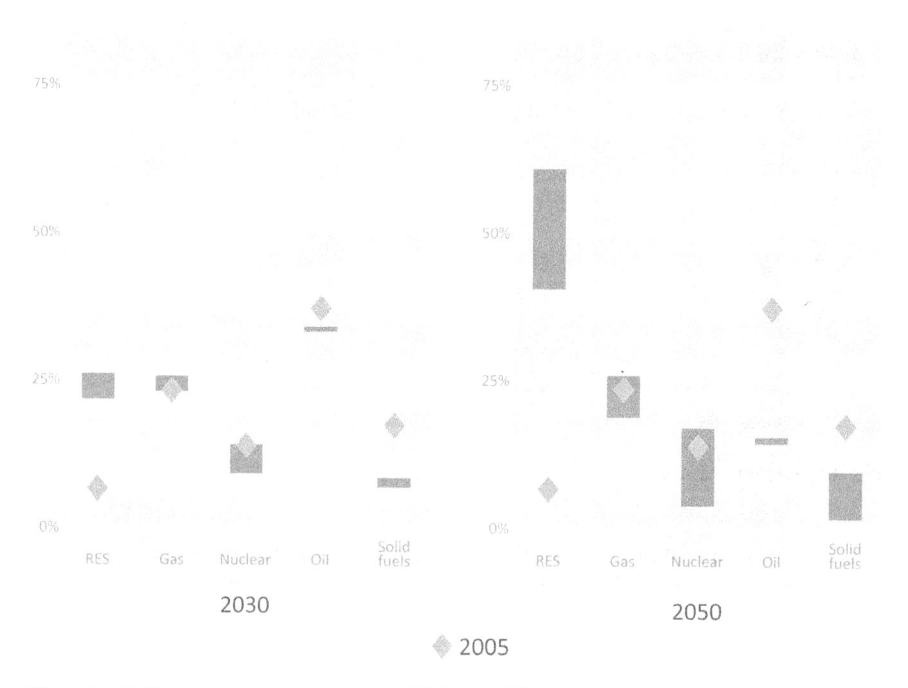

Fig. 4.6 EU Decarbonization scenarios—2030 and 2050 range of fuel shares in primary energy consumption compared with 2005 outcome (in %) (European Commission 2012)

set of 36 agreed indicators. In a nutshell (for a further breakdown, see Tables A.1–A.3 in Appendix A), these are the following:

- 11 *environmental* covering energy and mineral resources, climate change, ecosystem impacts from normal operation and severe accidents, and special chemical and medium and high level radioactive wastes;
- 9 *economic* including impacts on customers (electricity price), overall economy (employment and autonomy of electricity generation), and utility (financial risk and operation);
- 16 *social* addressing security/reliability of energy provision, political stability and legitimacy, social and individual risks, both expert based (normal operation and accidents) and perceived, terrorist threat, and quality of residential environment (landscape and noise).

A broad spectrum of a total of 26 possible advanced power generation technologies for fossil (coal, lignite and gas), nuclear (thermal pressurized water and fast sodium-cooled breeder reactors) and renewable (biomass, solar, wind, water) energy sources was considered, with a view to their medium-term (2030) and long-term operational readiness (2050), taking into account conditions in four countries (France, Germany, Italy, Switzerland). The starting point was today's (2005) technologies. The highly uncertain expectations for the future were categorized into three scenarios:

"pessimistic", "realistic-optimistic" and "very optimistic". The characteristics of the technologies for 2050 are summarized in Table A.4 in Appendix A.

A variety of methods were used to quantify the indicators such as life cycle assessment (LCA) and impact pathway approach (IPA). "Severe accident risk assessment" is based on worldwide historical data from a database (ENSAD) and trend analyses specially set up for major accidents in the energy sector; probabilistic analyses (PSA) were used for hypothetical nuclear accidents. The "social indicators", which are difficult to quantify, are based on an opinion poll among experts from a wide variety of fields up to a "Delphi" among the "experts". Some results are summarized below.

Renewables and nuclear have GHG-emissions that are one to two orders of magnitude lower than those from fossil technologies without carbon capture and sequestration (CCS). CCS has the potential to reduce the emissions by one order of magnitude (Fig. 4.7).

Radioactive wastes are inherently associated with nuclear energy (Fig. 4.8). The volumes of critical non-radioactive wastes are today highest for solar PV but expected to get strongly reduced in the future.

Most remarkable capital cost reductions are anticipated for solar PV. Nuclear, solar PV and hydro capital costs are subject to large uncertainties, though their character is different. While for nuclear the circumstances in project implementation are decisive, the uncertainty of hydro costs depends highly on-site characteristics and for solar PV on the successful realization of potential advancements.

Current electricity generation costs of nuclear and hydro credit the fact that capital costs are partially amortized. Capital cost reductions and their uncertainties are reflected in generation costs (Fig. 4.9). As capital costs dominate the solar PV and future hydro costs, the uncertainties in generation costs are highest for these technologies.

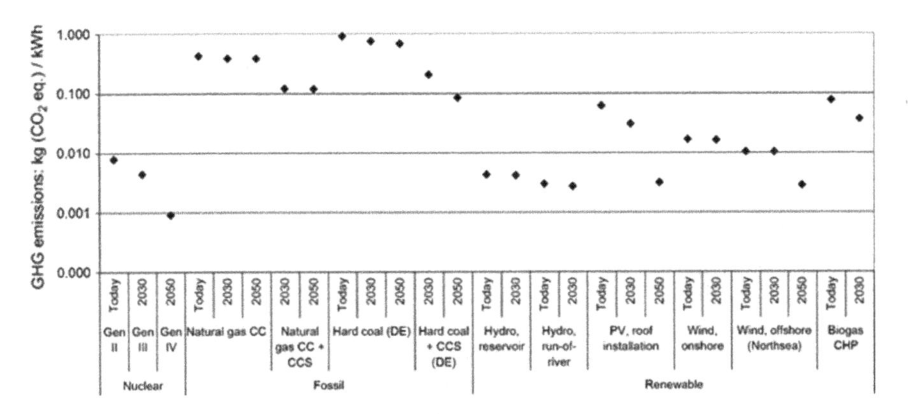

Fig. 4.7 Greenhouse gas (GHG) emissions of selected technologies (Hirschberg and Burgherr 2015). CC stands for combined cycle, CCS for carbon capture and sequestration, PV for photovoltaic, CHP for combined heat and power

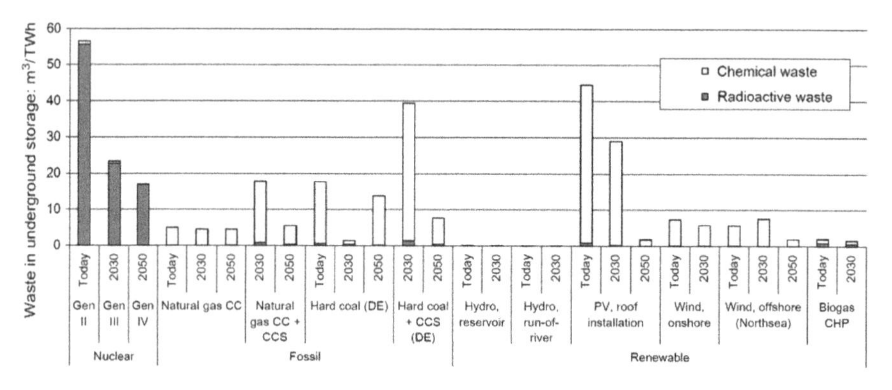

Fig. 4.8 Production of medium and high-level radioactive wastes and special chemical wastes stored in underground repositories (Hirschberg and Burgherr 2015)

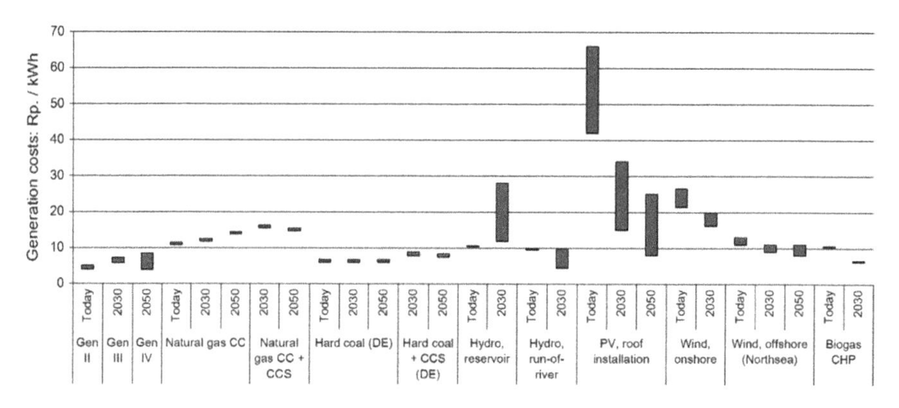

Fig. 4.9 Electricity generation costs of selected technologies (Hirschberg and Burgherr 2015)

Fossil technologies, in particular combined cycle gas, exhibit a very high sensitivity to fuel costs. The corresponding dependence of nuclear is very low or negligible in the case of fast breeder. Renewables apart from biomass are practically independent of fuel costs. Concerning health effects, hydro has the lowest impact from normal operation followed by nuclear and wind (Fig. 4.10). Coal and current biogas have the highest estimated health effect but there is a high potential for reduction in the case of biogas.

Fossil technologies have the highest expected risks due to severe accidents (Fig. 4.11). New generations of nuclear reactors exhibit very substantial reduction of the risk level compared with the current one. However, the maximum credible consequences of nuclear accidents are clearly highest for nuclear, which leads to high level of risk aversion. Also, large hydro depending on the location has the potential for very large consequences of accidents but without the component of very long-term land contamination, a feature unique for nuclear. In the case of most severe nuclear

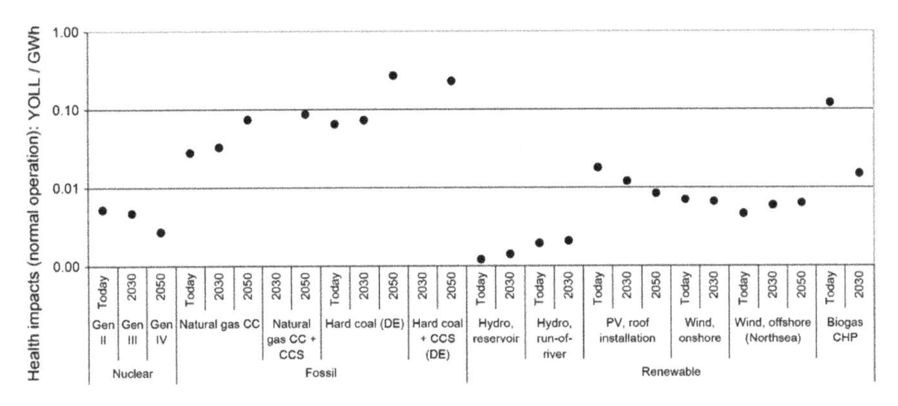

Fig. 4.10 Health effects of normal operation in terms of mortality measured in years of life lost (YOLL) per GWh (Hirschberg and Burgherr 2015)

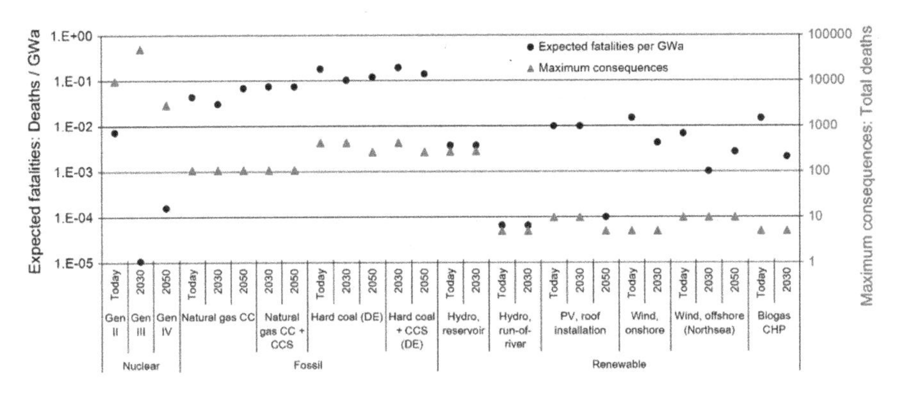

Fig. 4.11 Expected fatality rates due to severe accidents and maximum consequences per accident (Hirschberg and Burgherr 2015)

accidents apart from evacuation of large number of people, a long-term relocation may be necessary.

If desired, aggregation can be based on the total (internal plus external) cost approach or on Multi-Criteria Decision Analysis (MCDA). Costs are called *external*, if they are not born by the party that causes them, but rather by society as a whole. They include the costs of health damage that result from air pollution. Such damages are monetized, that is, are measured in or converted to monetary units, and also include those resulting from future climate change. These are very uncertain today and can vary over a large range. Further aspects are reduced harvests and damages to buildings caused by air pollution. Not all factors that play a role in the evaluation of a technology are amenable to expressing them in monetary units. Doing it may be controversial; above all what concerns are subjective aspects like perceived risks or visual disturbances to the landscape.

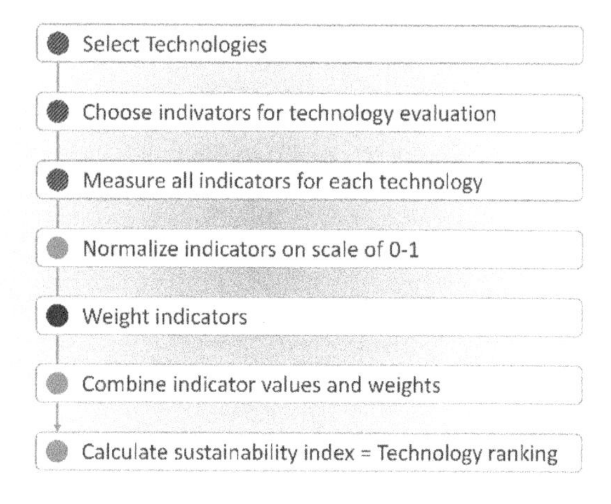

Fig. 4.12 Chart of the multi-criteria analysis process (subjective elements in red and objective steps in blue) (Hirschberg and Burgherr 2015)

In spite of these limitations, external costs are very valuable for cost-benefit analysis. Total cost of nuclear are lowest with the following reservations (mentioned before): (i) current internal costs are based on partially amortized capital costs (this applied also to hydro); (ii) future internal costs depend on smooth implementation of projects without delays caused, for example, by major safety-related modifications introduced during construction; and (iii) the scope of external costs assessment does not include aspects that are difficult or not meaningful to monetize.

MCDA has the capability to explicitly reflect subjective social acceptance issues. The approach builds on the steps shown in Fig. 4.12. Single indicators can already be used individually for technology comparisons. And from them, a single comprehensive index value can be calculated, which reflects how sustainable the individual technologies are compared to each other. Then the indicators are each weighted, based on the individual user preferences. Figure 4.13 shows the average indicator weights assigned by individual European stakeholders in the NEEDS project.

The overview of the results based on all stakeholder responses as elucidated in the NEEDS project is shown in Fig. 4.14 along with total costs.

While within the external cost estimation framework applied in NEEDS nuclear energy exhibits the lowest total costs, its ranking in the MCDA framework tends to be lower, mainly due to consideration of a variety of social aspects not reflected in external costs. Thus, nuclear energy ranks in MCDA mostly lower than renewables, which benefit from much improved economic performance. Coal technologies have mostly lower total costs than natural gas. In the MCDA framework, coal on the other hand performs worse than centralized natural gas options; the latter are in the midfield and have thus ranking comparable to nuclear. The performance of CCS is mixed.

The individual preference profiles have a decisive influence on the MCDA ranking of technologies. Given equal weighting of the environmental, economic, and social dimensions and emphasis on the protection of climate and ecosystems, and minimization of objective risks and affordability for customers, the nuclear options are top ranked.

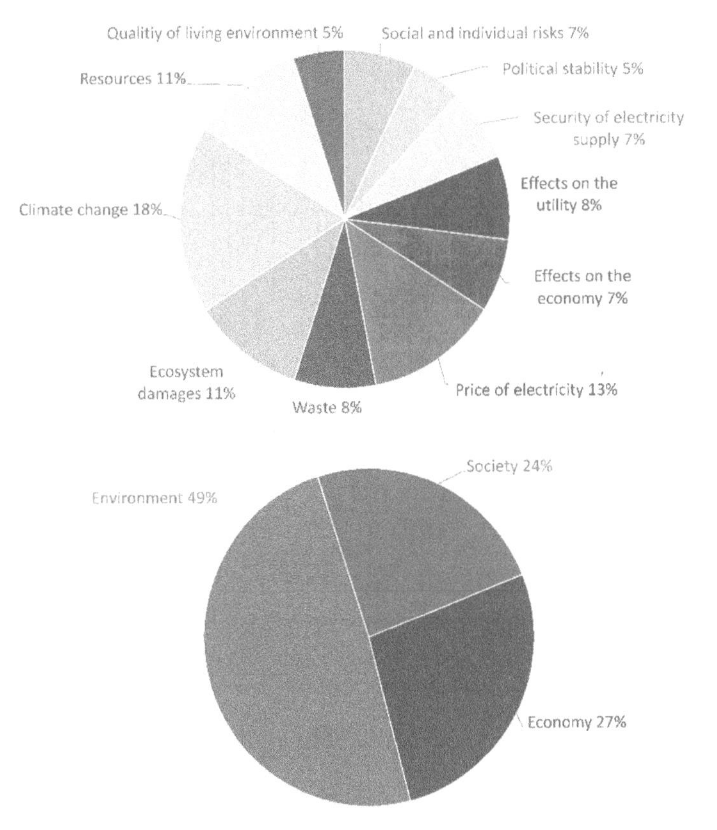

Fig. 4.13 Average indicator weights for technology assessment, obtained via online survey from stakeholders engaged in the European energy sector (not representative of the overall population) (Hirschberg and Burgherr 2015)

On the other side, focusing on radioactive wastes, land contamination due to hypothetical accidents, risk aversion and perception issues, terrorist threat and conflict potential, the ranking changes to the disadvantage of nuclear energy. The ranking of fossil technologies highly depends on the degree of emphasis on the environmental performance, which in relative terms remains to be a weakness, more pronounced for coal than for gas. Renewables show mostly a stable very good performance in terms of relatively low sensitivity to changes in preference profiles, based on highly improved economics.

An important result is that no energy technology is superior to the others in all aspects and meets all sustainability criteria ("no silver bullet solution"). The MCDA approach favors renewables with an almost equal weighting of the three dimensions of sustainability and challenges nuclear technology to make improvements, especially in the social sphere, while fossil fuels are poorly replaced by a stronger emphasis on environmental friendliness. Energy strategies should be based on a "good" mix that can take into account the respective (national) preferences (although taking global

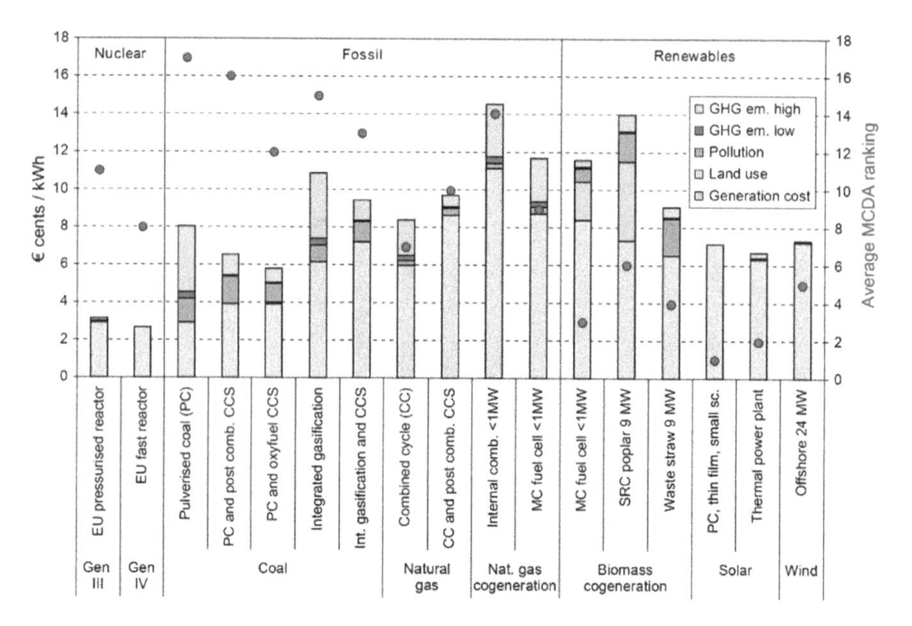

Fig. 4.14 Average MCDA ranking of future (year 2050) technologies compared with total costs. The figure shows a subset of the 26 subsystems evaluated. GHG low/high values represent low and high estimates of damage costs due to climate capture and storage; MC, molten carbonate; PV, photovoltaics (Hirschberg and Burgherr 2015)

aspects into account); in view of considerable technological, political and economic uncertainties, they should follow the imperative of diversification.

In recent years, however, climate protection and the necessity of "mitigation", human intervention to reduce the sources of GHGs or strengthen sinks, have largely asserted themselves as an urgent collective task on a global scale (IPCC 2014b). Due to the dominant role of energy supply, and emissions from the use of fossil fuels in particular, future energy supply strategies, including the transport, industrial heat and housing sectors, should follow the principle of decarbonization.

4.3 Scenario Analysis, Consideration of Technologies

Scenario analyses[3] have been provided by various organizations (OECD-IEA *World Energy Outlook* (WEO) and *Energy Technology Perspectives* (ETP), EU DG Energy (TREN, ENV), European Climate Foundation (ECF), Greenpeace *Energy [R]evolution* (ER), Eurelectric, IPCC etc.) with different time horizons (2030, 2050

[3]Collection of possible developments conditioned by basic conditions and assumptions, not to be equaled with predictions.

and 2100). They differ in regard to their base models (bottom-up, top-down; optimization of given market shares), assumptions about available technologies and their commercial deployment and other standards (costs, increase of energy requirements and the Gross Domestic Product (GDP), availability of infrastructures, market structures, etc.); whereas political framework conditions and goals are assumed to remain unchanged.

Aim of almost all scenario analyses was to demonstrate whether and how climate protection targets (max. increase of global temperature of 2 °C by 2100) and intermediate targets (in 2020, 2030 and 2050) can be reached. The results were in part severely divergent, if not confusing, which is understandable considering the diversity of approaches mentioned above.

Aim of one of these projects (Deutsch et al. 2011) was to compare selected scenario analyses and draw some generalized conclusions. Among the considered analyses (see Table 4.3, left column), the following energy carriers and their associated technologies were given special emphasis[4]:

- *RES* for electricity production and heat provision (Wind offshore, onshore; biomass; solar PV, solar thermal (CSP); geothermics)
- *fossil* [modern coal- and gas power plants with *Carbon Capture & Sequestration* (CCS)]
- *nuclear* [reactors and fuel cycles of generations III or III+ (e.g. EPR)]
- *associated infrastructures* (*smart grids*, *super grids*; new storage technologies).

Table 4.2 summarizes the universalizable results regarding energy requirements (increasing), the role of sectors of consumption (increasing relevance of electricity, including for the domain of transportation) and of different energy carriers/technologies and other aspects, in an overview.

The assumptions about technologies on which prospective strategies of energy (electricity) supply can be built, are consistently conservative. Only sufficiently mature, commercially viable ("deployable") technologies are being considered; "all changing technological quantum leaps" are dismissed even in long term studies. This approach is being shared in this paper.

Tables 4.3, 4.4 and 4.5 include information to technologies considered "available", including the grid, their development and share in demand coverage. These information are being used as well to infer key technologies that are portrayed first in an overview and then in detail in Sect. 4.4.

[4]Greenpeace ER, however, excludes the use of nuclear power and CCS-technology by principal.

4.4 Status and Future Prospects of Key Technologies

4.4.1 Overview

Development trends and challenges to technological development vary from one energy source to another; experience often points to challenges and barriers to the implementation of technically mature developments (e.g. IEA 2015c), which can have a significant impact on the pace of use of new technologies and the speed of change (see also Table 4.1). A prominent example of this is the increase in efficiency (for example in the housing sector), whose greatest potential for saving energy is undisputed, but experience has shown it to be difficult to exploit.

It is also of overriding importance that

- the decarbonization imperative forces the abandonment of fossil fuels, which, together with the provision of modern forms of energy for those who have so far been excluded, cannot be achieved by renewable energies alone (IEA 2016a);
- the transformations in the electricity sector towards large proportions of intermittent renewable energy sources (RES) require corresponding storage capacities that are not yet secured in any way and must be accompanied by an expansion or conversion of the transmission and distribution networks, with a trend towards a decentralized, even cellular structure being observed; furthermore, our thinking and actions must take place in "systems", which has seldom been successful so far;
- all estimates are subject to huge uncertainties (previous statements of the IEA do not agree with reality because the oil price has not risen but fallen, "*shale gas*" turns the USA from a gas importer into a gas exporter with corresponding geo-political effects, the expansion potential of renewable energies has been systematically underestimated (SRU 2011), Fukushima Daiichi has at least considerably thrown back nuclear technology, etc.).

Before individual energy sources and associated technologies are examined in detail with regard to the state of the art and current use as well as their development potential, rough trends are briefly outlined as follows:

- *Solar and wind*: Increasing use and falling costs per unit of energy will continue, albeit at decreasing rates. Increased efficiencies (load factors) and cost reductions are regarded as technological challenges. The dependence on an adequate adaptation of the electricity transmission network and the need for "complementary technologies" (e.g. reserve gas-fired power plants) are generally acknowledged; projects to use solar energy in the sunbelt and to overcome the intermittent nature of electricity generation by *super grids* (such as Desertec) have stalled. The problem of its high demand on land and material, visual and acoustic "*land pollution*" and declining public acceptance is increasingly being addressed (see also Asafu-Adjaye 2015).

Table 4.1 Summary of progress according to 2 °C scenario (2DS) (IEA 2016b)

	Status against 2DS targets in 2025	Policy recommendations
Renewable Power ⇨	With the improving competitiveness of a portfolio of renewable technologies and after the recent policy changes around COP21, the outlook for renewable power is more optimistic. Onshore wind and solar PV are on track to meet their 2DS targets but more effort is needed for other technologies to ensure renewable power is fully on track with 2DS	■ Maintain or introduce further policy support and appropriate market design that enhance competition while sending clear and consistent signals to investors, notably long-term arrangements needed to de-risk investment in capital-intensive technologies ■ Avoid any policy uncertainties – especially retroactive change – that can create higher risk premiums, undermining the competitiveness of renewables ■ Take a holistic approach that maximises the value of a renewables portfolio to the overall system. Countries beginning to deploy variable power plants should implement well-established best practices to avoid integration challenges
Nuclear power ⇳	Long-term policy and financial uncertainty remain for nuclear power, but significant increases in both construction starts and grid connections in 2015 helped make progress towards meeting 2DS targets	■ Policy support is needed to encourage long-term operation of the existing fleet and construction of new plants, given their vital contribution to GHG emissions reductions, as well as their contribution to energy security ■ Incentives such as carbon taxes or electricity market designs providing stable revenues may be required in liberalized markets, which otherwise favor lower-fixed-cost technologies
Natural gas-fired power ⇳	The recent fall in coal prices relative to natural gas and falling electricity demand have curtailed growth in capacity and generation, making it difficult to gauge if natural gas-fired power would reach its 2DS potential	■ To encourage coal-to-gas switching, electricity market incentives, such as carbon prices and more stringent regulation of plant emissions, are needed ■ Electricity market mechanisms are needed that recognize natural gas-fired power's operational flexibility to support the integration of variable renewables

(continued)

- *Water, biomass*: Their use has fallen short of expectations for a variety of reasons; there are limits in principle, e.g. nature conservation and public resistance.
- *Geothermal energy:* Is in the development stage with high exploration and drilling risks (4–6 km deep drilling is necessary in unfavorable regions); high investments in *Research, Development & Demonstration (RD&D)* are necessary and currently

Table 4.1 (continued)

Coal-fired power ⇔	Despite a slowdown in coal consumption globally, the projected trajectory of emissions reduction from coal is not on track to meet 2DS projections	■ Where coal-fired capacity is expanding, policy measures are required to ensure assessment of the full range of lower-carbon generation options to satisfy new capacity. ■ Generation from the less-efficient subcritical coal units should be phased out and new coal-fired units should have efficiencies consistent with global best practice – currently supercritical or ultra-supercritical technology
CCS ⇙	Moderate progress in CCS was made in 2015. Significant investment in projects and technology development by industry and governments are needed to get CCS on track to meet the 2025 target of 541 million tons of carbon dioxide (CO2) stored per year	■ New projects need to be proposed and supported from development to operation to ensure a growing stream of projects through the development pipeline. ■ Investment in storage resource development will de-risk projects and shorten the development time. Storage characterization and assessment are often the longest aspect of project development and outside the skill base of CO2 capture project developers. ■ Continued research and development of CO2 capture technologies are needed, including innovative technologies to reduce the costs and operational penalty of CO2 capture

Recent trends: ⇙ negative developments, ⇔ limited developments, ⇗ positive developments

Improvement, but more effort needed

not on track

questionable. Induced light earthquakes have set back drilling and development projects in some countries, e.g. Switzerland.

- *Coal*: The share of electricity generation (and related CO_2 emissions) has increased (currently 70% in China). For technical and economic reasons, coal has made a major contribution to meeting the (increased) demand for electricity; in some countries it has again displaced the more environmentally friendly gas from the reserve and base load range. Plants can ("*clean technology*") and will have to become even more efficient and less CO_2-emitting in the future. The equipment with so-called Carbon Capture and Sequestration (CCS) technologies would comply with the requirement of decarbonization; although the individual elements have been developed, CCS use has fallen short of expectations/plans; demonstration plants that have been decided upon have been abandoned or stagnated; breakthroughs in the use of CCS are questionable at least in the near future due to high costs, lack of acceptance, limited storage space or doubts about the achievable necessary tightness (HIS CERA 2011), etc.

Table 4.2 Interpretation of the results of the comparative scenario study (excerpt from Deutsch et al. 2011)

- Without new policy measures, energy demand will increase due to *GDP growth*
- *Electrification* occurs in (almost) all scenarios. Electricity is estimated to gain higher shares in final energy demand, especially in scenarios confronted with ambitious GHG-targets
- In *road transport*, the use of hybrid cars and electric vehicles is generally increasing towards 2050
- The *role of nuclear power* is generally an outcome of the bounds and investment cost assumptions set by the scenario developers. Without (major) restrictions, nuclear power tends to expand and gain increasing shares (especially worldwide
- Nuclear power is estimated to gain higher *importance in scenarios with ambitious GHG-targets* (unless the development is restricted exogenously)
- Scenarios expect future nuclear power plants up to 2050 to be dominated by *generation III or III+ reactors*
- Absolute and relative *increase of renewables (RES)* in the power sector
- Investment *costs for RES decrease* significantly in the scenarios, especially in the long term
- *Wind power, solar heat and solar PV* as well as *biomass* show the main contributions in the deployment of RES
- *CSS* plays an increasing role in several scenarios (from 2030) and can be seen as the second main power generation technology to provide base load power
- In scenarios applying optimization models, development of *carbon prices* is found to be crucial for the emergence of CCS
- Special attention is generally given to *gas-fired power plants*, which are estimated to serve as bridging technology and *dispatchable capacity* in a range of the considered scenarios, but with far lower utilization rates than currently observed
- Studies with ambitious emission reduction targets assume technological progress in *grid technology and management* and estimate that large increases in transmission capacities are needed
- Transmission extension induces large *additional capital costs* in the power sector
- Almost all studies emphasize the relevance of technologically advanced *smart grids and smart metering frameworks*, especially those confronted with ambitious emission reduction
- The *linkage of grid (sub-) models* and the development of generation technologies remains rather unclear in the studies analyzed

- *Gas*: Due to new gas deposits ("*shale gas*" in the USA), the flexibility of the power plants and the greater environmental friendliness compared to coal, the use of gas is constantly highly attractive and already accounts for a large share of the energy/electricity supply. Further technological developments to increase efficiency, adaptability and CCS equipment are necessary and recognized as bridge technology towards decarbonization (OPCC 2014) or in the pipeline.
- *Nuclear*: The use of nuclear energy has stagnated or, since 1993, accounted for a slightly declining share of global electricity generation, recently reaching around 11%. Attitudes towards the use of nuclear energy vary greatly from country to country: Germany, Belgium and Switzerland with phaseout decisions, some countries (re-)entered—Great Britain as a prominent example, France and Japan with

Table 4.3 Development estimated in selected technologies and sectors (Deutsch et al. 2011)

Study and scenario	CCS	Road transport	Other new power/energy technologies
WEO 450 ppm	Significant share of CCS in coal plants (depending on political framework)	EV and plug-in hybrids 2nd gen. biofuels from 2015	CSP in Northern Africa and Southern Europe No breakthrough
ETP Blue Map	CCS in power generation, fuel transformation and industrial production Large shares of coal plants use CCS (global: 90%)	By 2050: 70% of new vehicles are electric 2nd gen. biofuels for trucks, ships, aircraft Hydrogen after 2030	CSP-plants slightly develop in Europe
EU DG TREN Reference	Demonstrational CCS plants, but no large deployment after 2020	Hybrid vehicles and biofuels, but no EV	n/a
EU DG-ENV NSAT	CCS-technology mature until 2030	Biofuels in transportation increases towards 2030	Geothermal, tide/wave develop at a slow pace
ECF Pathways	CCS for coal/gas plants beyond 2020, CCS-retrofit	Mix of EV, biofuels and hydrogen: 20% EV (2020)	No breakthroughs
E[R] Advanced	CCS power plants not considered	Final energy share of EV: 14% (2030), 62% (2050) Increasing role of hydrogen Biomass in stat. applications	Hydrogen for industrial heat CSP, geothermal, ocean energy
Euroelectriv Power Choices	CCS: pilot projects by 2020, rising storage and transport costs	80–90% of road transport electrified by 2050 (PIHV and EV)	No consideration of new power technologies

reoriented energy strategies, China, Russia and India with massive expansion programs, etc. At the end of 2016, 449 reactors were in operation in 31 countries, there were 60 new construction projects (most of them in China); many existing plants are striving toward extending their operating lives (87 of the 99 plants in the USA have been granted an extension) or are about to be shut down, partly for political or economic reasons or because of expensive safety upgrades.

Newer third generation plants (e.g. EPR) with improved protection against catastrophic accidents are available and, in Europe with considerable and in China with lesser difficulties, are being built and put into operation with considerable delays; further improved plant designs and fuel cycles (fourth generation plants)

Table 4.4 Comparison of grid-development estimated in the studies (Deutsch et al. 2011)

Study and scenario	Grid development	Characteristics of future power grids	Smart grids
WEO 450 ppm	Lower global grid development compared to the reference (lower power demand)	n/a	Smart-grids seen as important for EV-deployment
ETP Blue Map	n/a	Crucial role for new grid technology (e.g. HVDC-lines)	Smart grids seen as an important factor
EU DG TREN Reference	n/a	Development in grid technology and management	n/a
EU DG-ENV NSAT	Reinforcement of power grids needed	n/a	n/a
ECF Pathways	Up to 170 GW of interregional capacity (significant expansion between FR and SP) Expansion required in the next 5–10 years	Future transmission mix: 73% AC, 27% DC, 67% overhead, 33% underground	Smart grids seen as a critical technology
E[R] Advanced	58.3 GW strengthened HVAC interconnection; 27.56 GW strengthened HVDC connections	Combination of smart grids, micro grids and efficient large-scale super grids (HVDC) High importance for transmission lines from north and south to Central Europe	Crucial role of smart grids
Euroelectriv Power Choices	C 40% interconnection capacity needed 40 (of 241 lines overall) new transmission lines	n/a	Crucial role for the timeline of the implementation of smart grids

are under development. In addition to increased plant and proliferation safety and cost-effectiveness, *public acceptance and financing* and the solution of the repository problem are regarded as the greatest challenges. Radically different concepts, including fusion reactors, are being researched, but are unlikely to play a commercial role in the coming decades. Recently, there has been a noticeable increase in interest in small to medium power modular reactors (Sornette et al. 2019).

Table 4.5 Main electricity and heat technologies available in the alternative scenarios (Deutsch et al. 2011)

Study and scenario	Conventional (heat and power)	RES-E	RES-H	Focus
WEO 450 ppm	Supercritical coal plants from 2020 Natural gas (power 2025) Nuclear Power, Gen. III	Wind (offshore after 2020, onshore before 2020)	Biomass, solar thermal, geothermal	Wind Nuclear Natural gas Biomass
ETP Blue Map	Coal-plants (CCS) Gas CHP, NGCC and NGOC (incl. CCS) Nuclear Power, Gen. III	Wind (first onshore, then offshore), biomass, solar PV/CSP (after 2020)	Heat pumps, solar thermal, biomass (CHP)	Wind Nuclear Coal CCS
EU DG TREN Reference	Growth of NGCC until 2015 Supercritical coal plants Nuclear Power, Gen. III	Wind (offshore after 2020), biomass and solar PV	Biomass, heat pumps, solar	Natural gas Nuclear Wind Oil
EU DG-ENV NSAT	New clean coal technologies by 2030 Nuclear Power, Gen. III	Wind (30% offshore) and biomass	Biomass/waste	Wind Nuclear Biomass Natural gas
ECF Pathways	Nuclear Power, Gen. III Equal generation shares for gas and coal (CCS)	Wind (offshore after 2020) and biomass	Biomass, heat pumps	Wind Solar Nuclear CCS
E[R] Advanced	Gas (as transition fuel) Coal-fired plants with lifetimes of 20a	Decentralized power generation: wind, solar PV, CSP, geothermal	Biomass, solar thermal, geothermal	Wind Solar Geothermal
Euroelectriv Power Choices	Nuclear Power, Gen. III Peak for gas-fired power generation: 2025	Wind (offshore after 2020), biomass, solar PV	Conventional biomass, solar, geothermal	Nuclear Wind CCS Natural gas
FEEM-WITCH	Nuclear Power, Gen. III Coal IGCC-CCS Natural gas	Wind and solar power + breakthrough technologies	Breakthrough technology	Nuclear Wind Breakthrough

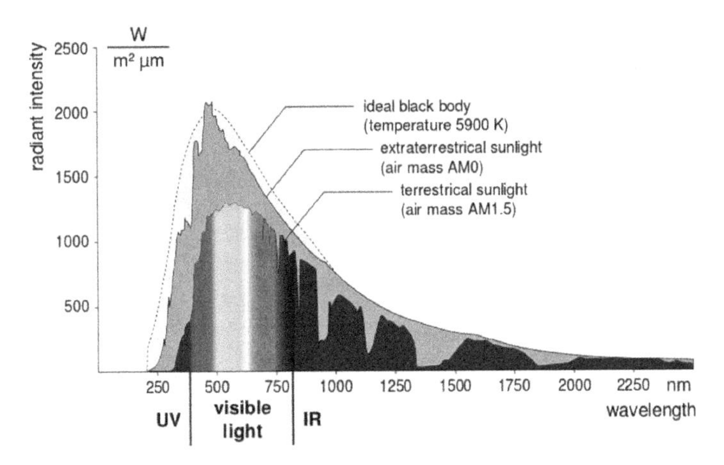

Fig. 4.15 Intensity at direct solar radiation (solar spectrum) as a function of wavelength. https://commons.wikimedia.org/wiki/File:Sonne_Strahlungsintensitaet.svg. The original uploader was Degreen at German Wikipedia. Improved Baba66 (opt Perhelion) on request; En. translation Locusta Fr. translation Eric Bajart Nl. translation BoH (https://commons.wikimedia.org/wiki/File:Sonne_Strahlungsintensitaet.svg), "Sonne Strahlungsintensitaet", https://creativecommons.org/licenses/by-sa/2.0/de/legalcode

4.4.2 Solar Power

Solar energy is a virtually unlimited source of energy in the long term and is available to such an extent that the total solar radiation of 1 h could cover global energy consumption for a whole year.[5]

Figure 4.15 shows the spectrum of the sunlight, the area under the curves corresponds to the intensities.[6] This shows that the majority of the energy is contained in the visible range (40%) and in infrared radiation (50%). Since the radiation available on earth fluctuates strongly throughout the day and the whole year, averaged values are used. Without consideration of the atmosphere and meteorological effects, the average intensity corresponds to 342 W per m^2 Earth's surface. In reality, this mean value is even smaller, since not all available radiation reaches the earth's surface because of shielding, absorption and scattering effects in the atmosphere.

Depending on the technology, different measured values are used to calculate the resulting yield as accurately as possible: For photovoltaics (PV) the *Global Horizontal Irradiance* and for solar thermal (CSP, ibid.) the *Direct Normal Irradiance*.

[5]The theoretical potential of global solar radiation in general corresponds to 3.9×10^6 EJ/a. The technological potential (depending on the available surface area and efficiency) is converted into 1338–14,778 EJ/a for PV and 248–10,791 EJ/a for CSP. Global annual energy consumption in 2014 was around 570 EJ (IEA 2015).

[6]The intensity (power per area) and thus the energy of the solar radiation depends on the wavelength (the smaller the higher the energy) as well as the number of photons incident per area.

- *Global Horizontal Irradiance* (GHI) is the total incidence of radiation on a surface horizontal to the Earth's surface. This is made up of the proportion of radiation that strikes directly without atmospheric losses due to scattering or absorption and the proportion of indirect radiation that results from scattering by air molecules, aerosols or other particles.
- The *Direct Normal Irradiance* (DNI) corresponds to the component of radiation that directly hits the earth without atmospheric losses. In contrast to the GHI, the irradiation on a surface is measured perpendicular to the radiation, which is why the DNI can assume higher maximum values compared to the GHI.

However, the influence of the geographical location on the possible energy production must not be neglected, since the yield increases proportionally to the incoming radiation. This can vary up to a factor of 3 for photovoltaics and even up to a factor of 10 for CSP (see Figs. 4.16 and 4.17). This is why in northern Germany, for example, twice the area of PV cells is needed to generate the same output as in southern Spain. The graphs also show that the DNI, in contrast to the GHI, also depends strongly on the meteorological conditions at the same latitude; evident in China or the USA, where this results in larger differences between East and West.

Solar thermal (Concentrated Solar Power)

Concentrated solar energy provides a virtually unlimited source of clean, non-polluting, high-temperature heat. Solar thermal and thermochemical approaches utilize the entire solar spectrum, and as such provide a favorable path to solar power and fuels production with high energy conversion efficiencies and, consequently, economic competitiveness. The industrial implementation of concentrated

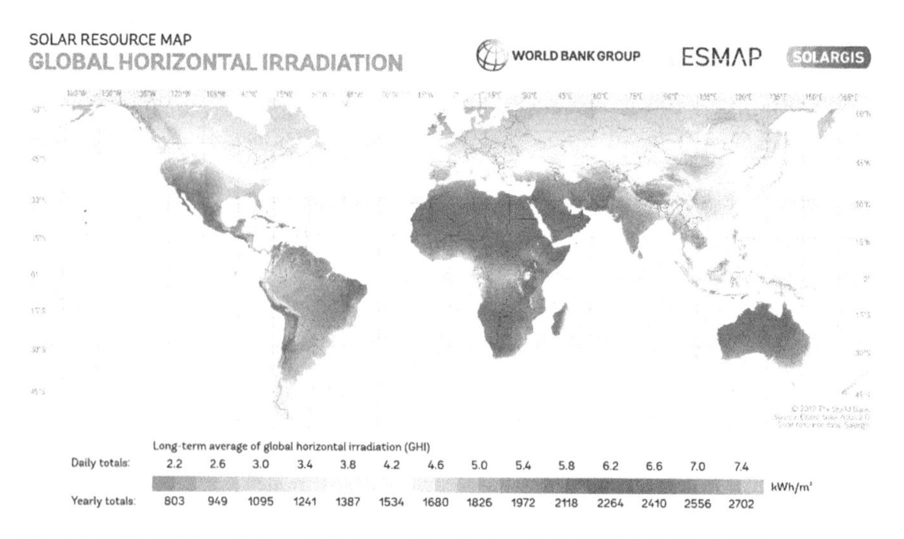

Fig. 4.16 World Map of Global Horizontal Irradiation. Period 1994–2015. © 2017 The World Bank, solar resource data: Solargis. URL: https://globalsolaratlas.info/download/world (accessed 13-Dec-2019)

solar power (CSP) is rapidly increasing, with commercial projects already totalizing several GW. Because CSP is restricted to direct normal solar irradiation, the sunbelt region (±40°) is mainly considered for its application. Four solar concentrating technologies are currently applied at commercial CSP plants, see Fig. 4.18. Current technologies are based on solar receivers that operate with thermal oil or water/steam at temperatures usually below 500 °C, coupled to steam-based "Rankine cycles" for power generation at 20% peak efficiency. Representative solar-to-electric conversion efficiencies and annual capacity factors are listed in Table 4.6. Industry roadmaps anticipate a 60% cost reduction of CSP by 2025.

CSP plants can dispatch power round-the-clock by incorporating a thermal storage system and over-sizing the solar field accordingly. Thermal energy storage systems store excess sensible heat collected by the solar field and, alone or in combination with fossil fuel backup, keep the plant running under full-load conditions. This storage capability enables penetration into the bulk electricity market where substitution of intermediate-load power plants of about 4000–5000 h per year is achieved.

Novel receiver concepts based on volumetric absorption of directly irradiated porous structures, particles, and alternative thermal fluids, operate at high temperatures/high fluxes and promise more efficient solar energy capture and conversions. Moreover, these advance concepts can be applied for the thermochemical production of solar fuels (Romero and Steinfeld 2012) the evolution of CSP is depicted in Fig. 4.19.

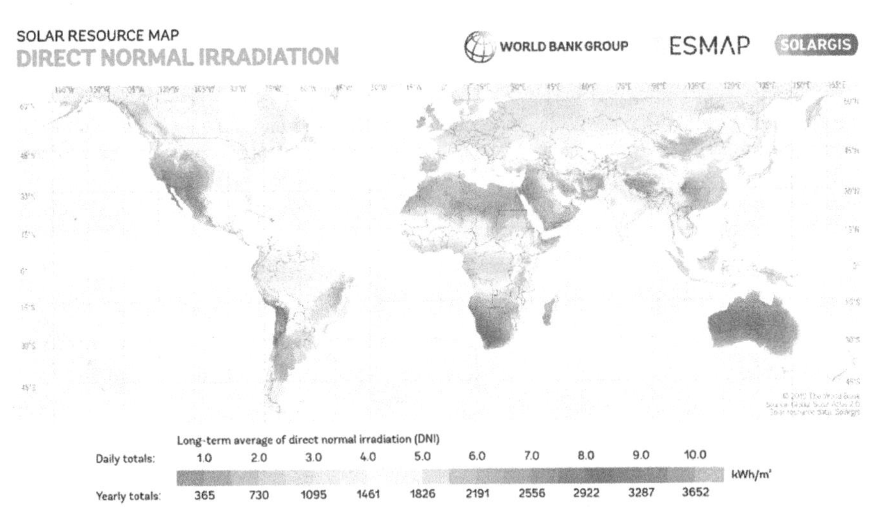

Fig. 4.17 World Map of Direct Normal Irradiation. Period 1994–2015. © 2017 The World Bank, solar resource data: Solargis. URL: https://globalsolaratlas.info/download/world (accessed 13-Dec-2019)

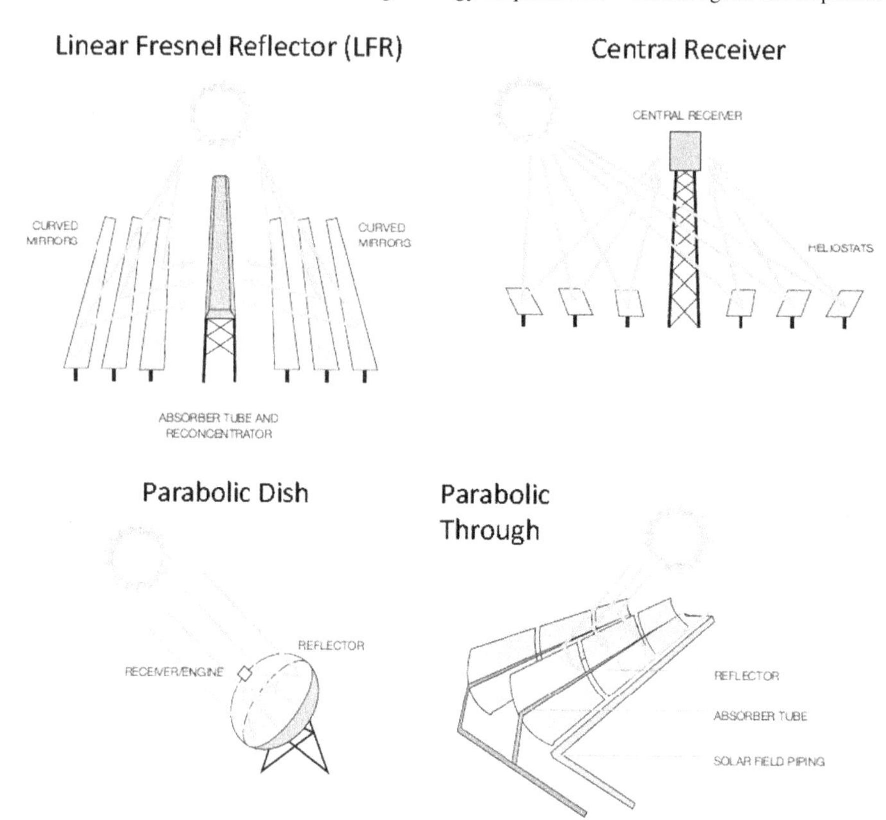

Fig. 4.18 Schematics of the four solar concentrating technologies currently applied at commercial concentrated solar power (CSP) plants. © Greenpeace International. https://energypedia.info/wiki/Concentrating_Solar_Power_(CSP)_-_Technology (accessed 13-Dec-2019)

Photovoltaics

Photovoltaics (PV) is regarded as one of the most promising sustainable energies (IPCC 2011) and has the potential to contribute significantly to energy supply in virtually every country, although the influence of the geographical location and local meteorological conditions must not be neglected.

In contrast to traditional power generation technologies, PV causes costs and CO_2 emissions during the production and installation of the plants, apart from the costs for cleaning, maintenance and disassembly (waste disposal/recycling). In order to generate economically and ecologically competitive electricity across the board using PV, however, there are still hurdles to overcome when it comes to implementation.

In photovoltaic cells, solar radiation is converted directly into electrical direct current by means of a photoelectric effect. Semiconductors (p-n-doped) are usually used as the basic material. If the p-n junction (space charge zone) of a semiconductor is hit by photons (whose energy is greater than the bandgap energy), electrons are

Table 4.6 Characteristics of CSP systems (Romero and Steinfeld 2012)

	Parabolic troughs	Central receiver	Dish/engine
Power unit	30–80 MW[a]	10–200 MW[a]	5–25 kW
Temperature operation (°C)	390	565	750
Annual capacity factor (%)	23–50[a]	20–77[a]	25
Peak efficiency (%)	20	23	29.4
Net annual efficiency (%)	11–16[a]	7–20[a]	12–25[a]
Commercial status	Mature	Early projects	Prototypes demos
Technology risk	Low	Medium	High
Thermal storage	Limited	Yes	Batteries
Hybrid schemes	Yes	Yes	Yes
Cost W installed			
$/W	3.49–2.34[a]	3.83–2.16[a]	11.00–1.14[a]
$/Wpeak[b]	3.49–1.13[a]	2.09–0.78[a]	11.00–0.96[a]

[a]Data interval for the period 2010–2025
[b]Without thermal storage

excited from the valence band into the conduction band.[7] This creates free charge carriers (electrons and holes), which are separated by the electric field prevailing in the space charge zone and dissipated via contacts to the load. A PV module is a serial or parallel connection of individual PV cells that are mechanically protected from environmental influences and together with other modules, any inverters and feed controllers, form the photovoltaic system (see Fig. 4.20).

Thanks to rapidly falling costs, PV systems have begun to play an important role in power generation in many countries. Meanwhile, even non-subsidized PV systems are becoming increasingly competitive, with electricity production costs in various parts of the world already at the level of fossil fuels. These developments have led to massive growth in the photovoltaic market in recent years (see Fig. 4.21). In 2015 alone, global capacity was increased by 50 GWp[8] to a total of 228 GWp, equivalent

[7]A PV cell with a p-n junction can only use a certain part of the spectrum of sunlight, i.e. not all occurring photons have a suitable energy to excite a valence electron. If one wants to use as large a part of the spectrum as possible (and thus increase the efficiency), several semiconductor layers with different band gaps can be stacked on top of each other (see Multijunction Cells).

[8]W_p (Watt peak) refers to the peak output of a solar system under standard test conditions (STC) and maximum solar radiation.

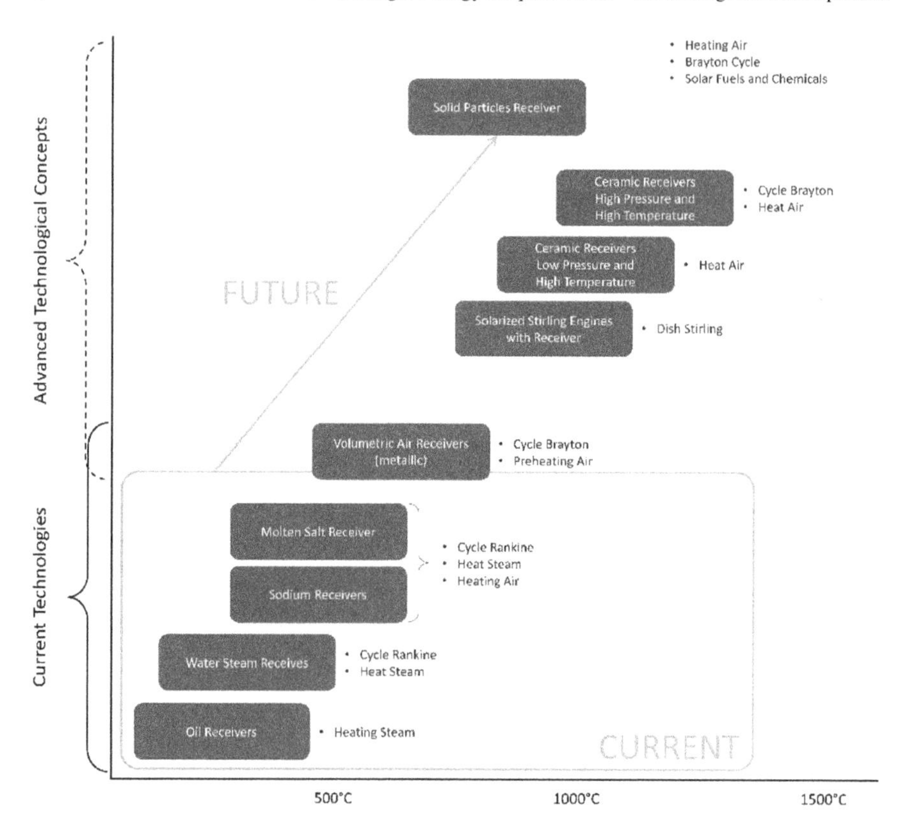

Fig. 4.19 Evolution of the use of CSP. The next generation of technologies allows surpassing 1000 °C and enables higher efficiencies via Brayton and combined cycles, as well as the thermochemical production of solar fuels (Romero and Steinfeld 2012)

to growth of just under 30%. This increase was mainly attributable to individual countries that have more than doubled their capacity in recent years,[9] led by China (+13 GWp) and Japan (+11 GWp), followed by the USA (+7 GWp) and the UK (+4 GWp). In terms of total capacity, China (43 GWp) has overtaken Germany (39.7 GWp), which is practically stagnating, in terms of total capacity and, together with Japan (34 GWp) and probably the USA (26 GWp), will clearly differentiate itself.

PV systems can be roughly divided into feed-in types (see Box 1). Only *on-grid systems* are of economic relevance for PV development in Western countries with good electricity connections. The systems can either be *distributed* or *centralized and* connected to the public power grid. In heavily populated countries such as Germany, Japan or Switzerland, *distributed systems* dominate, which are *rooftopped* or *structure-integrated. The* construction of *centralized systems is made more* difficult because their land requirements are in conflict with other uses, or because reasons

[9]Growth in Asia, America and Africa is offset by a decline in growth in Europe, "due to reductions in policy support and retroactive taxes" (REN21 2015:60).

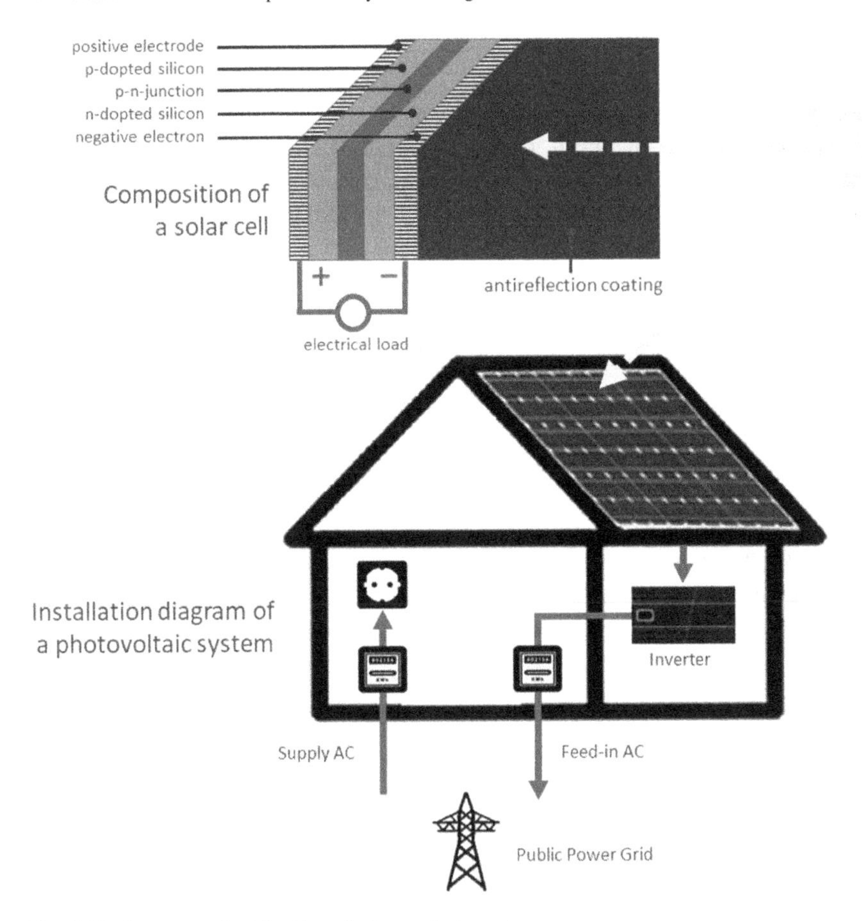

Fig. 4.20 Structure of a (distributed) photovoltaic system

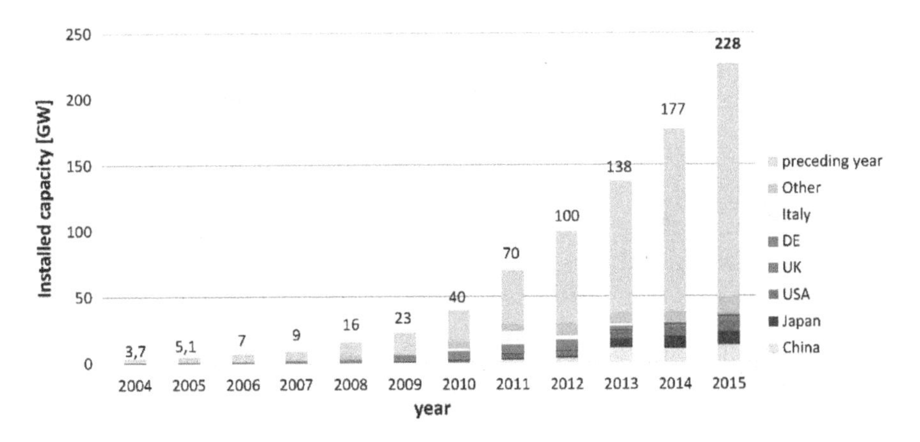

Fig. 4.21 Worldwide installed PV capacity (2004–2014) (REN21 2015)

for landscape conservation might argue against it. In countries such as the USA or China, which have a large proportion of deserts or unpopulated barren areas, *centralized systems are* already important and have great potential; in 2014 in China alone a capacity of over 8 GWp (20% of the worldwide capacity and 80% of the capacity installed by China) was installed using *large-scale power plants. In* terms of plant size, China (two plants above 1000 MWp) and the USA (four plants above 500 MWp) are the leaders.

In contrast, off-grid plants have great potential in developing countries without comprehensive electricity supply and access to electricity for the population. Estimates predict that more than three billion people will be supplied with electricity via *off-grid* systems by 2030 (Greenpeace/EPIA 2008).

In addition to the manufacturing and installation costs for a PV module, the efficiency of the PV cell is an important factor when comparing different technologies. The degree of efficiency indicates which proportion of the incoming solar energy is used or converted into electricity. A distinction must be made between the realized efficiency of commercial cells and that of laboratory cells under standard test conditions (STC). The latter can be up to 10% higher than series production. In general, a theoretical upper limit is set for the efficiency of cells with only one p-n junction.[10] Figure A.1 in Appendix A shows the development of efficiency rates for selected technologies over the last 40 years.

Box 1 Definitions from (Mearns 2014)

Off-grid PV systems are power supply systems without a connection to the public grid, also known as stand-alone operation; the electrical power generated must be temporarily stored and reused on site using an extra medium.

On-grid PV systems are connected to the public grid. A distinction is made between two types:

Distributed PV systems (rooftop, structure-integrated) supply either a connected customer or the public grid directly. The system can be installed on a private property (residential), on public or commercial buildings or on infrastructure elements such as motorway noise barriers. Typical sizes of these plants range from a few kW (residential) to a few MW (rooftop).

Centralized PV systems (power plants) generate electricity in the style of a power plant, which is fed directly into the public grid. They are rich area installations, therefore they are area intensive. The installed capacity of a plant normally exceeds 1 MWp.

The most important technologies today *are wafer-based solar cells* and *thin-film cells*; Table 4.7 contains a list of the most important cell types.

---

[10]The "*Shockley-Queisser limit*", which depends on the band gap energy of the semiconductor (maximum 34% at 1.34 eV).

Table 4.7 List of key features of wafer-based and thin film cells

	Commercial efficiency[a] (%)	Laboratory efficiency[b] (%)	Market share production 2014[c] (GWP)	Energy payback time[c] (a)	Degradation rate modules median[d] (%/a)	Lifecycle GHG emission median[e] CO_2-eq/kWh (g)
m-Si	15–18	25	16.9 (35.6%)	1.8/3.3	0.36	75
p-Si	13–16	20.8	26.2 (55.1%)	1.2/2.1	0.64	55
a-Si	5–7	13.6	0.8 (1.7%)	1.3/2.4	0.87	25
CIGS	12–14	21.7	1.7 (3.6%)	0.8/1.5	0.96	–
CdTe	11–12	21.5	1.9 (4.0%)	1/1.8	0.4	20

[a]Buechler (2012)
[b]NREL (cf. Fig. A.1 in Appendix A)
[c]Fh-ISE (2015)
[d]Jordan and Kurtz (2013)
[e]Edenhofer et al. (2011)
The measuring points of the GHI data are Italy (Sicily; 1925 kWh/m^2) and Germany (1000 kWh/m^2) (de Wild-Scholten 2013)

Wafer-based solar cells made of crystalline silicon have the highest market shares (more than 90% of production capacity by 2014 (Fh-ISE 2015) as a long-established and advanced technology. As is usual in the semiconductor industry, the silicon is melted into ingots, sawn into slices (wafers) and then further processed into solar cells. The technology is characterized by a high commercial efficiency, a good availability of the raw material Si and longevity. However, the minimum thickness of crystalline Si cells is limited to approx. 160 μm for reasons of mechanical stability and production limits. Therefore, a large amount of Si is required as the basic material, which in combination with high process temperatures leads to high energy consumption and considerable production costs. Further disadvantages resulting from the use of wafers are the fragility/inflexibility as well as a limited size of the individual cells.

With a cell efficiency of 15–18% in series production, *monocrystalline Si* (m-Si) is the leader in this segment. The laboratory efficiency has remained practically unchanged at 25% for 20 years and, limited by the theoretical upper limit, can no longer increase significantly. The challenge is to increase the quality (efficiency) of series production to a similar level as in research laboratories. The production of m-Si cells is very energy- and cost-intensive, which is reflected in a high energy recovery time and lifecycle GHG emission.

The production of *polycrystalline Si* (p-Si) is simpler compared to monocrystalline Si and requires less energy and therefore also less financial expenditure. However, the crystal lattice of p-Si has more impurities and irregularities, so the efficiency of 13–16% (commercial) and 20.8% (laboratory) is lower than that of m-Si. This disadvantage is offset by reduced module costs, which is why polycrystalline cells now have the largest market share. Another advantage of more efficient production is the shorter energy recovery time and lifecycle GHG emission.

Thin film solar cells differ from crystalline Si cells in a completely different manufacturing process. In contrast to crystalline silicon, the functional layers of amorphous silicon or compound semiconductors are applied directly onto a (often transparent) low-cost carrier material such as glass or plastic film. The thickness of the photon-absorbing semiconductor layer can be reduced to approx. 5 μm which greatly reduces the amount of semiconductor material required. In contrast to wafer-based cells, no high-temperature processes are required ($T_{max} < 700$ °C), which further reduces energy consumption and thus CO_2 emissions during production. The new production technology compared to wafer cells allows a simpler and more extensive production and installation of thin film cells as well as smaller power losses at high operating temperatures. Due to the possibility of plastic foils as carrier material, the individual cells can be made lighter, more flexible and larger in area, which can lead to many new application possibilities for *rooftop* or *structure-integrated* systems. The most striking disadvantage of thin-film cells is their reduced efficiency compared to wafer-based cells.

Compared to crystalline Si cells, PV cells made of *amorphous silicon* (a-Si) have a low-cost and raw material-saving manufacturing method and correspondingly low module prices, lifecycle GHG emissions and energy return times. They are characterized by improved low-light behavior in diffuse light. The major drawback of this technology is the low efficiency of commercial (5–7%) and laboratory cells (13.6%) and a light-induced reduction in efficiency after commissioning.

Cells made of *copper indium gallium diselenide* (CIGS) have the highest commercial (12–14%) and research (21.7%) efficiencies but also the highest production costs of all thin-film cells (costs comparable to m-Si). On the other hand, the energy recovery time is significantly lower than that of wafer-based cells. In the future, the shortage of the heavy metal indium could have an impact on production volumes and prices. CIGS-based cells are very interesting structurally, as they are applied not only to glass substrates but also to plastic films, resulting in flexible ultra-light PV cells.

Compared to CIGS, *cadmium telluride* (CdTe) cells have a lower commercial efficiency (11–12%) with virtually identical laboratory values (21.5%). This is mainly due to the fact that CdTe cells have shown a strong growth in laboratory efficiency over the last 5 years of about 5%, which has not yet been transferred to commercial cells. Advantages of this technology lie in the production. Disadvantages are possible bottlenecks in tellurium extraction and the toxicity of cadmium.

In addition to the dominant technologies, there are cell types that are already commercially available but are only used in niche markets and *emerging technologies* that could become important for the PV market in the future. As a parallel development to the PV cells, research is being conducted on the *Concentrated Photovoltaic*, a combination of CSP and PV.

In order to increase the amount of photons absorbed and thus the efficiency of a solar cell, *Multijunction Cells (stack cells)* layer different p-n semiconductor transitions on top of each other with decreasing bandgap energy. This allows a larger part of the radiation spectrum to be captured (see Fig. 4.22) and the theoretical upper limit to be overcome. These cells therefore have very high efficiencies (30–40% in

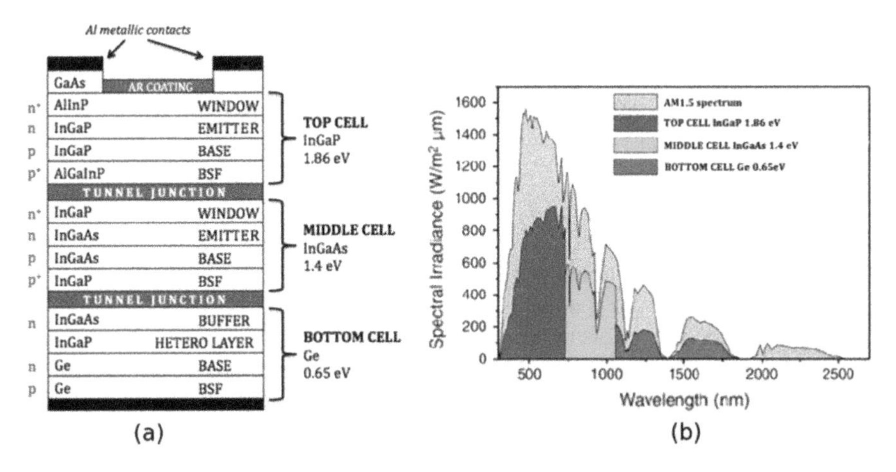

Fig. 4.22 Structure and absorbed spectrum of a multijunction cell. Fh-ISE; Ncouniot (https://commons.wikimedia.org/wiki/File:StructureMJetspectre.png), "StructureMJetspectre", https://creativecommons.org/licenses/by-sa/3.0/legalcode

the laboratory); due to the correspondingly high costs, these cells have so far only been used where costs are of secondary importance (e.g. in space flight).

One *emerging technology* is *Organic PV Cells* (OPV), which consist of high-molecular crystalline hydrocarbon compounds with electrical properties similar to those of inorganic semiconductors. The efficiency for commercial cells is currently still very low at less than 3%; in the laboratory 11.5% is achieved. Furthermore, the inadequate long-term stability is a sticking point of this technology. Possible advantages would be low material costs and energy consumption during production as well as the mechanical properties of the plastic (flexibility, transparency and easy handling), which provide possibilities for easy integration in building technology or in everyday objects such as clothes or backpacks.

The so-called *Pervoskite Cells* rely on a simple and resource-poor production and have been achieving new record laboratory efficiencies for almost 3 years (currently at 20.1%; growth limited by the theoretical upper limit). However, there are still a number of obstacles in the way of market viability (cell stability, environmental compatibility of the materials).

The *Concentrated PV* (CPV) combines PV cells with mirrors or lenses to concentrate a large amount of radiation on a smaller area (up to 500 times the intensity of the sun) to save the required cell area and thus semiconductor material. Unlike normal PV cells, the CPV can only use direct solar radiation (DNI) to focus them at the focal point of the optics, where a highly efficient (often multijunction) PV cell is located. Therefore, the range of application of this technology (as with the CSP) is reduced to regions with high direct solar radiation and little cloud coverage. The PV module must also be equipped with a good *tracking system* that aligns the CPV cells to the position of the sun. As shown in Fig. A.1 in Appendix A, CPV can massively increase the efficiency of individual cells. The CPV is making continuous progress in cell and module efficiency but is struggling to keep up with the falling prices of conventional PV.

Even in conservative scenarios, a rapidly rising growth trend in PV is assumed for the future outlook. Depending on the scenario, total installed PV capacity is expected to reach 540 GWp by 2020 and up to 4.5 TWp by 2050 (approx. 6000 TWh of generated energy and 16% of global electricity supply), depending on the scenario (see Fig. 4.23).

Despite the great potential and impressive progress in the development of technologies, there are a number of hurdles to overcome in order for photovoltaics to reach its projected potential by 2050.

Although photovoltaics is no longer the most expensive form of electricity generation, the decisive factor for further growth will be the extent to which *module and system costs* decrease. This reduction depends on research into new and optimization of old production and PV technologies, the demand for PV modules (installed capacity) and the cost development for the overall system (especially inverters).

The lifecycle GHG emission (between 25 and 75 g CO_2-eq/kWh), which is already one order of magnitude smaller than for fossil fuels, must be further reduced in order to improve the "*attractiveness*" of PV. This can be achieved by reducing energy

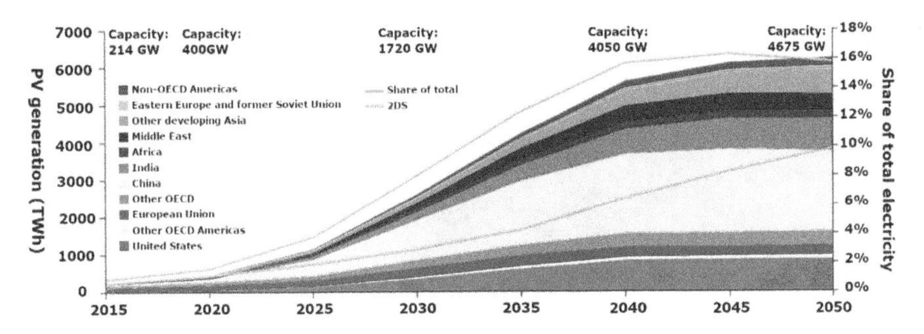

Fig. 4.23 Outlook for regional electricity production (TWh) and installed capacity from solar PV (IEA 2014a: 60)

consumption during production and increasing the sustainability of[11] the energy required. Other aspects are the environmental compatibility and rarity of the materials required and the problem of recycling PV cells, the impact of which can only be assessed in the next decade, when the first generation of large-scale solar cells will have to be disposed of.

With the growth of PV, there are also new demands on the *infrastructure of* the public electricity grid (see Sect. 4.4.13). In order not to slow down the progress of PV unnecessarily, the development of the remaining elements of electricity supply (*grid, storage, control*) must keep pace. Furthermore, the public acceptance (keyword: land pollution) of PV is a topic that can be mitigated by focusing on buildings or structure-integrated systems (away from open space systems).

It is not assumed that—despite the enormous progress made, especially in the last decade—developments will occur in the next 50 years that could completely turn the whole concept upside down. In general, the actual costs of each technology will continue to determine market share. It can be assumed, however, that thin-film cells, despite current stagnation, will generate a larger market share in the future because energy savings in production and good integration in building structures represent a major advantage over wafer-based technologies; in addition to cost development, the successful development and implementation of *emerging technologies* will also depend on the extent to which the production steps from the laboratory can be scaled up to series production.

4.4.3 Wind Power

The use of the kinetic energy of the wind for various purposes has a long tradition in the history of mankind; the first windmills for electricity supply date back to the end

[11]The majority of the PV modules are produced in Asia. First and foremost is China (60%), which provides the large amounts of energy largely through coal-fired power plants.

Top 10 Cumulative Capacity Dec 2015

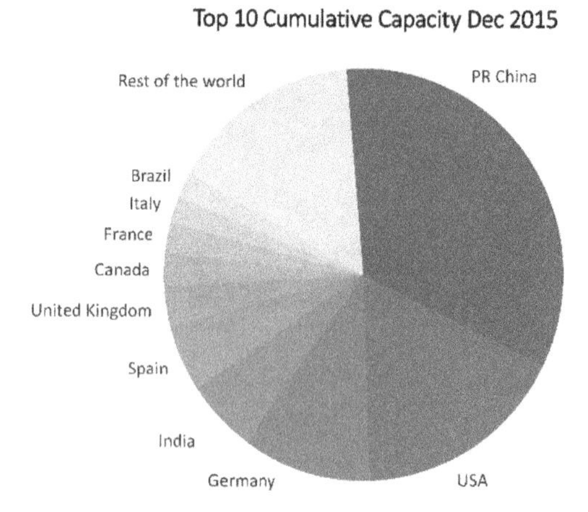

Country	MW	% Share
PR China	145,362	33.6
USA	74,471	17.2
Germany	44,947	10.4
India	25,088	5.8
Spain	23,025	5.3
United Kingdom	13,603	3.1
Canada	11,205	2.6
France	10,358	2.4
Italy	8,958	2.1
Brazil	8,715	2.0
Rest of the world	67,151	15.5
Total Top 10	**365,732**	**84,5**
World Total	**432,883**	**100**

Fig. 4.24 Top 10 cumulative capacity as of December 2015 (GWEC 2016)

of the 1880s.[12] The massive expansion of wind power began at the beginning of our century, fueled by political measures including a legal obligation to purchase wind power produced and fixed feed-in tariffs.[13] At the end of 2015, wind turbines with a rated output of just under 433 GW were installed worldwide, with an increase of 63 GW in 2015, equivalent to an increase of around 22%. China is the undisputed leader in terms of new capacity and total capacity, the EU states together rank third, with Germany contributing about half of all installed systems (GWEC 2016), see also Fig. 4.24). Worldwide, wind turbines supplied about 706 TWh of electrical energy, corresponding to about 3% of the electricity demand with an average capacity utilization of just under 26%.[14]

Land based (onshore) wind power plants at favorable locations can now produce electricity at production costs[15] below those of new hard coal, gas and nuclear power plants; only lignite-fired power plants are still cheaper; offshore plants are significantly more expensive, despite higher full load hours, by a factor of about 1.7–2 (Fh-ISE 2018).

From a technical point of view, wind turbines can be used in all climate zones, at sea and at all land locations.[16] Today, onshore locations dominate; offshore wind farms, on the other hand, lag well behind, accounting for just under 3% of installed

[12]James Blyth, a Scot from Scotland, built the first wind turbine in 1887 to generate electricity to charge the batteries used to light his holiday home.

[13]In Germany in 2013 according to the "Erneuerbare Energien Gesetz" (EEG §29 Abs. 2) for at least 5 years 8.8 ct/kWh for land-based plants; thereafter the basic remuneration is reduced to 4.8 ct/kWh, both decrease annually by 1.5%.

[14]In 2019, in the EU with 192 GW installed and an electricity production of 192 GWh, wind energy accounted for 15% of total electricity production, corresponding to 26% in Germany.

[15]These fluctuate between 4.5 and 10.7 Euro cent/kWh (as of 2013 (Fh-ISE 2015), depending on the location.

[16]For wind resource data cf. e.g. www.globalwindatlas.com/map.html (accessed 13-Dec-2019).

capacity. Worldwide, ground-based wind energy offers a physical potential that is probably higher than the current global primary energy demand of about 570 EJ (equivalent to an average output of 18 TW). Today, only a fraction of this potential is used. Geophysical and ecological repercussions of the use of wind energy to completely cover global demand are not to be expected, but effects are documented on scales of up to approx. 100 km in the atmospheric state as well as in the maritime milieu (Carpenter et al. 2016).

Wind power gives variable power which is very consistent from year to year, but which has significant variation over shorter time scales. It is therefore used in conjunction with other electric power sources to give a reliable supply. As the proportion of wind power in a region increases, a need to upgrade the grid, and a lowered ability to supplant conventional production can occur. Power management techniques such as having excess capacity, geographically distributed turbines, dispatchable backing sources (gas powered plants), sufficient hydroelectric power, exporting and importing power to neighboring areas, using vehicle-to-grid strategies or reducing demand when wind production is low, can in many cases overcome these problems.

The dominant design by far is the three-bladed wind turbine with horizontal axis and rotor on the windward side, the so-called resistance rotor, whose nacelle is mounted on a tower and actively follows the wind direction. Other designs, so-called Leerunner, in which the rotor runs behind the tower and active wind tracking can be omitted, or "buoyancy rotors" with vertical rotation axis and also without active wind tracking, have not been able to assert themselves at least as large turbines.

The achievable output of a wind turbine increases with the third power of the wind speed and also depends on flow losses and technical efficiencies. Accordingly, there is a strong dependence on the weather both for the planning and for the actually achievable yield, the most[17] accurate forecast of which is also indispensable for the integration of the generated electricity into the electrical grid. The typical output of current turbines is 2–5 MW for onshore and 3.6–8 MW for offshore. In an effort to achieve ever lower electricity production costs, wind turbines are gradually growing in size (see Fig. 4.25): since the rotor blade length is squarely integrated into the power-determining rotor surface, it has meanwhile risen to over 50 m and with it the hub (tower) height up to 130 m; modern low wind turbines now have rotor diameters of 130 m and hub heights of up to 150 m.

For cost reasons, manufacturers are increasingly focusing on series production, standardization and modularization as well as the transportation of complete plant components to the site. The turbines switch on at a typical wind speed of 3–4 m/s and switch off at a speed of 15 m/s (storm); the high variability of the electricity generated by the wind is important in terms of control and grid technology.

[17] A distinction has to be made between the electrical construction-related nominal output (GW installed) multiplied by the theoretical maximum running time (8760 h) and the actual output and running time, i.e. the electricity generated (GWh); the resulting "capacity utilization" in the EU was 25.3% in Germany at 22.3%.

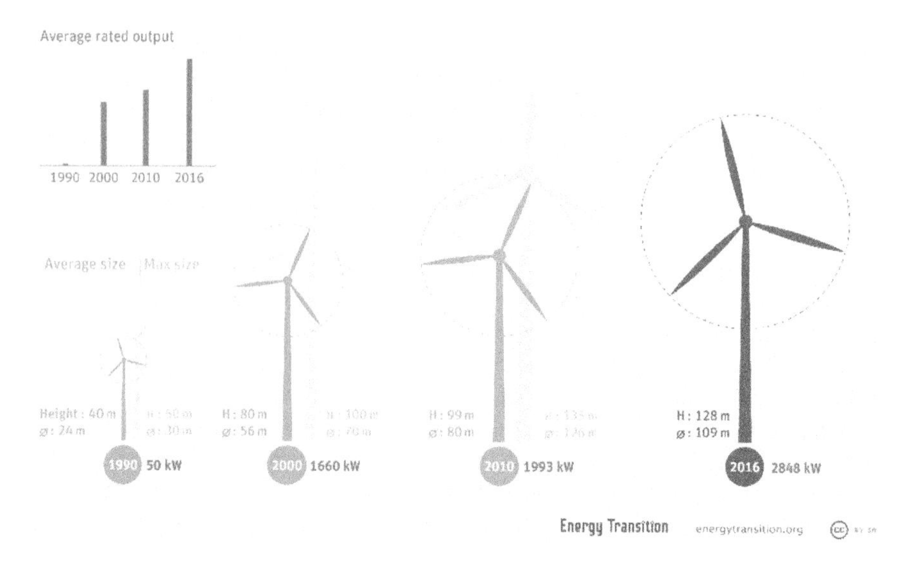

Fig. 4.25 Development in size and power of wind turbines, 1990–2016. © energy-transition.org. URL: http://wiki.energytransition.org/the-book/technology-for-sustainability/wind-power/ (accessed: 13-Dec-2019)

Wind energy is considered to be environmentally friendly, since the emissions of greenhouse gases and other pollutants associated with its technical use, as well as the demand for water, are low; the area occupied by the wind turbine is also advantageously small[18] at around 4000 m^2 (foundation area about 300–500 m^2). Disadvantages are noise emissions and the killing of birds caused as well as optical impairments and shadows. The use of wind energy is very safe in view of the occurrence of serious accidents. Accidents, usually far away from settlements during storm phases, do not result in personal injury; an exception are accidents at work during assembly and maintenance.

Wind turbines have recently developed rapidly in terms of technology and can be used competitively in good locations. Nevertheless, it is a relatively new technology, which still has considerable uncertainties, but which can be expected to make great progress (Herbert et al. 2014). The central starting point for the associated research and development is the further reduction of power generation costs through (1) more efficient production techniques, (2) rotor blades with ever longer blade lengths using lighter and more stable materials, with new profiles and "intelligent" design, and (3) turbines with only two rotor blades. In addition, the development of less windy regions through special weak wind turbines is moving into the focus of manufacturers and research.

[18] According to the Potential Atlas of Germany, wind energy can cover 20% of Germany's electricity requirements in 2020 on 0.75% of the land area (AEE 2010).

4.4.4 Geothermal Energy

The use of extracted geothermal energy can take place either directly, for example in the heat market, or indirectly, to generate electricity or in a combined heat and power system. Geothermal energy is a practically unlimited energy source in the long term[19]; its classification depends on the available temperature levels:

- *High enthalpy deposits*: Geological heat anomalies with several hundred degrees Celsius hot fluids (water/steam), strongly correlated with volcanic regions. The hot fluid can be used to provide industrial steam, to feed heat networks or to generate electricity using a steam turbine. The techniques are existing and mature and subject to further improvement. The oldest power plant with a steam turbine and a capacity of 750 MWel, built in 1913, is located in Tuscany.
- *Low enthalpy-deposits*: As a rule, deep boreholes are necessary in order to enable economically viable uses (T > 100 °C). There are three different types of heat extraction: hydrothermal systems, petrothermal (HDR) systems and deep geothermal probes (see also Sect. 4.4.8). While hydrothermal systems are being planned, for example in Germany, HDR processes are being tested in pilot projects. Geothermal probes consist of a borehole more than 1000 m deep with a coaxial tube in which the fluid (mostly water) circulates. The output of such closed systems is several hundred kilowatts; they are in the testing phase. Alternatively, probes with direct evaporators (heat pipes or *heat pipes*) have been proposed, which use liquids or mixtures with a correspondingly low boiling point as the working medium and enable higher extraction rates.
- *Near-surface geothermal energy*: This includes the use of geothermal energy up to a depth of approx. 400 m, for which properties would in principle be suitable. The geothermal heat is used by means of collectors, geothermal probes, energy piles or heat well systems; the extracted geothermal heat is then transported via piping systems with a circulating liquid, which is usually connected to a heat pump. The described system can also be used inexpensively (without heat pump) for cooling purposes.
- *Geothermal energy from tunnels and mining facilities*: To generate thermal energy from tunnel structures, escaping tunnel water is also used, which would otherwise have to be temporarily stored in cooling basins for environmental reasons before it can be discharged into local waters. The first such known facility was put into operation in Switzerland in 1979 at the south portal of the Gotthard road tunnel. It supplies the Airolo motorway works yard with heat and cold. In the meantime, further systems have been added which mainly use hot water from railway tunnels. At the north portal of the Gotthard Base Tunnel, tunnel water with temperatures between 30 and 34 °C is leaking out and is to be used via a district heating network. Closed mines and depleted natural gas deposits are conceivable projects for deep geothermal energy. This also applies to a limited extent to deep tunnels.

[19] With the reserves stored in the upper 3 km of the Earth's crust, the current global energy demand could in principle be met over 100,000 years.

Table 4.8 Correlation between volcanic activity and theoretically usable geothermal energy in volcanic regions (Stefansson 2005)

Country	Number of volcanoes	Theoretical continuous power (MWel)
USA	133	23,000
Japan	100	20,000
Indonesia	126	16,000
Philippines	53	6000
Mexico	35	6000
Iceland	33	5800
New Zealand	19	3650
Italy (Tuscany)	3	700

Depending on the depth of the deposit, the formation water there is between 60 and 120 °C hot; the boreholes or shafts are often still present and could be reused to supply the warm deposit water for geothermal purposes. Pilot plants are located in Heerlen, Czeladź, Zagorje ob Savi, Burgas, Novoshaktinsk in Russia and Hunosa near Oviedo.

Electricity generation from geothermal energy is traditionally concentrated in countries with near-surface high-enthalpy deposits (mostly volcanic or *hot-spot areas*, see Table 4.8). In countries that do not have this, the electricity must be generated at a comparatively low temperature level (low enthalpy deposit with about 100–150 °C), or it must be drilled correspondingly deeper. Newly developed *Organic Rankine* (ORC) *plants* enable the use of temperatures from 80 °C for power generation. They work with an organic medium (e.g. pentane) that evaporates at relatively low temperatures.

Geothermal energy is a renewable energy with a high global potential, but it has to contend with development risks (deep drilling), major uncertainties (exploration risks), lack of competitiveness and environmental risks; it is on the rise worldwide, Table A.6 in Appendix A gives an overview of projects in Central Europe.

4.4.5 Hydropower

Hydroelectric power stations convert the kinetic energy of water into mechanical or electrical energy using turbines on watercourses, dams or the sea, the latter when using currents and tides. In 2018, the installed capacity of hydropower plants worldwide was 1292 GW. With 4200 TWh of electrical energy produced, hydropower covers 16.5% of the world's electricity needs (IHA 2019). This corresponds to ¾ electricity generation from renewable sources. The share of electricity production varies from country to country (see Table 4.9) and reaches its highest levels in

Table 4.9 Ten of the largest hydropower producers in 2019

Country	Annual Production (TWh)	Installed capacity (GW)	Capacity factor	Percentage of total electricity production in that country
PR China	1064	311	0.37	18.7
Canada	383	76	0.59	58.3
Brazil	373	89	0.56	63.2
USA	282	102	0.42	6.5
Russia	177	51	0.42	16.7
India	132	40	0.43	10.2
Norway	129	31	0.49	96
Japan	87	50	0.37	8.4
Venezuela	87	15	0.67	68.3
France	69	25	0.46	12.2

Data from Wikipedia, https://en.wikipedia.org/wiki/Hydroelectricity?oldformat=true (access: 13-Dec-2019)

Norway and Brazil; in the People's Republic of China alone, hydropower plants of around 32 GW are under construction or planned.

The type designation depends on the height difference of the water level in front of and behind the turbine:

- *Low-pressure power plants* with drop heights of approx. 15 m are often located in the middle course of rivers or as tidal or wave power plants by the sea.
- *Medium pressure power plants* with heads between 25 and 400 m require low dams (dams or higher weirs on rivers).
- *High-pressure power plants* with heads of more than 250–400 m with dams in low and high mountain ranges.

The turbines (Pelton or Francis turbines at high pressures, see Fig. 4.26) and generators enable efficiencies of up to 90% and thus have only a small improvement potential. The largest hydropower plants have nominal capacities of up to 12.6 GW (Itaipú) and 18.2 GW (Three Gorges); 18 plants alone have capacities of more than 5 GW (see Table A.7 in Appendix A) (Fig. 4.27).

Dam hydropower plants can also be equipped and operated as pumped storage plants and serve as electricity storage facilities (see Sect. 4.4.2.11). The potentials of hydropower result from topographical conditions and the justifiability of considerable environmental interventions as well as the high investment requirements; in countries such as Switzerland and Germany it is estimated to be rather low (Akademien der Wissenschaften Schweiz 2012).

A special type of hydroelectric power plant is a *tidal power plant that* functions according to the dam principle. They are built at sea bays and in estuaries (river estuaries) which have a particularly high tidal range (at least 5 m difference between

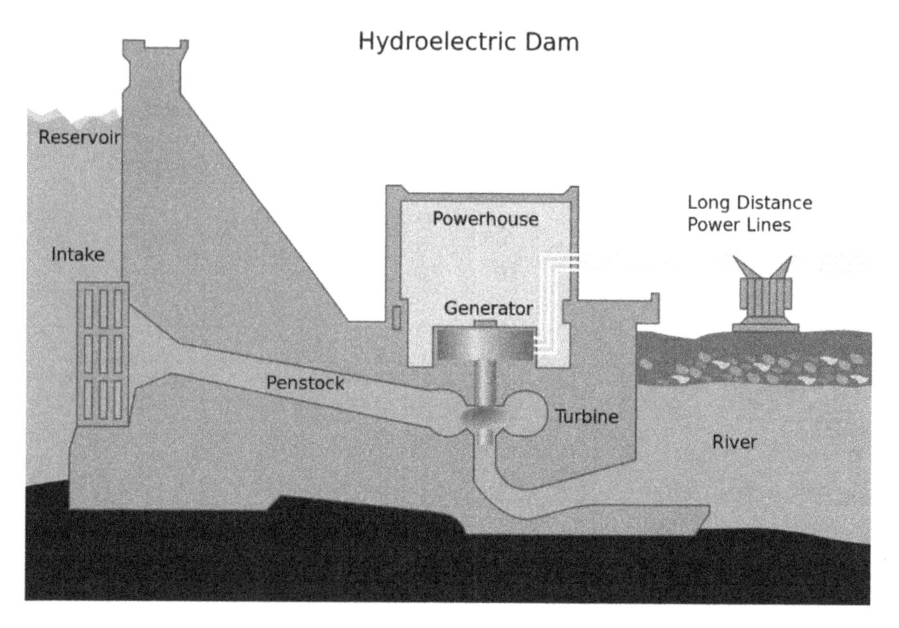

Fig. 4.26 Schematic view of a hydropower plant with Francis turbine. © Tennessee Valley Authority; SVG version by Tomia (https://commons.wikimedia.org/wiki/File:Hydroelectric_dam.svg), "Hydroelectric dam", https://creativecommons.org/licenses/by-sa/3.0/legalcode

high and low water level). In order for this to be effective, the bay in question is dammed up with a dike in which water turbines are operated, which at high tide are operated by the incoming water, at low tide by the outgoing water. This is achieved by changing the rotor blades. So that even a small water gradient can be used to generate electricity, so-called bulb turbines are used, the best known of which is the Kaplan turbine. A tidal power plant of this type can also use excess electricity from other power plants to pump seawater into the reservoir, and generate additional electricity during the subsequent backflow, thus acting as a pumped storage power plant. It is estimated that there is a potential for technologically harvestable tidal energy resource at about 1 Terawatt worldwide (IRENA 2014).

There is a total of seven, most smaller, tidal power plants worldwide with a total capacity of just over 520 MW. For a long time, the largest, with 240 MW, was built in 1961 on the Atlantic coast in the mouth of the Rance in France and opened in 1966. The tidal range in the bay near St. Malo is normally 12 m, sometimes 16 m. The concrete dam is 750 m long, creating a reservoir with a surface area of 22 km^2 and a useful volume of 184 million m^3. The entire plant supplies around 600 GWh of electrical energy annually (corresponding to a load factor of approx. 13%) and also operates as a pumped storage power plant. In 2011, Sihwa-ho, South Korea's largest tidal power plant, was completed 40 km southwest of Seoul with a total of 254 MW, including a 13 km dam separating a natural bay from the Yellow Sea. Of the plants seriously being planned, the successor to the Severn Barrage project with 320 MW, which was abandoned for economic and ecological reasons, the much

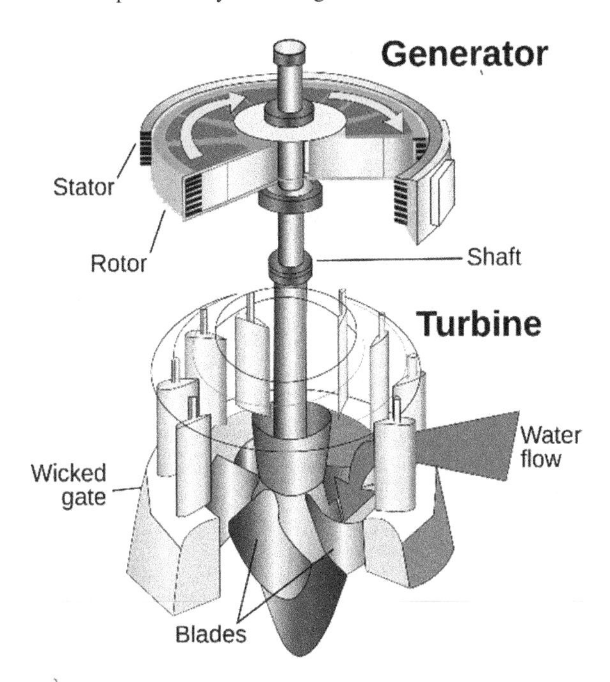

Fig. 4.27 Structure of a turbine. © U.S. Army Corps of Engineers (Vector image: Gothika, Edit: Bammesk) (https://commons.wikimedia.org/wiki/File:Water_turbine_-_edit1.svg), "Water turbine—edit1", marked as public domain, more details on Wikimedia Commons: https://commons.wikimedia.org/wiki/Template:PD-US

smaller Swansea Bay tidal power plant is worth mentioning. It is to be built in a bay on the north coast of the Bristol Canal; after approval by the Ministry of Energy, construction planning began in 2015.

4.4.6 Biomass

Biomass is defined as organic, renewable raw materials in which solar energy has been chemically bound by photosynthesis and all resulting secondary and by-products, residues or waste. Mainly used are renewable raw materials such as wood, agricultural products or organic residues. Depending on the use, processing operations of varying complexity are carried out and the intermediate products are either directly converted (in solid, liquid or gaseous form) into heat for buildings or industry, used for electricity generation or converted into gaseous or liquid fuels for transport (see Fig. 4.28 for an overview). This diversity and variable plant sizes make biomass a form of renewable energy with the highest fuel and usage flexibility. Depending on the technology and materials used, the increased pollution of the air during combustion and the conflict over land use during the cultivation of raw materials have a negative impact.

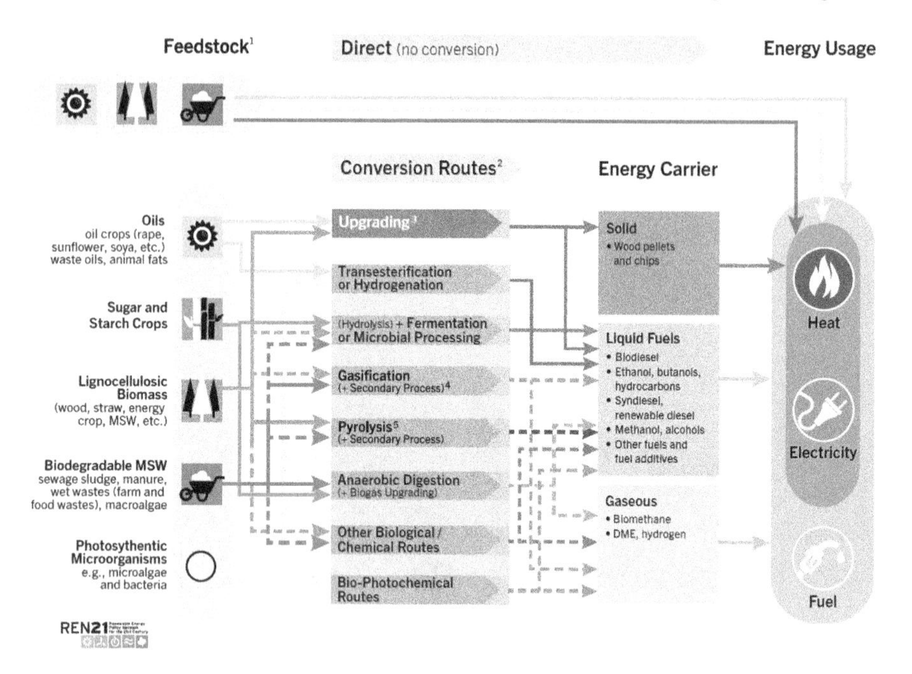

Fig. 4.28 Bioenergy conversion pathways (REN21 2015)

In 2014, energy produced from biomass accounted for almost 14% of global primary energy consumption and is therefore the form of renewable energy with the largest market share, with the main share of this energy (approx. 90%) being used for heating, as shown in Fig. 4.29. Solid fuels account for the largest share of the biomass used for heat and electricity, while liquid fuel is the most important source

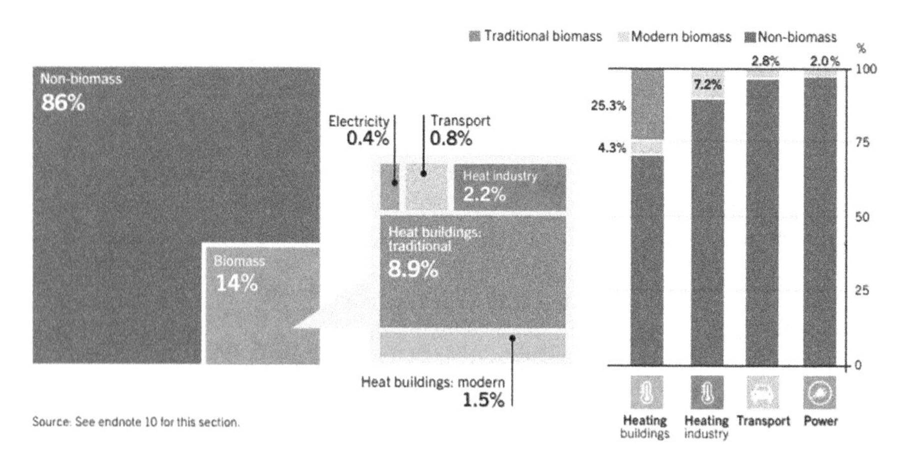

Fig. 4.29 Shares of biomass in total final energy consumption and in final energy consumption by end-use sector in 2014 (REN21 2016)

of biomass in the transport sector. There is a total of 106 GW of installed capacity worldwide for electricity production and 305 GW for heat generation, with Europe, the USA, Brazil and China accounting for the largest share. The technical potential of biomass to meet global energy demand is seen as considerable, but also limited. Depending on the scenario, a technical potential of 300–500 EJ/a is estimated, of which probably 100–300 EJ could actually be used in 2050 (AEE 2014).

Since biomass is the only resource that can be used to generate energy that is available locally virtually anywhere in the world, it is of great importance to the population of poor and technologically poorly developed countries. More than 60% [31 EJ in 2015 (REN21 2015)] of the global biomass energy turnover takes place in developing countries. This *traditional* use of biomass involves burning firewood or charcoal in simple, often inefficient devices to provide heat for cooking or heating. Over 2.5 billion people worldwide (80% of the population in sub-Saharan Africa, 50% in Asian developing countries and 14% in Latin America (REN21 2016) are currently dependent on the traditional use of biomass for cooking, with a slight upward trend. The poor, incomplete combustion in the existing primitive furnaces can cause strong pollutant emissions with health damages for the users directly on site and also has a very low efficiency of only 10–20%. Technological solutions to remedy this situation exist, but have not yet been implemented on a large scale.

In contrast, the *modern* use of biomass—often in thermal plants, which are usually operated in identical or similar form for fossil fuels—is the opposite. This includes the combustion of solid, liquid or gaseous biomass to produce heat for domestic or industrial use and/or electricity. In addition, there is the use of liquid or gaseous biomass as fuel.

For an efficient and low-emission use of biomass in modern plants, it first undergoes a number of different processing steps, which can be subdivided as follows according to Kaltschmitt and Thrän (2009):

- *Thermal conversion*: Drying and *palletization* compresses and dewatered biomass, thus increasing combustion efficiency and energy density. *Torrefaction* heats woody biomass in the absence of an oxidizing agent, further increasing energy density compared to pellets and approaching the quality of coal.
- *Thermo-chemical conversion*: Solid biofuels are sub-stoichiometrically mixed with an oxidizing agent (e.g. air or water) and heated under controlled conditions. These include processes such as gasification (gaseous), pyrolysis (liquid) or carbonization (solid), which differ methodically mainly in the composition of the resulting secondary biofuel.
- *Physico-Chemical Conversion*: This process is used to obtain liquid biofuels from biomass containing oil or fat. The oil is mechanically separated from the solid phase for this purpose.
- *Bio-Chemical Conversion*: In this method, the biomass is broken down into its components by microorganisms, e.g. by fermentation.

A general evaluation and comparison of the different conversion methods is difficult, since the suitability of a method often depends on the specific application, as can be seen from the *life cycle* GHG emissions for biofuels. *This mainly*

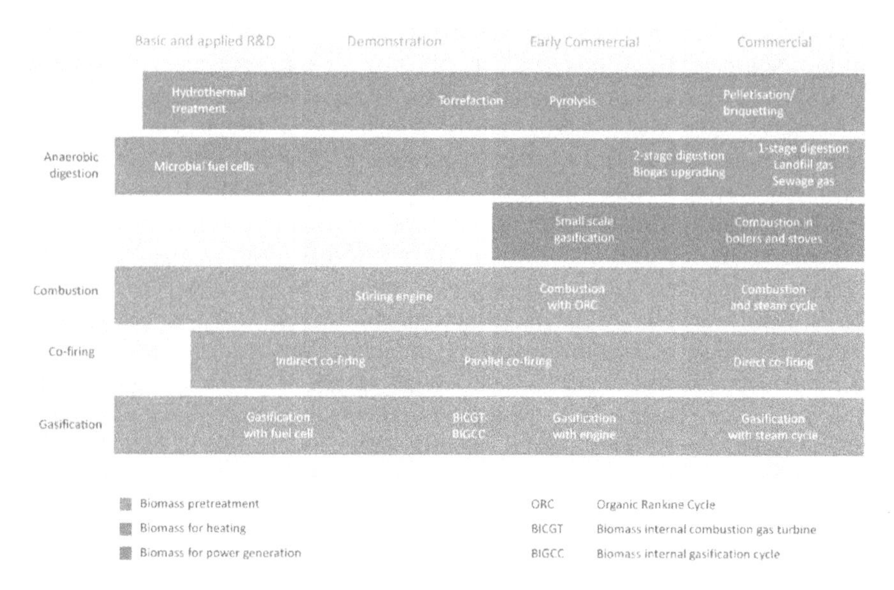

Fig. 4.30 Overview of current status of conversion and combustion technologies (IEA 2012)

depends on how energy-intensive the cultivation of the raw material is and how complex the transport and conversion are.[20] *Therefore, closed, decentralized cycles in which domestic energy crops are used efficiently are optimal, which favors different processes depending on regional conditions* (AEE 2009).

The efficient *use of biomass to generate heat and electricity is* experiencing strong growth worldwide [almost 4% capacity growth in 2015 (REN21 2016)]. However, the long-term development potential is strongly dependent on the availability of biomass, on political decisions and on crude oil and commodity prices, which account for 40–50% of energy costs (IEA-ETSAP and IRENA 2015). The further development and commercialization of existing technologies and the development of new processes for efficient combustion and environmentally friendly conversion are also decisive for the future generation of heat and electricity from biomass. An overview of the status of today's (2012) technologies can be found in Fig. 4.30.

Biofuels, i.e. liquid and gaseous fuels derived from biomass, can play an important role in reducing CO_2 emissions in the transport sector, and enhancing energy security. By 2050, biofuels could provide 27% of total transport fuel and contribute in particular to the replacement of diesel, kerosene and jet fuel. The projected use of biofuels could avoid around 2.1 Gt of CO_2 emissions per year when produced sustainably. To meet this vision, most conventional biofuel technologies need to improve conversion efficiency, cost and overall sustainability. With a sound policy framework in place,

[20] In contrast to fossil fuels, GHG emissions from biogas plants are about one order of magnitude smaller but still significantly higher than those from other renewable energies or nuclear power; however, in combination with CCS (see Sect. 4.4.9), biomass power plants could even act as CO_2 sinks.

it should be possible to source the required biomass from residues and wastes, along with sustainably grown energy crops. Most biofuels are supposed to be competitive by 2030, unless their production costs are coupled to strongly to oil prices (IEA 2011).

Independent of the specific application of biomass, it plays a multi-dimensional role which is one of its unique characteristics. It is currently utilized for food, feed, fiber and energy supply, all using the same land for its production. It also supports different types of ecological aims, including biodiversity, greenhouse gas (GHG) emission reductions and landscape development. When evaluating the feasibility of a certain application for energy production, the whole set of criteria has to be considered and weighted accordingly since biomass utilization can dynamically change those relationships and produce either positive or negative impacts, both locally and globally. Therefore, one of the main challenges related to use of biomass is the management and optimization of potential benefits and trade-offs, such as GHG savings, biodiversity, employment opportunities and energy/food security. A comprehensive approach which covers all the sustainability issues from their economic, environmental and social perspectives is needed. In Sect. 4.2 is shown how such an evaluation pattern is applied. Some of the aspects of biomass, biogas and *combined heat and power generation* are being discussed there at length (IEA ETSAP and IRENA 2015).

4.4.7 Coal and Gas Power Plants

Fossil fuels accounted for 90% of global primary energy consumption in 2018; their share of electricity generation was 65.2%, with coal accounting for 38%, followed by gas at just over 23%, while oil was of secondary importance. In some countries the share of coal in electricity generation is significantly higher (China, USA, Poland); in Germany the share of fossil fuels (2018) was about 45% (lignite 24%, hard coal 13%, natural gas 8%), currently with a slightly decreasing tendency and in the long run with the goal of complete renunciation. Advantages of fossil power plants are the availability of fuels and mature available technologies as well as relatively low plant costs and short construction times; disadvantages are (differently) high emissions of GHG and air pollutants, which can be reduced with modern technology, however with efficiency losses and cost increases.

From the point of view of climate protection, the switch to more modern plants and from coal to gas makes[21] sense in the short term; in the longer term, fossil-fired power plants are only acceptable if CO_2 is separated from the flue gases and subsequently stored or used (CCS/CCU, see Sect. 4.4.9). Thermal power plants are diverse in terms of supply technology and unit size. Coal-fired steam power plants

[21]Specific CO_2 emissions [gCO_2/kWh]; lignite 850–1200, hard coal 750–1200 (best technology: 740), natural gas CCGT 400–550 (best technology: 330) (VDI 2013).

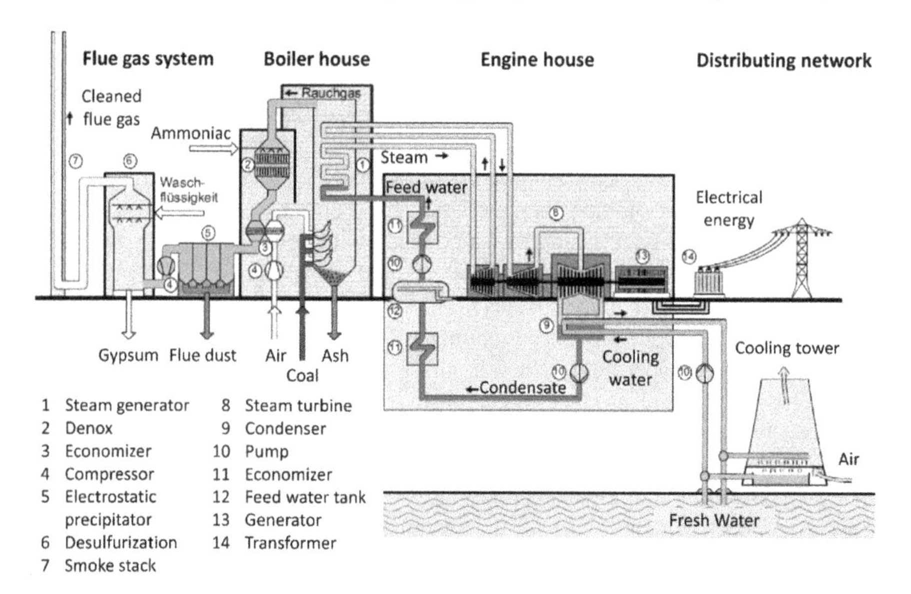

Fig. 4.31 Coal-fired steam power plant, schematic (Bennauer et al. 2009)

and gas-fired combined cycle power plants with typical "electrical" unit outputs of 300–1000 MW are available for large-scale power generation.

Steam-fired power plants fired with hard coal or lignite consist of essential components, which are shown schematically in Fig. 4.31. Typical steam parameters today, with live steam pressures of 285 bar and temperatures of 600 °C, are in the "ultra-supercritical" range and allow efficiencies of approx. 46%, the electrical own consumption is approx. 5.5–8.5%. Compared with the average global power plant efficiency of 33%, progress in technology development is clearly evident and can be expected to continue in the future. These include further improvements in steam parameters and individual component efficiencies as well as a reduction in electrical power needs. Material developments play an important role in this context. In addition, there are innovations to reduce starting times and increase load change rates. Both are important in order to integrate coal-fired power plants into modern energy systems; these properties are secondary when they are used in the base load range, for example in countries with strongly increasing electricity demand.

Even for gas-fired power plants, important evolutionary developments are more likely than radical new developments. Table 4.10 illustrates the development from gas turbine and combined cycle power plants up to the world's most modern combined cycle power plant Irsching with an advanced gas turbine of the so-called H-class, with which an efficiency of just over 60% could be achieved with a net output of 578 MWe.

Gas-fired power plants most frequently use natural gas, but also other gases, including biogas from fermentation or gasification of biomass (see Sect. 4.4.6). Combined cycle power plants (see Fig. 4.32[22]), which are characterized by short

Table 4.10 Typical characteristics of gas turbines and combined cycle power plants (VDI 2013)

	Unit	E-Class	F-Class	H-Class
Gas turbine	MW	172	295	375
Gross output[a]	–	12.1	18.8	19.2
Pressure ratio	kg/s	531	692	829
Exhaust gas mass flow	°C	537	586	627
Exhaust gas temperature	%	35.3	40	40
Gross efficiency[b]	–			
Combined cycle power plant				
Number GT × ST[c]	–	1 × 1/2 × 1	1S/2 × 1	1S/2 × 1
Net output	MW	253/512	431/862	570/1140
Net efficiency[b]	%	52.5/53.1	58.7	>60

[a]ISO conditions
[b]Related to the lower thermal value
[c]Single shaft arrangement of gas turbines (GT) and steam turbines (ST)

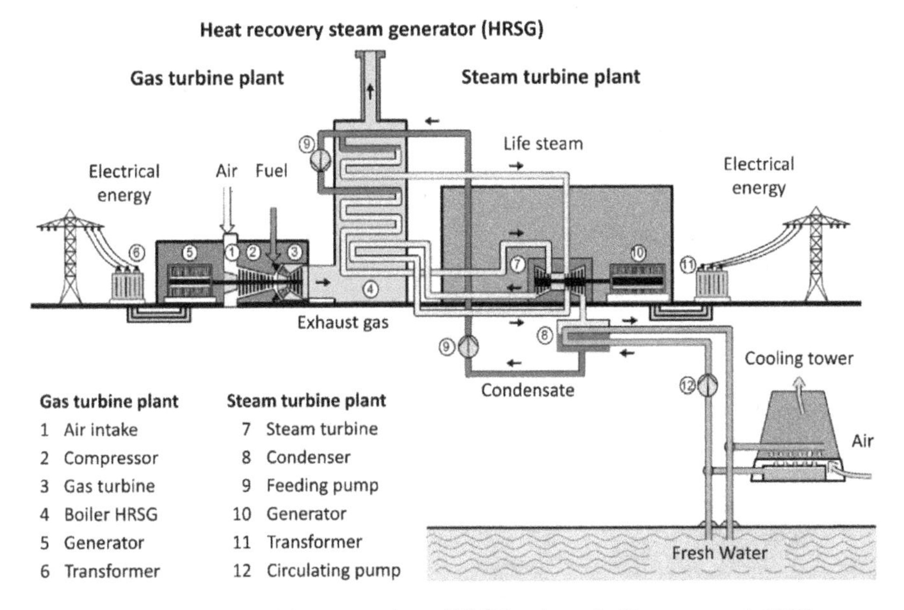

Fig. 4.32 Gas and steam turbine power plants (CCGT), schematic (Bennauer et al. 2009)

start-up and load change times, have gained acceptance. Blocks of small services that can be interconnected in modules provide a high degree of flexibility, which should be particularly attractive for developing countries.

[22]E.g. load gradients of 35 MW per minute between minimum and base load; start up in less than half an hour after 6–8 h standstill.

4.4.8 Fracking

Hydraulic fracturing or short fracking (Meiners et al. 2012) is a method of creating, widening and stabilizing cracks in the rock of a deep underground deposit with the aim of increasing its permeability. This makes it easier and more stable for gases or liquids to flow to the well and be recovered. A liquid ("fracfluid") is pressed into the geological horizon from which it is to be extracted through a borehole, typically under high pressure of several hundred bar. When the process of crack formation is complete, the overpressure created in the depth is relieved and the fracfluid is brought back to the surface. The fracfluid is water, which is usually mixed with chemical additives (below 1%) and supporting agents (approx. 5%), such as quartz sand, in order to stabilize cracks ("pathways") as long as possible and guarantee permeability.

Since the end of the 1940s, fracking has been used primarily in oil and gas production for better exploitation of deposits, as well as in the development of deep aquifers for water extraction and the improvement of heat transport in deep geothermal energy. In the latter cases no supporting agents or chemical additives are required. Since the beginning of the 1990s and especially in the USA from about the year 2000, production has focused on so-called unconventional crude oil and natural gas (including "shale gas", see Fig. 4.33). The fracking boom there significantly changed the US energy market and caused prices to fall. Since around 2013, the US government has

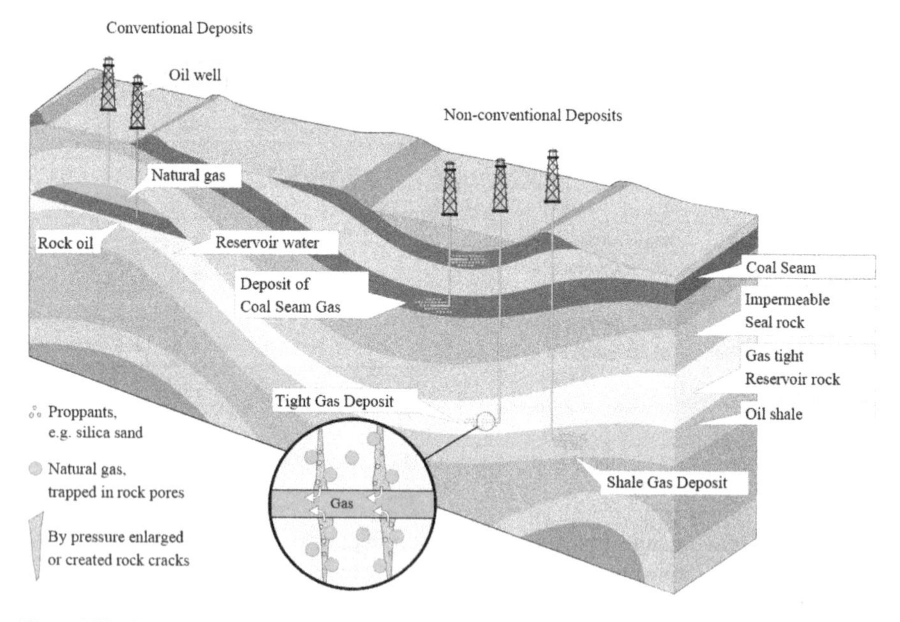

Fig. 4.33 Schematic representation of conventional (left) and unconventional deposits (right). MagentaGreen (https://commons.wikimedia.org/wiki/File:(Non)_Conventional_Deposits.svg), https://creativecommons.org/licenses/by-sa/4.0/legalcode

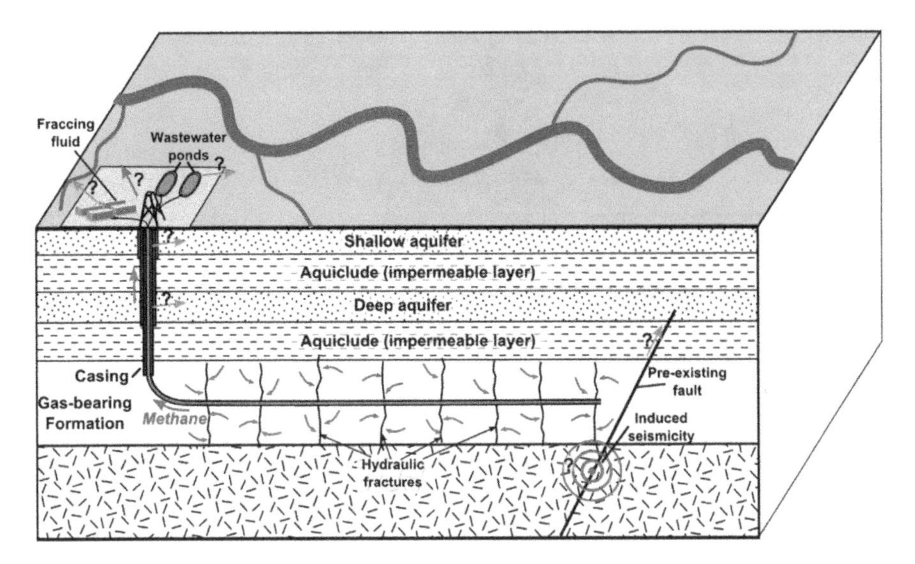

Fig. 4.34 Schematic representation of the potential environmental risks of a well. Mikenorton (https://commons.wikimedia.org/wiki/File:HydroFrac2.svg), "HydroFrac2", https://creativecommons.org/licenses/by-sa/3.0/legalcode

therefore been supporting efforts to achieve energy independence and to increase exports of liquefied natural gas to Europe and Japan.

Exploration of unconventional gas deposits and the use of fracking are being promoted in some countries (in Europe, for example, in Poland), while other countries have banned or are considering banning them (e.g. France, Netherlands or Germany).[23] South Africa imposed a moratorium on fracking in 2011, but lifted it again in 2012 and allowed the exploitation of shale gas deposits on about 20% of the country's surface area. In 2011, China developed a shale gas source with fracking for the first time and has the world's largest unbalanced gas reserves.

Gas fracking technology is well established and is being further developed or optimized; its use is often countered by inadequate data on deposits and, above all, feared environmental risks. These include (see also Figs. 4.34 and 4.35):

- The danger of water pollution by fracfluids and the chemicals contained therein, by the gas conveyed and the partly toxic reservoir water[24];
- the high land and water consumption and lack of climate neutrality of the gas produced;
- migration of substances from the deposit to other layers;
- vibrations during deep drilling and triggering of artificial earthquakes.

[23] For an overview cf. https://en.wikipedia.org/wiki/Hydraulic_fracturing_by_country.

[24] At the end of the borehole, this can escape into the groundwater through leaks in the piping or into the surface water due to improper disposal of the frac fluids.

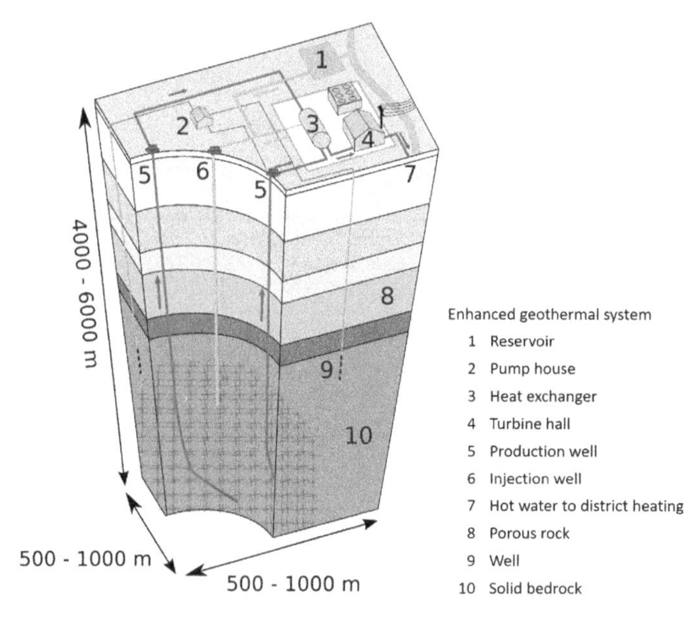

Fig. 4.35 Principle of geothermal use from hot, dense rock (HDR). Geothermie_Prinzip.svg: *Geothermie_Prinzip01.jpg: "Siemens Pressebild" http://www.siemens.com derivative work: FischX (talk) Geothermie_Prinzip01.jpg: "Siemens Pressebild" http://www.siemens.com derivative work: Ytrottier (talk) (https://commons.wikimedia.org/wiki/File:EGS_diagram.svg), "EGS diagram", https://creativecommons.org/licenses/by-sa/3.0/legalcode

Fracking for the *use of deep geothermal energy*, which enables temperatures of the fluid to exceed 80–100 °C and thus—in addition to heat utilization—electricity generation, is similar to the process for extracting unconventional gas deposits. In the case of hydrothermal geothermal energy, i.e. the presence of a layer of rock carrying water, one borehole is sufficient; fracking is not necessary in every case, but may improve the connection of the borehole to a reservoir. Petrothermal (so-called *Hot Dry Rock*, HDR) systems (see Fig. 4.35), which contain hot but less permeable rock from which no water can be extracted, require the creation of an artificial crack and fissure system by pressing water under high pressure, i.e. fracking. In order to circulate water, at least two deep boreholes are required; to create a petrothermal reservoir, large quantities of water are required (10,000–20,000 m^3),[25] but the use of supporting agents and chemicals is not mandatory. Accordingly, "pollution risks" are eliminated, but the potential for earthquakes is increasing due to drilling in crystalline rock.

Depending on the geological database and the local conditions, the use of HDR systems is also associated with considerable drilling, exploration and operating risks as well as competing uses. The technique with circulation of water, which must

[25]According to Akademien der Wissenschaften Schweiz (2012).

be continuously supplemented, is in the phase of large-scale testing and further development.

4.4.9 *Carbon Capture and Sequestration (CCS)/Carbon Capture and Utilization (CCU)*

CCS is the process of capturing CO_2 from large point sources such as fossil power plants, transporting it to an underground storage facility and storing it there without returning it to the atmosphere (see also Fig. 4.36). By equipping modern conventional coal-fired power plants, for example, with CCS, their overall CO_2 emissions could be reduced by 80–90% (the German Federal Environment Agency calls this an "actual" value of 70%), coming along with a major loss in efficiency (see Fig. 4.37). The resulting CO_2 emissions are still well above that of renewable energy sources (RES) or nuclear power plants, but makes coal use compatible with the achievement of

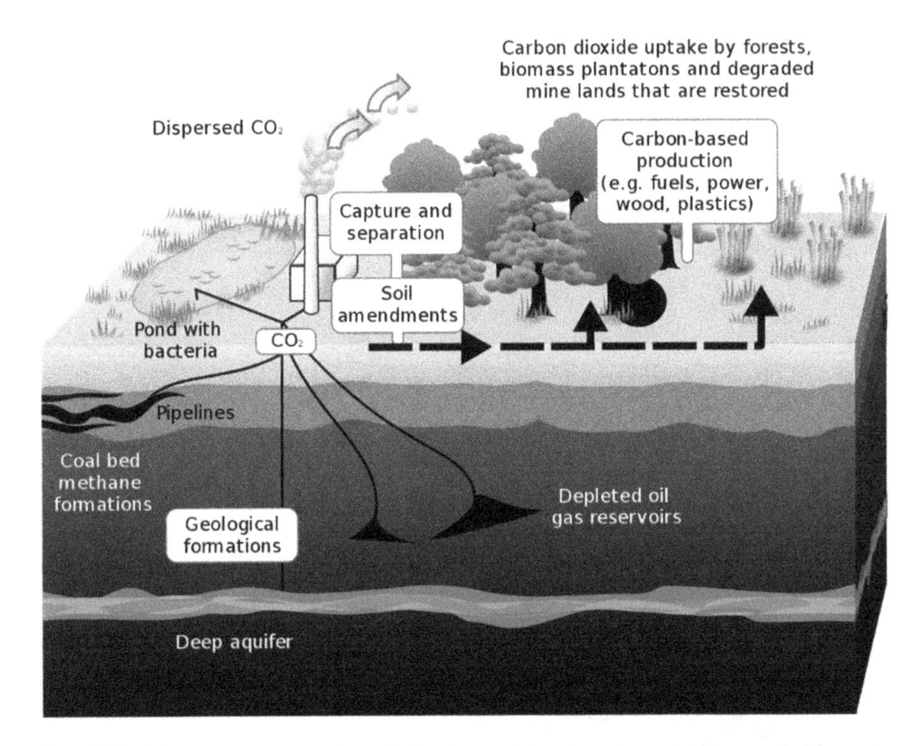

Fig. 4.36 Schematic representation of the terrestrial and geological storage of CO_2 emissions from a fossil power plant. LeJean Hardin and Jamie Payne derivative work: Jarl Arntzen (talk) (https://commons.wikimedia.org/wiki/File:Carbon_sequestration-2009-10-07.svg), "Carbon sequestration-2009-10-07", https://creativecommons.org/licenses/by-sa/3.0/legalcode

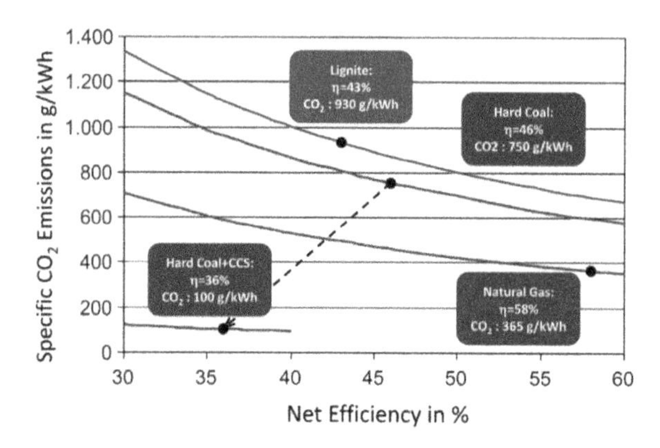

Fig. 4.37 Dependence of specific CO_2 emissions on electrical efficiency in power supply with various fossil fuels (VDI 2013)

climate protection targets (IPCC 2011). The latter applies all the more to modern gas-fired power plants; biomass power plants with CCS even function as CO_2 sinks.

By separating and compressing CO_2, the fuel requirement of a new coal-fired power plant increases by up to 40%, the capital costs by 70% (HIS CERA 2011) and the land requirement by up to 50%; in addition, there are facilities and costs for transport and storage.

A distinction is made between "*post-combustion*" (scrubbing from the exhaust gas), "*pre-combustion*" (separation after coal gasification) and combustion in an oxygen atmosphere (*oxyfuel*), see Fig. 4.38. Each technology has advantages and

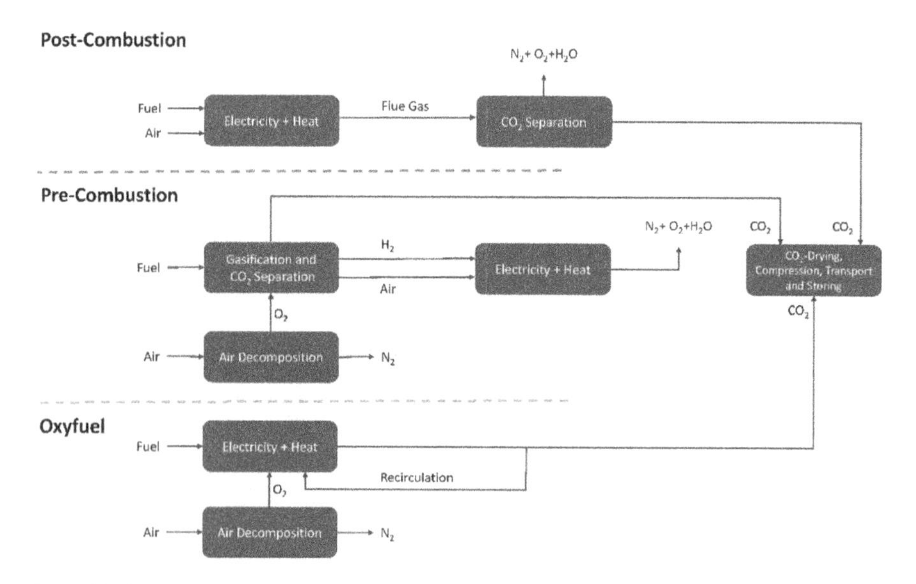

Fig. 4.38 Subdivision of the different process routes (VDI 2013)

disadvantages with regard to critical factors such as efficiency losses, separation rate, purity of the separated CO_2, further environmental effects, costs and inertia of the process in load following operation. Which technology (and whether one at all) will prevail on a large scale is completely open (Grünwald 2008).

For the permanent storage of gaseous CO_2 in geological formations, deep sedimentary layers whose pores are filled with saltwater and exploited oil and gas reservoirs are favored. Deep disposal in the sea has now been abandoned due to insufficient retention times. In sediment layers at depths of 800 m and above, the CO_2 introduced remains in a supercritical state, so that a new outcrop is practically impossible, provided that these layers are covered by an impermeable top layer and boreholes are securely closed. While CO_2 has already been injected into geological formations for several decades for various purposes, the "final" fate of the gas poses a new challenge. From a climate policy perspective, the German government stated in 2007 that a maximum annual leakage rate of 0.01% would be acceptable, i.e. after 1000 years 90% of the CO_2 would still be in the repository, while IPCC considers a retention of 99% over more than 1000 years to be probable; even with higher leakage rates at least a time saving could be achieved.

In addition to this risk of outgassing, which is difficult to rule out, the question of what becomes of the salt water displaced from the aquifers, whether mixing with the groundwater after ascending into fault zones can be ruled out, up to and including the possibility of sudden emergence with locally high (dangerous) CO_2 concentrations, plays a decisive role in assessing the future prospects of this variant of sequestration.

CO_2 could also be stored in the form of carbonates, which could be deposited openly and without safety concerns. Silicates of the alkaline earth metals, which can be converted with dissolved carbonic acid, could be considered as starting materials. Research in this area is in its infancy; problems include slow reaction rates.

The future prospects of CCS for further decarbonization of fossil fuels are unclear:

- The potential is high, the individual elements of the process chain are present, but still completely untested; the most far-reaching project, "*Schwarze Pumpe*" in East Germany, was stopped in 2014 (after Vattenfall AB had invested about 70 million euros in the project).
- Large-scale research programs are aimed at improvements and optimizations as well as demonstration plants; the risks are still unexplored (SRU 2011).
- The crux of the matter is the considerable additional costs and thus questionable competitiveness of the electricity generated, the social acceptance problem of land-based CO_2 storage facilities and CO_2 pipelines in or through densely populated areas and the limited availability of storage capacities.[26]
- The massive use of CCS, if at all, is not considered possible until 2030.

[26]For Germany, the Federal Agency for Raw Materials estimated the storage capacity at 12 ± 3 billion tonnes of CO_2, which would be sufficient for a few decades for the German power plant park; at the same time, reference was made to competition for use, for example with geothermal energy.

One way to avoid the problems associated with storage is to *use the* captured CO_2. Such concepts (CCUs) include *enhanced recovery of* oil and gas fields as a tried and tested method of initiating CCS and injecting CO_2 into deep, non-degradable coal seams, its sorption from coal and displacement of usable methane, and *a number of exciting niche opportunities* (HIS CERA 2011) such as CO_2 for the production of building materials (gypsum), feedstock for chemical processes and third generation biofuels.

Bio-energy with carbon capture and storage (BECCS) is favored as a technology because it could lead to negative overall CO_2 emissions and is therefore counted as an important climate change mitigation scenario, for example in the IPCC Fourth Assessment Report. The technical potential for negative emissions from BECCS is estimated at 10 Gt CO_2 per year, while the economic potential for additional costs of less than 100 Euro is estimated at only up to 3.9 Gt (by comparison, emissions of anthropogenic greenhouse gases were just under 50 Gt at the end of 2010). Accordingly, their use is currently limited to a few smaller facilities (e.g. plants for corn reprocessing into syrups and ethanol in Illinois, USA) and larger-scale facilities are not in sight.[27]

4.4.10 Nuclear Fission

Initial situation, light water reactors

As listed by the IAEA in its 2018 Reference Data Edition (IAEA 2018) there are 450 nuclear power plants with a total capacity of 391.7 GWe are in operation[28]; the share of nuclear power worldwide was almost 11%—with strong fluctuations from country to country, with a maximum value of 72% in France. A total of 55 units with a capacity of 56.8 GWe are under construction worldwide, 120 units with a capacity of about 125.3 GWe are planned. A special position for the future use of nuclear energy is taken by China, where 12 blocks are under construction and 43 are planned. This would result in an increase in the share of nuclear power from currently just over 2% to over 20% by 2030 (Höhener 2014), followed by Russia (10% → 29%) and India (6% → 22%) (Fig. 4.39).[29]

[27]For actual data cf. https://en.wikipedia.org/wiki/Bio-energy_with_carbon_capture_and_storage (accessed 13-Dec-2019).

[28]Of the 44 units that were shut down in Japan after "Fukushima" until further notice, 3 retrofitted units are now back in operation (as of Feb. 2017), two retrofitted units have been banned from operation; 20 more retrofitted units are awaiting approval for operation, 17 units have not yet been requested to be put back into operation; 8 units have definitely been shut down.

[29]For an overview of actual data cf. https://www.world-nuclear.org/information-library/current-and-future-generation/nuclear-power-in-the-world-today.aspx (accessed 13-Dec-2019).

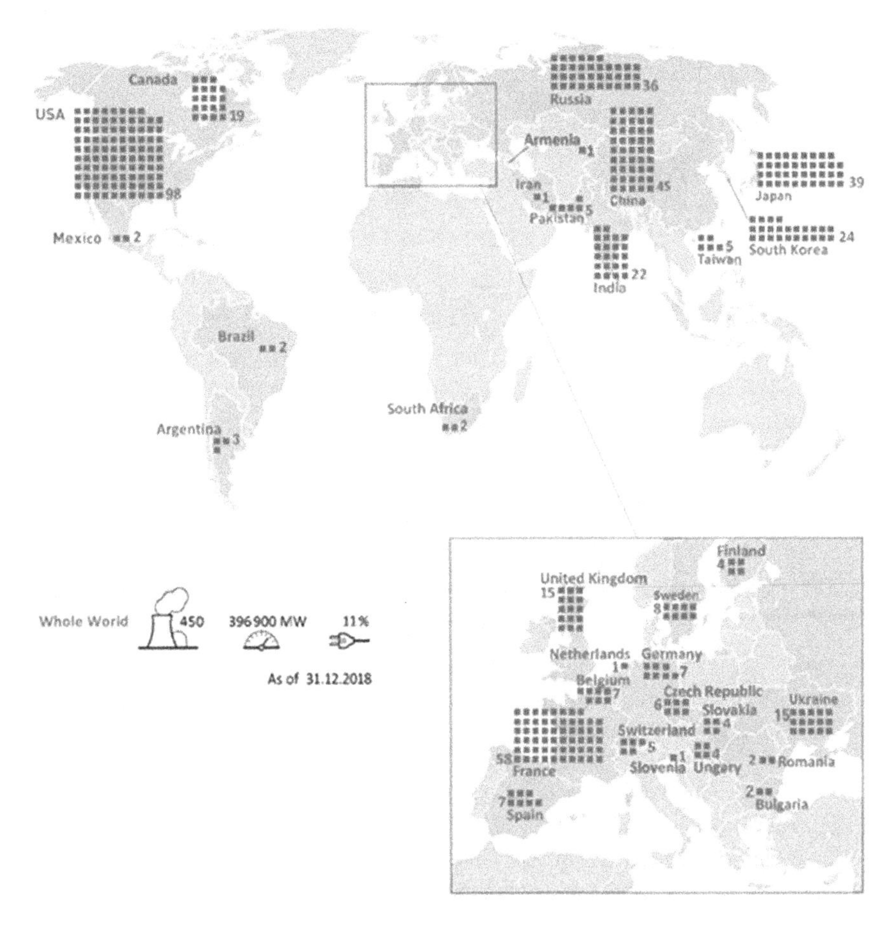

Fig. 4.39 Nuclear power plants worldwide, status 31.12.2018 (Nuklearforum Schweiz 2019)

With a share of almost 90%, reactors dominate which produce energy mainly by fission of the naturally occurring though enriched *uranium-235*[30] and use water as a coolant, which either evaporates in the reactor and is led to a turbine (*boiling water reactors, BWR*) or remains in the liquid phase thanks to high pressure and transfers its heat to a steam generator (pressurized water reactors, PWR); the downstream turbine drives a generator set for power generation, as in the BWR. The unit output of modern *light water reactors* (LWR) today—following the "*economy of scale*"—is more than 1315–1600 MWe, with a thermodynamic efficiency of about 33%.

Highly active fission products accumulate in the fuel rods of today's reactors, together with *plutonium* and other *transuranium elements* such as *americium* and

[30]The proportion of *U-235* in natural uranium is 0.7% and is enriched to 3–5% for use in light water reactors. The neutron capture produces *Pu-239*, whose fission contributes to the generation of energy.

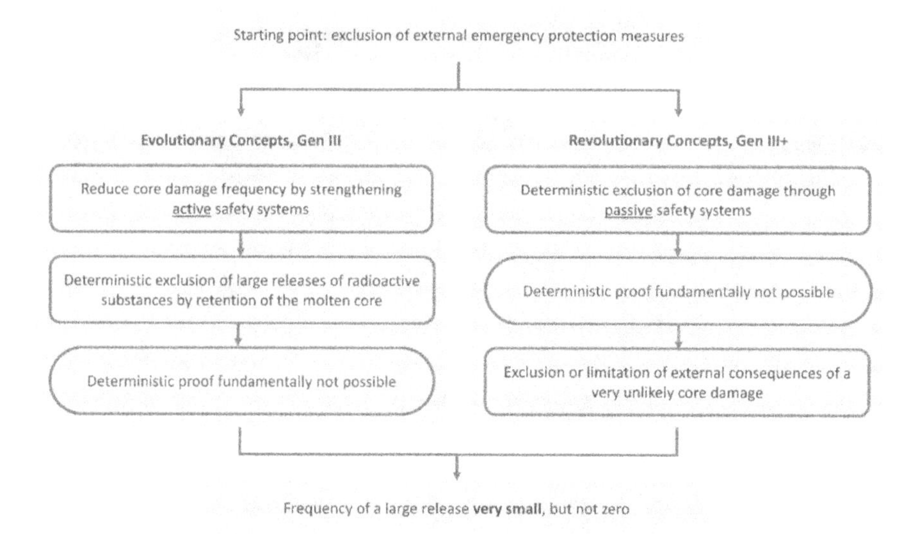

Fig. 4.40 Concept approaches for the further development of light water reactors (LWR) (Prasser 2014a)

curium, which are formed by the capture of neutrons in the non-fissile *uranium isotope-238. Zirconium* cladding tubes include these components. However, the fuel rods must be cooled constantly, otherwise they would heat up due to the decay heat and eventually melt, releasing the fission products, even if the chain reaction has already been stopped.

The overriding aim of the further development of LWR was to make emergency cooling even more reliable and to include the radioactive material contained in the reactor core in all "conceivable" accidents within the plant and to exclude large releases into the environment. Two general approaches have been followed (see Fig. 4.40):

- *Evolutionary concepts* with improvements of the known safety systems to further reduce the core damage frequency[31] and devices to retain the molten reactor core in the containment as well as to control hydrogen formation and its long-term cooling. A prominent example of these so-called Gen III reactors is the *European Pressurized Water Reactor* (EPR), which is currently under construction in Finland, France (in both countries with large construction time and cost overruns) and China (2 units)[32] (see Fig. 4.41).
- *Revolutionary concepts* are based on a transition to passive safety systems that manage without an external power supply and active triggering from the safety

[31] In the case of EPR, including plant-internal and external events, to close to 10^{-6} per reactor year.

[32] Taishan 1 and 2 have been taken into operation in June 2018 and June 2019. Construction work has started in the UK (Hinkley Point C) early 2019.

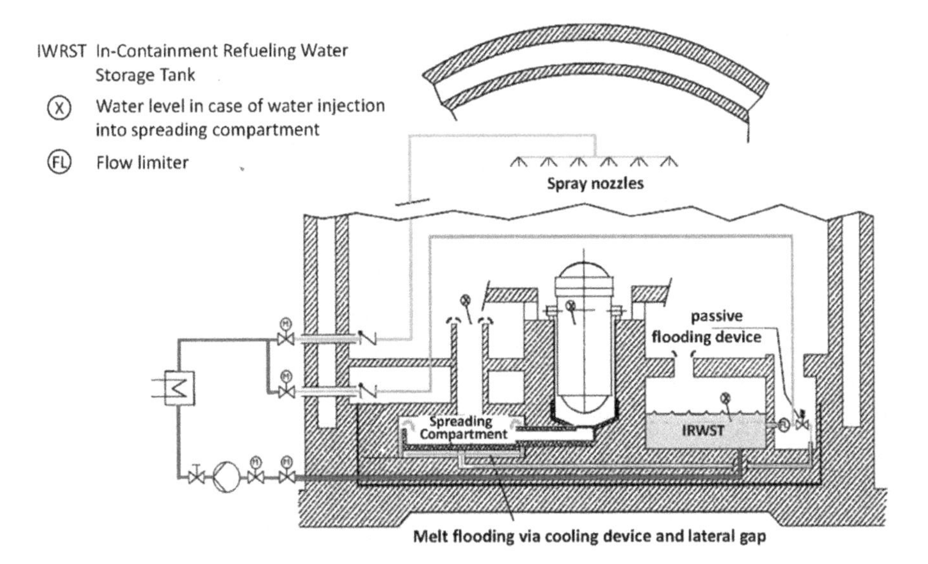

Fig. 4.41 Containment design of the EPR with core catcher and active cooling system. Areva NP (https://commons.wikimedia.org/wiki/File:CHRS_EPR_catcher_flooding.jpg), "CHRS EPR catcher flooding", marked as public domain, more details on Wikimedia Commons: https://commons.wikimedia.org/wiki/Template:PD-shape

system (IAEA 2009), see Fig. 4.48. A prominent example is the Advanced Pressurized Water Reactor (AP 1000/CAP 1400), which is being built in China[33] and is to become the new standard there (Fig. 4.42).

In the course of development, however, it became clear that severe accidents with large releases can be made "arbitrarily unlikely", but not deterministically excluded, even with purely passive safety systems.

Figures 4.43, 4.44, 4.45, 4.46, 4.47 and 4.48 show reactor concepts under development within the GIF-framework. China has begun the construction of a prototype high Temperature Reactor (HTR-PM) a first step towards the development of the VHTR. Both France and Russia are developing advanced sodium-fast reactor designs for near-term demonstrations. A prototype lead fast reactor is also expected to be built in Russia in the 2020 time frame.[34]

Radically innovative concepts with self-sustaining chain reaction

At present, however, innovative approaches beyond this are also being researched in order to make nuclear fission reactors and associated fuel cycles even safer (keyword: inherent safety properties), to make better use of the fuel (increase in burn-up from

[33]Yangjiang 5 has been taken into operation in July 2018.

[34]Figures 4.42, 4.43, 4.44, 4.45, 4.46 and 4.47 are provided by US Department of Energy Nuclear Energy Research Advisory Committee, marked as public domain. Cf. US DOE (2002). For more detailed and updated description cf. Pioro (2016).

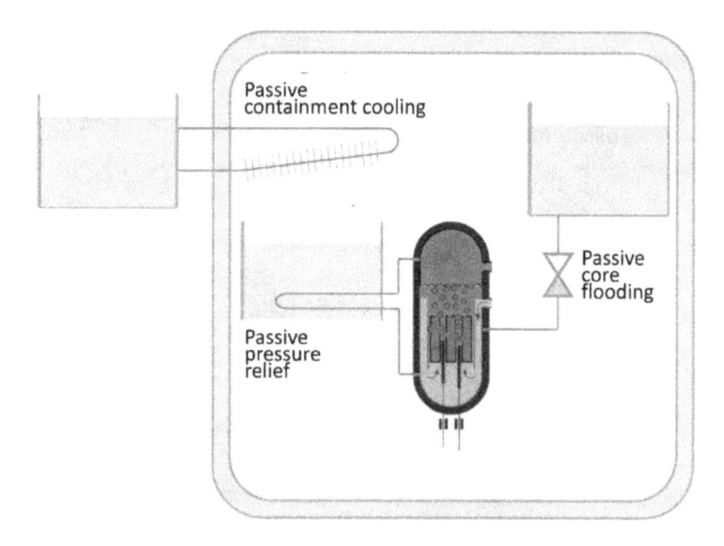

Fig. 4.42 Passive systems for reactor pressure relief, core flooding and containment cooling (Prasser 2014a)

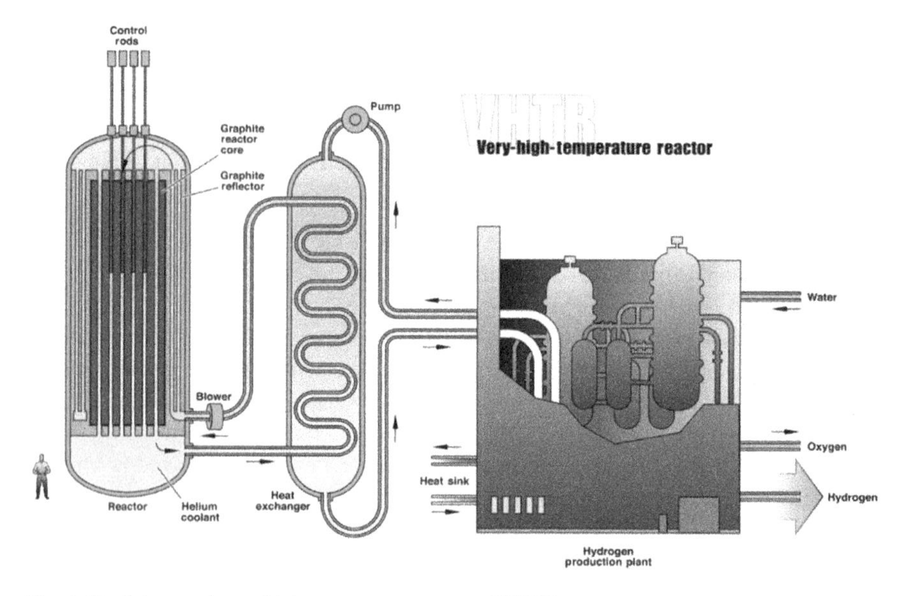

Fig. 4.43 Schema of very-high temperature reactor (VHTR)

approx. 50 (Gen III) to up to 100 MWd/kg natural uranium), to reduce the risk of proliferation and to relieve the disposal path [actinide separation and transmutation (P&T)] or even to largely avoid them (transition from natural uranium to thorium), both ultimately in order to increase sustainability. Such research programs run in countries that rely on nuclear power and/or in international collaborations such as

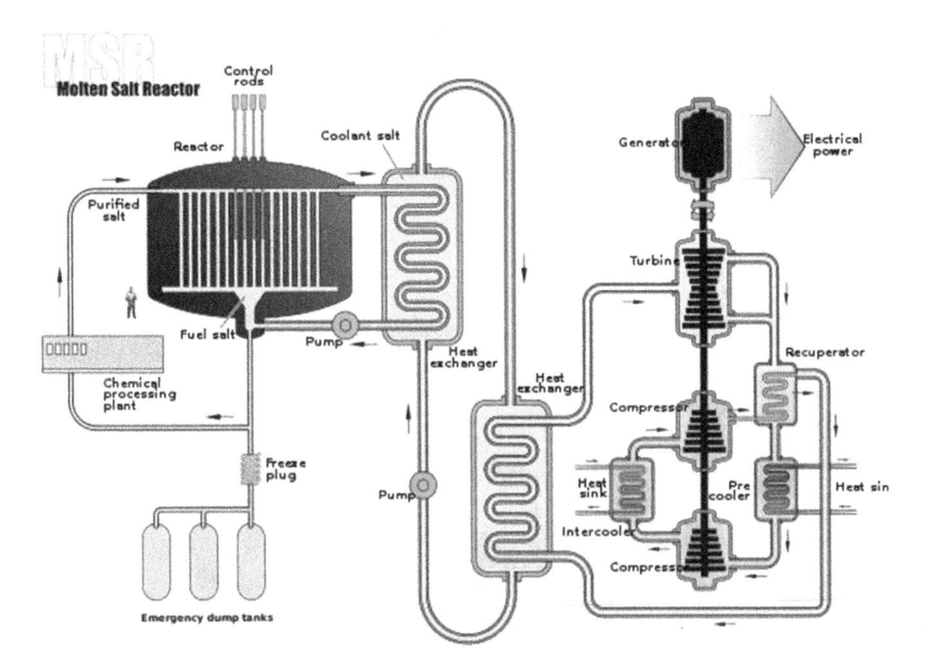

Fig. 4.44 Schema of a molten salt reactor (MSR)

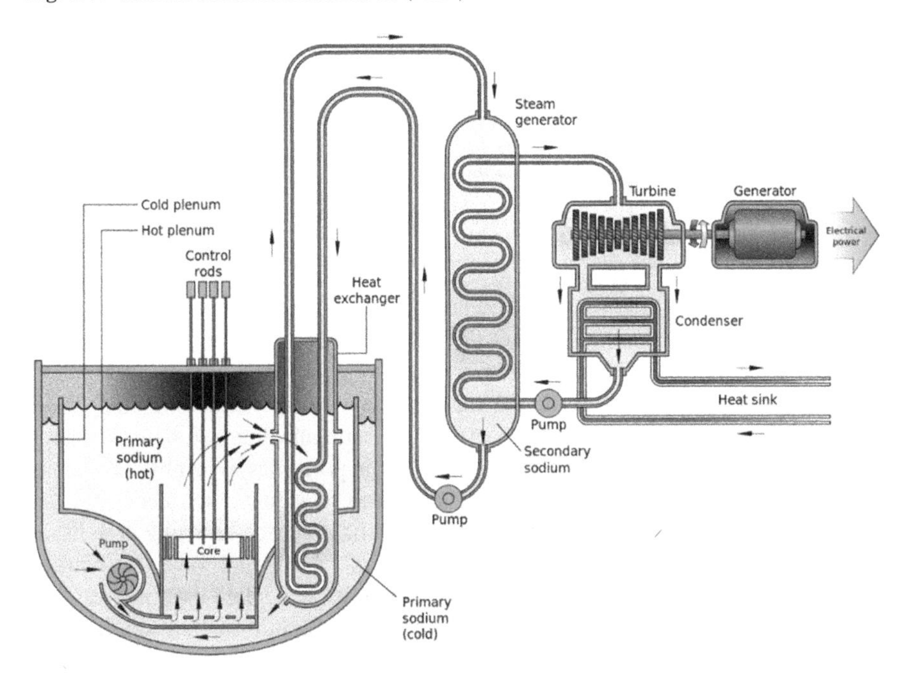

Fig. 4.45 Schema of a sodium-cooled fast reactor (SFR)

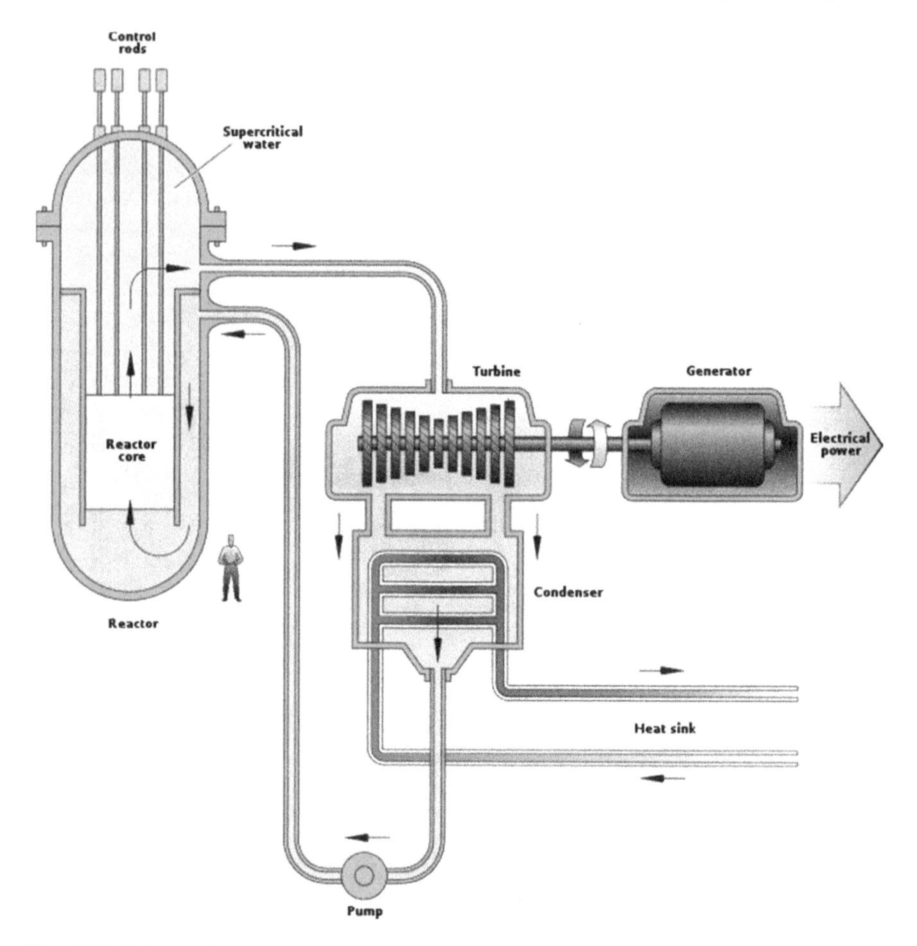

Fig. 4.46 Schema of a super-critical water-cooled reactor (SCWR)

the *Gen IV International Forum*[35] (GIF), which aims to research and develop six carefully selected concepts to market maturity.

These concepts represent a radical break with today's generation of light water reactors ["radical design break" (Kelly 2014)]. A distinguishing feature is the energy spectrum of the neutrons used for the nuclear fission processes, whereby fast spectra with non-thermalized fission neutrons ("fast reactors") dominate, since they are particularly suitable for better fission material utilization, for breeding new fissile material and/or actinide combustion ("waste burners") due to their favorable neutron economy. Another important distinguishing feature is the coolant used; the variants range from supercritical water via inert gas (helium) to exotic-looking single-phase

[35]GIF was set up to provide the basis for identifying and selecting six nuclear energy systems for further development. Gen IV systems for further development. GIF has 13 (11 active) members including EU, France, Japan, China, Russian Federation, Korea, USA and Switzerland.

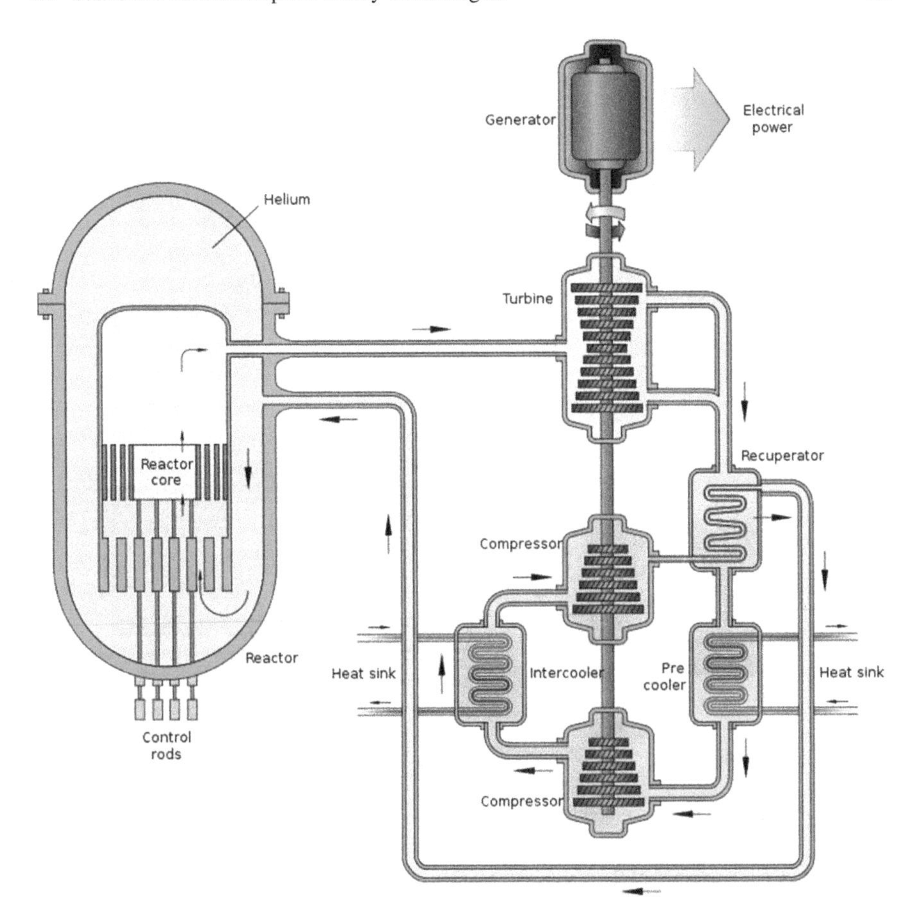

Fig. 4.47 Schema of a gas-cooled fast reactor (GFR)

fluids such as liquid metals (sodium, lead or lead/bismuth) and molten salts (fluorides or chlorides), the latter with partly excellent neutron-physical and thermodynamic properties. In addition, there are differences in size (very small to very large) and core outlet temperatures of over 500 to more than 800 °C, which, compared with today's LWR (around 325 °C), allow higher thermodynamic efficiencies or even the provision of heat for chemical processes (hydrogen generation) (see Table 4.11 with important design characteristics). Most concepts envisage a closed fuel cycle with reprocessing, whereby the development of proliferation-resistant processes and plants close to reactors is planned and, like the development of reactors, involves considerable effort and risk.

The early 2020s to the 2030s are named as the beginning of the "demonstration phase"; these currently extremely innovative concepts could be commercially applicable about ten years later, i.e. around 2030–2040, provided that "*show stoppers*" prove to be surmountable in the course of the development period.

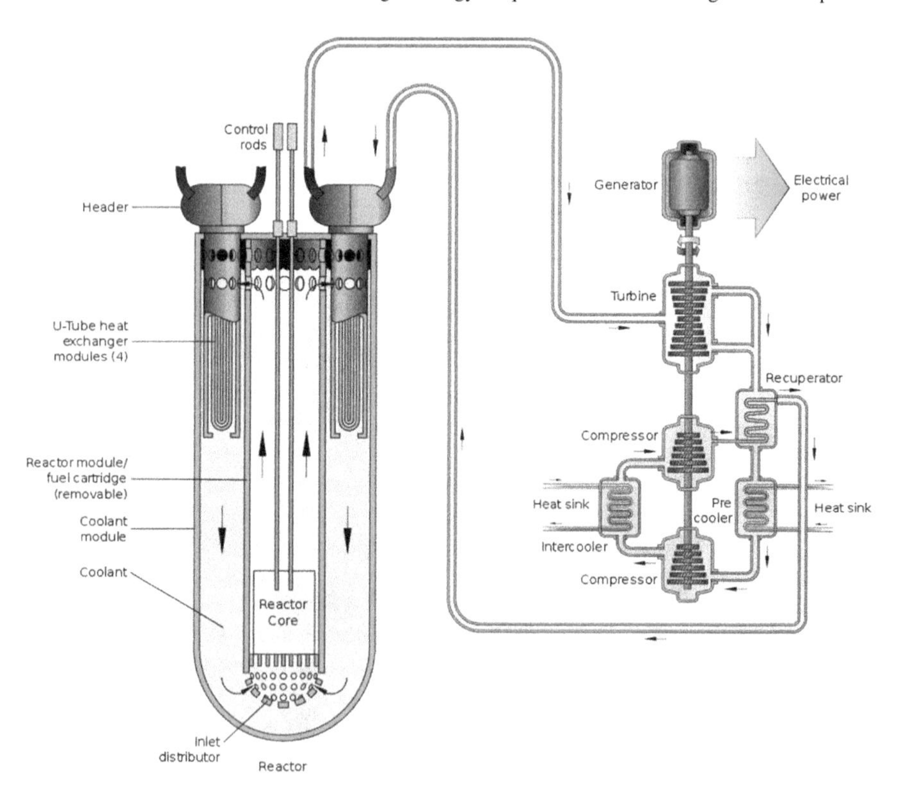

Fig. 4.48 Schema of a lead-cooled fast reactor (LFR)

Table 4.11 Characteristics of the six Generation IV Reactor Systems (Kelly 2014)

System	Neutron spectrum	System pressure (MPa)	Coolant	Outlet temperature (°C)	Nominal power density (MW/m^3)	Size (MWe)
VHTR	thermal	8	Helium	900–1000	8	100–300
SFR	Almost	0.3	Sodium	510–550	175	50–1500
SCWR	Thermal/fast	25	Water	510–625	100	1000–1600
GFR	Almost	7	Helium	850	100	1000
LFR	Almost	0.3	Lead, lead/bismuth	480–800	70	20–1200
MSR	Epithermal	0.6	Fluoride salts	700–800	170	1000
LWR	Thermal	8–16	Water	325	100	600–1600

VHTR very high temperature reactor; *SFR* sodium-cooled fast reactor; *SCWR* super-critical water-cooled reactor; *GFR* gas-cooled fast reactor; *LFR* lead-cooled fast reactor; *MSR* molten salt reactor; *LWR* light water reactor

In view of the early development stage, serious cost and construction time estimates are hardly possible at present; some concepts are characterized by a high degree of complexity and novelty of the safety approach, which can result in additional obstacles in the approval process.

Parallel to the development of reactors in today's common power class of up to 1600 MWe, interest in the development of *small to midsized modular reactors* (SMRs) has increased worldwide. The driving force behind this is the assumption that SMR, with outputs per module of up to around 300 MWe, can be better adapted to lower growth rates in energy consumption than large-scale plants and integrated into small supply networks, can incur lower investment costs, i.e. are easier to finance, and can penetrate the heating market in addition to electricity generation; they are also viewed as a possible replacement for old coal-fired power plants. In addition, SMR promises to better meet specific operator requirements, particularly in terms of simplicity through to series production in factories and transport to the site, longer operating times without fuel change and a new safety quality based on inherent safety features and passive safety systems, thus without active systems and rapid operator intervention.

Virtually all reactor development lines include SMR projects. At present, about 50 concepts for different applications advanced under development worldwide (see Fig. 4.49); three of them for power generation (with six units) are already under construction stage: a 25 MWe demonstration plant in Argentina and two floating 35 MWe reactors in Russia, each operating according to the PWR principle, as well as two gas-cooled HTR-PM with 200 MWe each in China as a construction of a non-coastal demonstration plant. Also, worth mentioning is the construction of a medium

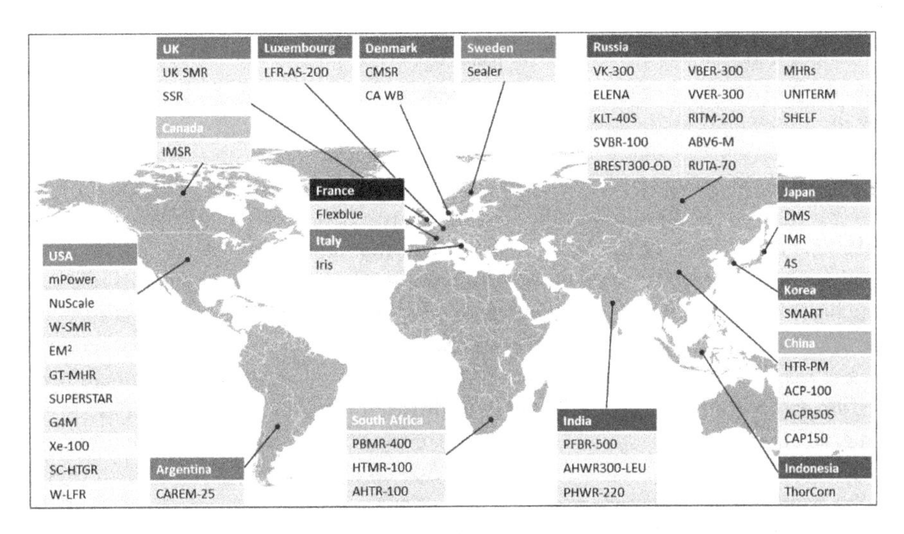

Fig. 4.49 Worldwide development of small modular reactors (following IAEO: Advances in Small Modular Reactors Technology Developments, September 2014, updated)

sized (500 MWe) fast neutron spectrum breeder reactor in India using liquid sodium as a coolant.

SMRs are often apostrophized as suitable for small countries or developing countries in particular, but their competitiveness in competitive markets is often questioned. Some "conventional" concepts, which are based on the technology of today's light water reactors, are already halfway mature and are associated with low development risks as "bridging technology"; for "more exotic" concepts this applies to the ongoing developments of the fourth generation. New, and perhaps attractive for newcomer countries, is the technology called floating nuclear power plant of small power, which is being deployed by Russia (starting from icebreakers) and now also by China. The reactor components are manufactured ashore in factories and assembled on a ship or lighter and then pulled to the intended location, moored and connected to the infrastructure there (electricity or heat network); the fuel is supplied and taken back by the manufacturer after use in the reactor ("full package nuclear service").

With a view to possible nuclear systems of the next (fourth) generation, four concepts are presented here that are often cited as bearers of hope: Helium-cooled, modular pebble bed reactors, molten salt reactors and thorium as substitutes for enriched natural uranium as well as actinide transmutation with accelerator-driven systems and lead/bismuth as target material and coolant (see also Sornette et al. 2019).

Modular pebble bed gas-cooled reactors (HTR-M) are designed in such a way that safety systems for rapid reactor shutdown and dissipation of decay heat can be dispensed with—they are considered inherently safe. The fuel is contained in spheres about 0.5 mm in size, which are surrounded by a durable, hermetic layer of silicon carbide. Several thousand of these so-called *coated particles* are sintered into graphite spheres of about 60 mm in size (see Fig. 4.50). The radioactive fission products, in particular the volatile radioisotopes of the noble gases *krypton* and *xenon* and the particularly problematic *caesium isotopes 137* and *134*, are reliably retained by the silicon carbide layer.

The spherical fuel elements are filled into the reactor vessel until the critical mass is reached at the desired operating temperature; they pass through the reactor core several times until the (high) target burn-up is reached. Even in the event of a total failure of the helium cooling system, there is no unacceptable overheating of the fuel, as the reactor power automatically drops to a level at which the heat still released can be dissipated via the wall of the reactive vessel. However, the thermal output of a reactor must be limited to about 200–300 MW, otherwise the maximum temperature in the reactor cannot be limited to 1600 °C. At significantly higher temperatures, *iodine* and *cesium* would slowly begin to diffuse through the silicon carbide layer, entering the helium cycle and from there into the environment. In addition, the high outlet temperature of the helium (≥850 °C) allow high efficiencies in power generation (>40%) and generation of process heat.

Prototypes of pebble bed reactors (AVR, THTR-300) were first built in Germany. Immature technological solutions and decreasing public acceptance have led to the project line being discontinued. In the meantime, the concept was taken up by South Africa, but then dropped again due to a lack of operator interest. The concept is now

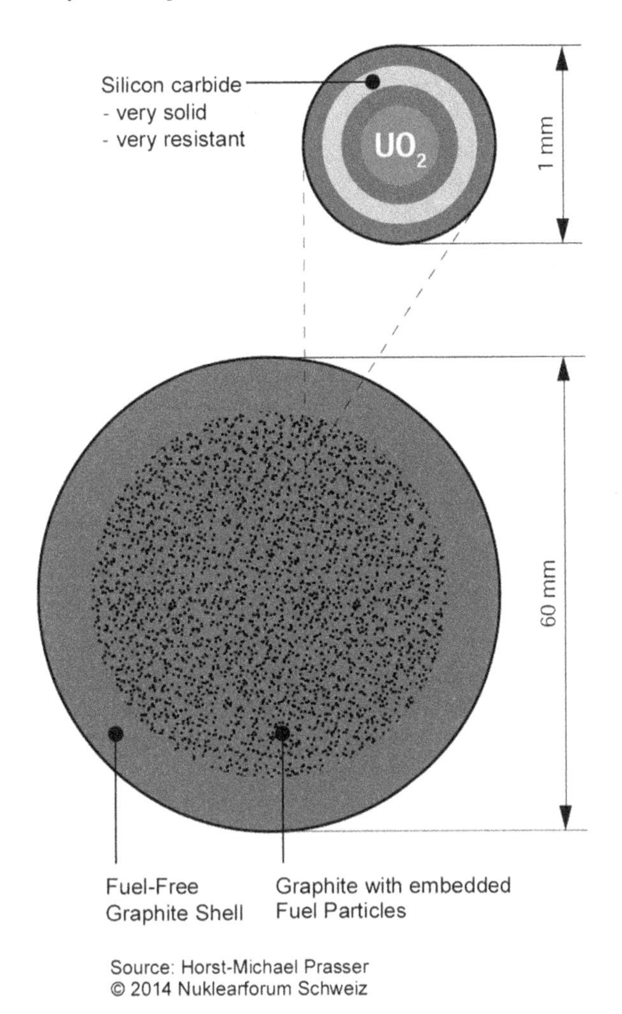

Fig. 4.50 Fuel assemblies for helium-cooled high-temperature reactors

being further developed in China: After successful tests on the test reactor HTR10, the construction of the HTR-PM with two reactors of 250 MW thermal power each was started in the province of Shangdong and is to mature into one of the two national standard designs; start of operation is scheduled for the beginning of 2020.

While the pebble bed reactor relies on the novel barrier concept of the *coated particles*, one tries to keep the radioactive inventory in the reactor core low in a variant of the molten salt reactor (MSR) (LeBlanc 2010) the quantities of radioactive substances that can be released in the event of an accident are limited in principle. In the reactor there is a pool of molten salt with a temperature between 500 and 700 °C or even above. The actual fissile material is dissolved or distributed in a molten mixture of various salts, such as *lithium*, *beryllium* or *zirconium fluoride*. These may be different fissionable nuclides, such as *uranium 235*, *plutonium 239* or *uranium 233*, which is obtained from thorium. If the corresponding starting nuclides

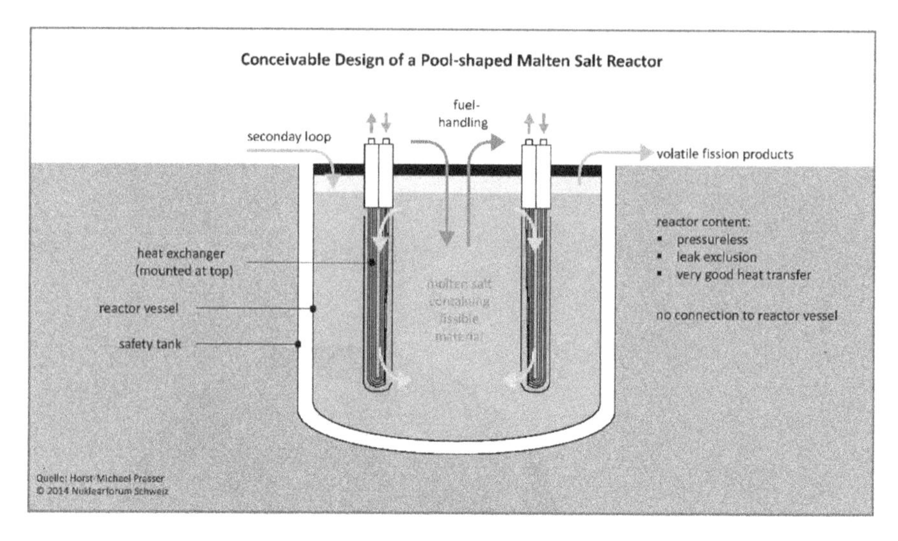

Fig. 4.51 Design of a salt smelting reactor in open pool design with continuous discharge of gaseous fission products and fuel preparation close to the reactor

are added to the molten salt, the reactor can serve as a breeder for *plutonium 239* or *uranium 233*. Similarly, transuranic elements such as *americium* and *curium can* be separated and transmutated from the waste of other reactor lines and thus eliminated when added (Fig. 4.51).

MSR are very flexible in terms of fuel composition. Ultimately, so much fissile material is always allotted that the chain reaction can be maintained. Therefore, they have no excess reactivity, which is a safety advantage, as is the pressureless reactor vessel. The reactor can be operated with thermal neutrons and high target burn-up, or with a fast neutron spectrum if breeding or transmutation processes are to be realized. Since there are no reactor internals due to construction, there are no neutron losses due to absorption in construction materials. The comparatively large neutron surplus permits a very good neutron balance and very efficient breeding and transmutation.

The high operating temperature permits power plant processes or the provision of process heat. Volatile fission products will already escape from the molten salt by themselves at these temperatures. This process can be intensified by bubbling an inert gas through the melt. Gaseous and volatile fission products can thus be continuously separated, separately conditioned and stored. Furthermore, the concepts provide for "on line fuel processing" and, if necessary, "re-processing", which has a positive effect on proliferation resistance.

The extremely long-lived, radiotoxic minor actinides (see Fig. 4.52) produced during the uranium-plutonium cycle largely determine the "*husbandry time*" *of* a repository; they could be converted into more short-lived fission products in transmutation systems—such as the molten salt reactor—or largely avoided by transition to the thorium-232/uranium-233 cycle.

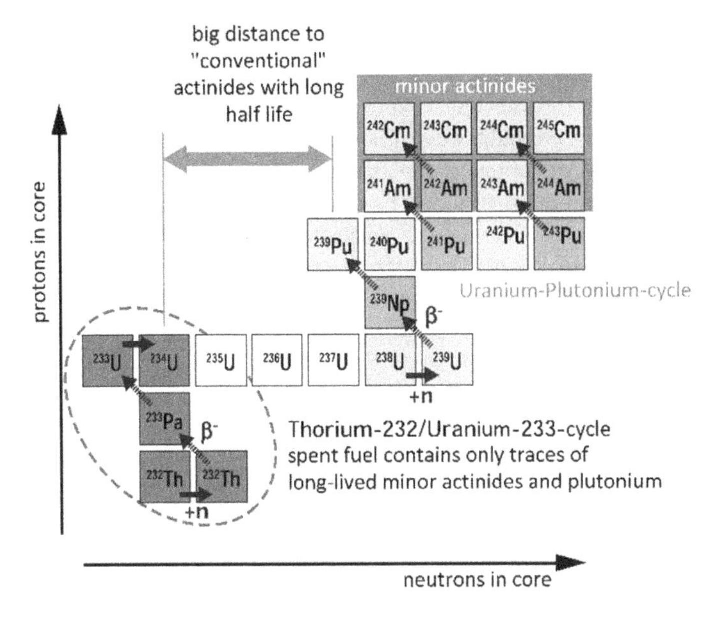

Fig. 4.52 Nuclide map with uranium-plutonium and thorium-uranium-233 cycle

This novel thorium-based fuel cycle reduced the required guarding times for highly active waste to less than 1000 years without transmutation and reprocessing. Another sustainability argument is the large potential range of thorium deposits, which can ensure fuel supply for much longer than the use of natural uranium in breeder reactors. However, a long transition period needs to be managed, as there is currently little *uranium 233* available. This means that we cannot put a large number of reactors into operation overnight with *thorium* and *uranium 233* alone but would have to "feed" other fissile material, for example plutonium, as a result of which the advantages in terms of waste would not initially become apparent.

Accelerator-driven systems; actinide transmutation

In today's reactors—with *U-238* and low enriched *U-235* as fuel and with a thermal neutron spectrum—besides fission products *plutonium* and transuranium elements are produced with even higher atomic numbers, such as *neptunium, americium* and *curium.* The latter, also known as minor actinides, are formed from the *U-238 isotope* by neutron capture and are largely characterized by strong radiotoxicity and extreme longevity[36]; they largely determine the confinement times of deep geological repositories to be secured. They can be transmuted into "harmless" nuclei of medium atomic numbers; however, as already explained, fast neutrons are usually required for

[36]The quantities of *neptunium/americium/ curium* are relatively small at 500/100/20 g per ton of spent nuclear fuel.

this. Depending on the effectiveness of partitioning and transmutation, the necessary husbandry times could be reduced to several thousand years.

Reactors suitable for transmutation must therefore refrain from moderating the fission neutrons (today's light water reactors are accordingly unsuitable) and use various liquid metals or helium as coolants; the fuel must have a higher fissile material content or have neutrons added from outside so that a self-sustaining chain reaction can be achieved. The latter connects the "subcritical" reactor with a strong proton accelerator and a target on which the proton beam impinges and becomes a neutron source via spallation processes. One such *accelerator-driven system* (ADS) for the transmutation of actinides and simultaneous power generation is the "*energy amplifier*" proposed by Nobel Prize winner *C. Rubbia,* also known as *Rubbiatron* (see Fig. 4.53), in which a lead-bismuth alloy is used both as a target and for its cooling and for cooling the actual reactor by means of natural circulation (without pumps).

The safety-related advantage of such a reactor is that so-called criticality accidents (RIA), i.e. core damage or destruction of the entire plant by a runaway chain reaction, are excluded; if the supply of neutrons from outside is interrupted, the function of the reactor would be immediately shutdown, but the problem of "residual heat removal" remains. The actinides would also have to be separated before ADS use, which would require reprocessing of spent fuel, with all its technical and safety difficulties.

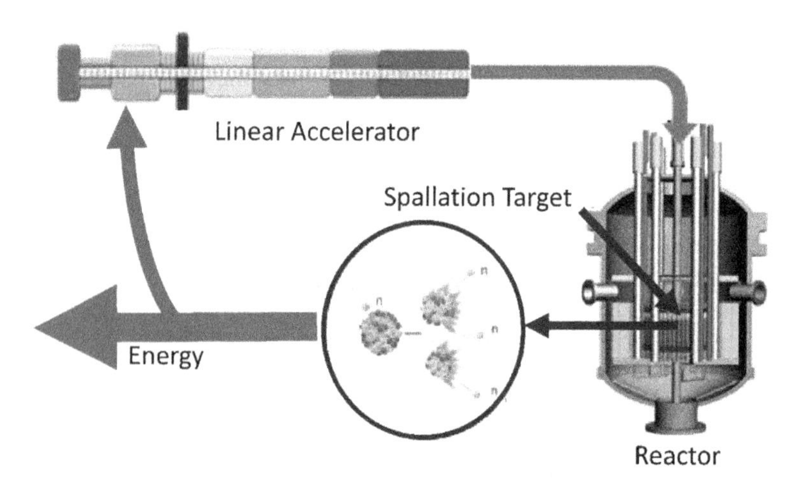

Fig. 4.53 Principle of an ADS ("Rubbiatron"). A strong proton beam is fired at a liquid metal target consisting of heavy elements. The spallation reaction produces about 20 high-energy neutrons per proton. They trigger fissions in the surrounding fuel elements, whereby actinide isotopes that cannot be fissioned with thermal neutrons are also burnt. The surrounding reactor core increases the neutron flux by a factor of 20–50, i.e. the majority of available neutrons are fast fission neutrons (Yan et al. 2017)

The advantages mentioned are probably countered by considerable additional costs and enormous technical challenges in its development. For example, a proton accelerator of an unprecedented performance class[37] and reliability is needed.

Another challenge is the spallation target. The enormous power density with the high necessary beam energy can probably only be mastered with a liquid metal target. This technique has been successfully tested at the Swiss Paul Scherrer Institut (PSI) with the *MEGAPIE* experiment. However, the target performance there must also be increased by a factor of 10 and the service life (previously 3 months) increased.

The most advanced research and development project for the construction of an ADS is *MYRRHA* at the Mol Research Centre in Belgium. With EU funding and strong international involvement, a 2.4 MW accelerator will be built and connected to a lead-bismuth cooled reactor. The reactor with a capacity of 50–100 MW would be operated in the highly subcritical range,[38] commissioning could take place in 2022–2024, the investment requirement is estimated at more than one billion euros.

Fast gas-cooled reactors with self-sustaining chain reactions are a competitor to ADS. They can "breed" new fuels parallel to actinide transmutation and are researched or developed within the framework of the Gen IV Initiative. This would avoid the considerable problems associated with the use of liquid metal, and would also eliminate the considerable electrical consumption caused by the operation of the accelerator.

Moreover, none of these innovative subcritical systems would be used on an industrial scale before 2030; plant costs, construction and approval times and ultimately their competitiveness are currently difficult to assess.

Conclusions

Nuclear power plants, which are essentially based on the technology of today's light water reactors, but which have been further developed in an "evolutionary" or "revolutionary" way, i.e. Generation III+ reactors, are partly under construction or even in operation which means that, if sufficiently mature, they can be used today. In terms of safety, they have been upgraded to such an extent that core damage events become even less likely and the consequences remain limited to the plant or can even be excluded. They use the fuel more efficiently because they are aiming for higher burnups and extend the service life of the plants to at least 60 years. High plant costs with corresponding financing requirements and long construction times represent major obstacles to implementation.

Generation IV reactors or even more "exotic" concepts go even further and would enable the avoidance of extremely long-lived fission products and actinides or their incineration. Some concepts promise a further increase of the inherent safety features, whereby the possibility of new concept-specific accidents (overcooling, freezing of

---

[37] A proton energy of approx. 1 GeV is required for beam currents around 10 mA, which corresponds to a beam power of 10 MW. One of the most powerful comparable plants is located at the Paul Scherrer Institute (www.psi.ch), but with 560 MeV and 2 mA it has a beam power of "only" about 1 MW.

[38] http://sckcen.be/en/Technology_future/MYRRHA (accessed 13-Dec-2019).

the coolant, violent chemical reactions when it escapes, etc.) must be taken into account, as well as possibly increased proliferation risks.[39]

The further development of these seemingly very complex systems requires considerable research efforts and successful experimental and demonstration phases, which are associated with correspondingly high technical and physical uncertainties and risks. It is not yet possible to make reliable statements about costs and construction times as well as licensing requirements and procedures to be changed. Leading projects such as MYRRHA and experience with large-scale technical developments suggest that these highly innovative or even "exotic systems" can only play a significant role in power generation in the long term (2040–2050), provided that considerable hurdles are overcome. These statements are to be reflected in the successes achieved with massive research and development in leading countries in this respect, above all China, but also Russia and perhaps also India and again the USA.

In the course of regional but global energy supply, smaller modular reactors (SMRs) deserve special attention. In terms of flexibility, safety properties, affordability and necessary infrastructure, they promise considerable advantages over reactors of today's size. Concepts based on the technology of today's light water reactors and possibly high-temperature gas-cooled reactors would be feasible in the next few years and would have a bridging function compared to the more "exotic" concepts.

4.4.11 Nuclear Fusion

The occurrence of a nuclear fusion, i.e. the fusion of two atomic nuclei into a new one, depends essentially on the degree of probability, the so-called cross section, with which colliding nuclei react with each other. The released energy results from the deficit of the masses before and after the nuclear reaction, occurs as kinetic energy of the new nucleus and produced neutral particles and is particularly large in the fusion of *deuterium* and *tritium* to *helium* (see Fig. 4.54).

Today's fusion experiments and concepts for fusion reactors rely on this process as the most promising candidate, in which the hydrogen gas is heated to about 100–250 million °C at an external pressure of about 1–2 bar; atomic shells and nuclei are separated from each other and form an electrically conductive plasma which is enclosed in a torus by means of strong magnetic fields with helical field lines. A self-supporting fusion reaction, a "burning" without external energy supply, is only possible if particle density, energy confinement time and temperature meet a minimum value,[40] the *Lawson criterion, which has* not yet been achieved.

Deuterium is present in the Earth's water in almost inexhaustible quantities, while *tritium* must be produced by "breeding" from the *lithium-6 isotope* in the blanket,

[39] See also a more recent publication Sornette et al. (2019).

[40] "Triple product" $n_e \times \tau \times T$ (electron density × Inclusion time as ratio of thermal plasma energy to power dissipation × temperature of the plasma) as a measure of the ignition condition; for tritium-deuterium reaction $2.8 \times 10^{21}\ m^{-3}$ s keV (Dolan 2012).

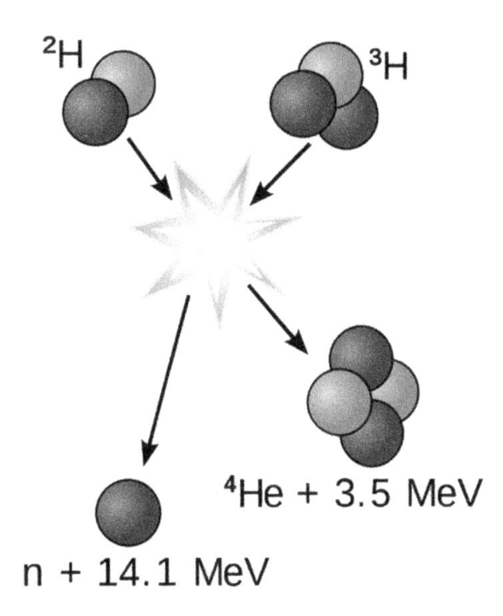

Fig. 4.54 Schematic representation of the fusion of tritium and deuterium; the kinetic energy of the neutron (80%) is used for energy generation (potential electricity generation), the charged helium nucleus (α-particles) remains in the plasma and heats the plasma with its kinetic energy. Wykis contribs (https://commons.wikimedia.org/wiki/File:Deuterium-tritium_fusion.svg), "Deuterium-tritium fusion", marked as public domain, more details on Wikimedia Commons: https://commons.wikimedia.org/wiki/Template:PD-self

the coating of the plasma vessel, by neutron capture. In addition to the necessary shielding of surrounding components against high-energy neutrons,[41] neutrons must be increased by 30–50% in the blanket, since the fusion provides exactly one neutron per tritium atom consumed; in principle, a new tritium atom could be generated from this, but losses due to absorption, escape and the transfer of the incubated *tritium* disturb the balance. Target draft designs are based on the use of the $(n, 2n)$—nuclear reaction to beryllium or lead. The heat generated by various processes is dissipated via cooling water. The necessary systems must also be accommodated in the blanket, which further increases its construction volume.

Experimental fusion research has been carried out jointly in Europe and the USA for several decades: The currently largest Tokamak, the *Joint European Torus* (JET), went into operation in 1983 (construction decision 1973), the *Tokamak Fusion Test Reactor* (TFTR) in the USA at about the same time, was operated until 1997 and replaced in 1999 by the *National Spherical Torus Experiment* (NSTX) (Fig. 4.55).

The primary goal was and is the generation and maintenance of hot hydrogen plasmas. A significant amount of energy was released via a deuterium-tritium reaction (1.8 MW over 2 s) in 1991 at the JET; in 1997 16 MW were reached, but with an external plasma heating of 24 MW.[42] In 1994 a plasma line of 10.7 MW was reached at the TFTR, three years later a plasma temperature of 510 million °C.[43]

[41] At 4.1 MeV, the energy of the neutrons produced during the fusion process is about one order of magnitude higher than that produced during the fission process.

[42] https://www.euro-fusion.org/fusion/history-of-fusion/ (accessed 13-Dec-2019).

[43] https://www.pppl.gov/Tokamak%20Fusion%20Test%20Reactor (accessed 13-Dec-2019).

Fig. 4.55 Section through the torus-shaped plasma vessel of ITER with an outer radius of 6.2 m and an inner radius of 2 m and a height of 6.7 m (with a person as scale in the lower right-hand corner); the volume is 837 m^3, which allows a thermal fusion power of 500 MW at a density of 1020 particles per m^3. The deuterium-tritium mixture is inductively heated to over 150 million °C and is to form a stable plasma which is generated by strong magnetic fields from the wall using superconducting coils. No machine-readable author provided. KentZilla assumed (based on copyright claims). (https://commons.wikimedia.org/wiki/File:ITER-img_0237_II.jpg), "ITER-img 0237 II", https://creativecommons.org/licenses/by-sa/3.0/legalcode

ITER,[44] which has been under construction since 2007, is expected to achieve a far greater fusion capacity than heating capacity[45] and a burning time of up to 1 h; 2035 is currently cited as the year for this (see Fig. 4.56). Research focuses on the testing of various methods for plasma heating, diagnostics and control as well as various concepts for the incubation of *tritium*. At the same time, materials research is being conducted at the *International Fusion Materials and Irradiation Facility* (IFMIF). ITER is still too small for a permanent fusion reaction when the *Lawson criterion is* reached or when the plasma is additionally heated, so that the plasma cools down too much. ITER is therefore not a direct precursor or prototype of a continuous fusion reactor with a closed tritium cycle for power generation.

This is to be achieved with the follow-up project DEMO (Maisonnier 2008) at a fusion power of 2–4 GW (electric power 1–1.5 GW), provided that ITER and accompanying materials research show that the Tokamak design can be extrapolated into

[44] *International Thermonuclear Experimental Reactor* (ITER, Latin for path) in which China, EURATOM, India, Japan, South Korea, Russia and the USA participate.

[45] A fusion power of 500 MW is to be achieved with a 10-fold increase in heating power; much higher values cannot be achieved with ITER technology for power plants (https://en.wikipedia.org/wiki/ITER).

ITER Timeline	
2005	Decision to site the project in France
2006	Signature of the ITER Agreement
2007	Formal creation of the ITER Organization
2007-2009	Land clearing and levelling
2010-2014	Ground support structure and seismic foundations for the Tokamak
2012	Nuclear licensing milestone: ITER becomes a Basic Nuclear Installation under French law
2014-2021	Construction of the Tokamak Building (access for assembly activities in 2019)
2010-2021	Construction of the ITER plant and auxiliary buildings for First Plasma
2008-2021	Manufacturing of principal First Plasma components
2015-2023	Largest components are transported along the ITER Itinerary
2020-2025	Main assembly phase I
2022	Torus completion
2024	Cryostat closure
2024-2025	Integrated commissioning phase (commissioning by system starts several years earlier)
Dec 2025	First Plasma
2026	Begin installation of in-vessel components
2035	Deuterium-Tritium Operation begins

Fig. 4.56 Phases of the ITER project in Cadarache, France. https://www.iter.org/proj/inafewlines#6 (accessed 13-Dec-2019)

this area. The cost of ITER, originally estimated at just under EUR 6 billion, was estimated at a good EUR 16 billion in 2010. Similar increases apply to the construction period and commissioning; the start of the deuterium-tritium process is scheduled for 2035[46] (see Fig. 4.56). The overcoming of the considerable technological obstacles and political decisions in favor of this novel technology are not expected before 2050 (Grunwald et al. 2002), a large-scale application not before the last quarter of this century.

Fusion reactors place high demands on the materials and construction concepts used:

- High flux density of high-energy neutrons leads to considerable radiation damage (power area density of about 2.2 MW m^{-2}, 100 times higher than for light water reactors).
- Formation of radioactive nuclides by activation of materials; austenitic chromium-nickel steel as blanket and structural material is not usable in the long run.
- Necessity of a periodic replacement of the innermost hull, since probably no material can withstand the high loads, primarily due to the high neutron fluxes of a commercial reactor, for years.

[46]"Under the new schedule, the start of deuterium-tritium operation is set for 2035" (http://www.world-nuclear-news.org/NN-New-schedule-agreed-for-Iter-fusion-project-2111164.html (accessed 13-Dec-2019).

Finally, it should be emphasized that thermonuclear fusion reactors with deuterium-tritium as fuel must be large-scale plants in the GW range. From a radiological and safety point of view, activated plant components represent the greatest challenge. It is required that the largest part must be stored in a controlled manner for only 100 years after the end of the service life, the small part for approx. 500 years—a final storage can therefore be omitted (EFDA 2001).

Reactivity accidents and core meltdown accidents after loss of residual heat removal, which characterize the risk of today's light water reactors, can be physically excluded.

In addition to *tokamaks, stellarators are* currently being pursued as an alternative reactor concept. In both types with toroidal arrangement of the reaction vessel a stable inclusion of the plasma can only be achieved if the magnetic field lines along the torus are not closed in a circle but twisted helically. In contrast to the tokamak, the so-called poloidal magnetic field component of the stellarator is not generated by an induced current flowing in the plasma, but by elaborately shaped coils optimized with high computer power from the outside, which has a positive effect on the stability of the plasma. A stellarator could later work as a power plant in continuous operation, but the permanent maintenance of a current in the plasma in a tokamak is by no means assured and the subject of current research. To prove the suitability of the stellarator concept for a possible fusion power plant, the *Wendelstein 7-X stellarator is* operated in Greifswald; after a 15-year construction period, commissioning and tests began in 2015, plasma was generated for the first time in February 2016 and enclosed at a power of 10 MW for one to two seconds. In addition to the realization of longer confinement times, the optimization of the magnetic field geometry and the complex coil system is the goal of the research project, in which more than one billion euros have been invested so far Fig. 4.57.

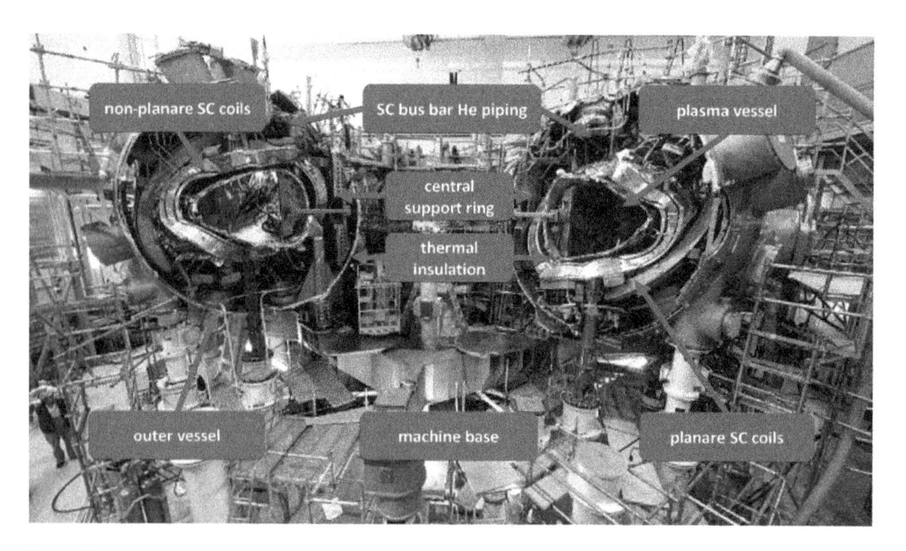

Fig. 4.57 Schematic representation of a fusion power plant according to the stellarator concept

Completely different concepts such as reactors with inertial confinement, cold fusion or the use of other fuels are being researched, but lag behind or aim at other applications and are not pursued here in the context of the topic and time horizons of this study.

4.4.12 *Power Storage*

In order to balance demand and supply at any time and any place, electricity storage is necessary, especially when variable wind and solar energy are increasingly used. The energy contained in the current can only be used for the moment of generation due to the conductor-bound charge movements at almost the speed of light. Later use is only possible by means of electrical storage (capacitors, coils) or conversion into another form of energy (mechanical, electro-chemical, thermal). For this purpose, technologies have been and are being developed that differ in the type of storage capacity and storage time, and thus according to the possible area of application (see Fig. 4.58).

When concentrating on long-term storage (days to weeks) for load balancing or the provision of reserve and control energy of large storage capacity (100–1000 MW), pumped storage, conventional and adiabatic compressed air storage, hydrogen storage and synthetic methane come to the fore.

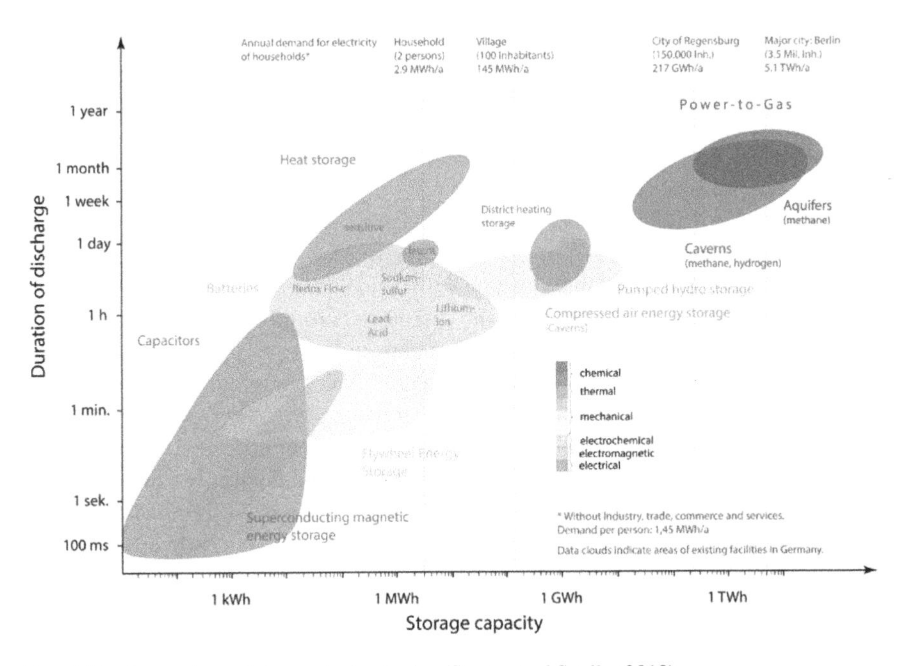

Fig. 4.58 Comparison of storage technologies (Sterner and Stadler 2018)

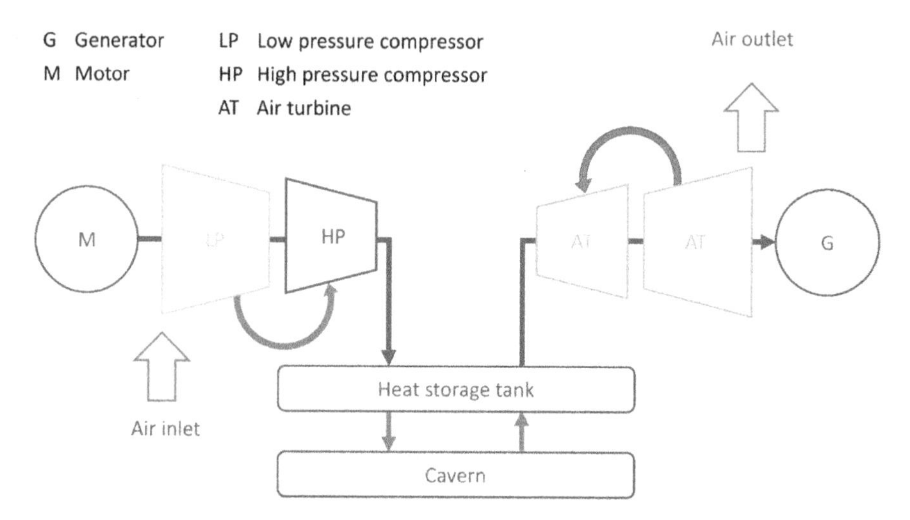

Fig. 4.59 Diagram of an adiabatic air reservoir (AA-CAES) (Meyer 2007)

Pumped storage facilities are hydropower plants that are designed not only for generator operation but also for pump operation. Today, it is the most cost-effective and most common form of electricity storage, but requires appropriate topographical conditions and public acceptance.

Conventional compressed air storage units (CAES) are gas turbine power plants that use excess electrical energy to compress and store ambient air in salt barracks and aquifers, which is then fed to a gas turbine at peak times after heating to several hundred degrees Celsius. To increase the efficiency from 42–54% to about 70%, adiabatic compressed air accumulators (AA CAES) are being developed, which do not release the compression heat into the ambient air like CAES, but use it to heat the stored compressed air (see Fig. 4.59).

In addition to storing heat ("power-to-heat"), *thermal storage systems* can also perform a function in the power supply system. Thermal energy storage devices exist in sensitive, latent and thermochemical versions, the last two of which are still largely in research and development and therefore currently have no relevance for the electricity system. Sensitive heat storage units in the form of buffer storage units in households, commerce or industry and district heat storage units in the form of district and local heating networks, on the other hand, play a very important role for the power generation system, since despite their very poor efficiency (due to a low exergy content) they represent one of the most favorable flexibility options for the power system. Heat accumulators can thus transfer the fluctuations in the electricity sector to the heat sector cost-effectively and with comparatively little effort (Agora Energiewende 2014: 137).

Excess electrical energy can also be converted into *hydrogen* ("power-to-hydrogen") by electrolysis and then compressed. It can then be stored, similar to compressed air, but due to the higher energy density with approx. 60 times higher

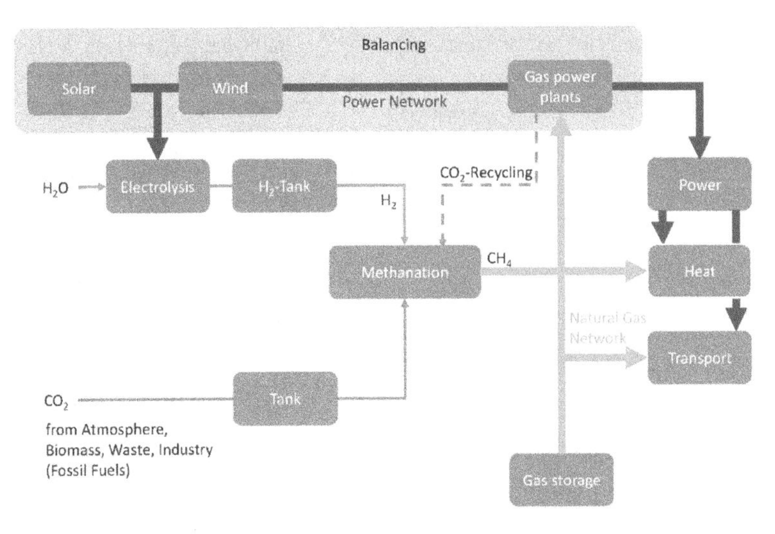

Fig. 4.60 Integrative *renewable power-methane concept* (following Sterner 2009)

useful energy quantity. An interesting alternative or further development of hydrogen storage is the sector-wide *power-methane concept*, which is based on the methanization of hydrogen and uses existing infrastructure (transport networks) (see Fig. 4.60). Of course, these "power-to-gas" concepts are only climate-neutral if the electricity is generated using CO_2-free or poor technologies (renewable or nuclear energy sources).

While thermal storage, pumped storage and conventional pressure storage have already reached market/implementation maturity, the other technologies are in the market preparation phase (AA-CAES) or in the stage of pilot/prototype plants. The costs vary from slightly more than 10 ct/kWh (pumped storage) and 15.5 ct/kWh (AA-CAES) to more than 50 ct/kWh (hydrogen storage) or are not yet foreseeable. Further characteristic values, including development potential, are listed in Table 4.12, see also Fig. A.2 and A.3 in Annex A for comparison; for information on capacities and efficiencies of various power storage devices, including direct-electric (coils, capacitors) and electro-chemical (rechargeable batteries) devices.

4.4.13 Power Transmission Grids/Intelligent Grids

Transmission grids have the task of transporting energy (primarily high-voltage electricity, gas and, in future, alternative fuels) from the place of generation to consumption and distribution centers and, together with storage facilities, of balancing usual

Table 4.12 Characteristic data for storage technologies with higher performance (Mahnke et al. 2014)

	Pump storage	Compressed air reservoir	Hydrogen storage	Synthetic methane
Field of application	Load balancing black start secondary/minute reserve	Load balancing black start minute reserve	Long-term storage load balancing, control energy, transport sector	Long-term storage, load balancing, control energy, transport sector
Efficiency (%)	56–77% (old plants), 70–80% (new plants)	42–54% (CAES), 70% (AA-CAES, future)	Approx. 40%	Approx. 20 to max. 36%
Power	From 2 MW, distributed 100–1000 MW	Expected in the future 100–1000 MW	100–1000 MW	100–1000 MW
Electricity storage capacity	Total 40 million kWh (Germany), approx. 80 billion kWh (Norway)	580,000 kWh to 2.86 million kWh, expected future approx. 1 million kWh	Theoretically usable up to approx. 200 billion kWh gas storage in Germany	Theoretically usable up to approx. 200 billion kWh gas storage in Germany
Energy density	0.27–1.5 Wh/l	Depending on pressure approx. 3–6 Wh/l (AA-CAES)	3 Wh/l (normal pressure), 750 Wh/l (250 bar), 2400 Wh/l (liquid)	Approx. 9 Wh/l (normal pressure), approx. 2.250 Wh/l (250 bar), 7200 Wh/l (liquid)
Specific. investment cost	5–20 Euro/kWh storage capacity, 500–1000 Euro/kW capacity	40–80 Euro/kWh storage capacity, 1000 Euro/kW output 2030 expected: 700 Euro/kW output (AA-CAES)	30–60 ct/kWh Storage capacity (cavern), 800–1500 Euro/kW Output (alkaline pressure electrolysis) 2000–6000 Euro/kW Output (PE electrolysis) <500 Euro/kW Output expected	No additional costs for storage in the gas network, 1000–2000 Euro/kW capacity

(continued)

fluctuations. It is generally assumed that enormous investments in transmission grids are necessary, especially in the electricity sector,[47] in order to

[47]In Europe alone several hundred billion euros; the EU estimates the financial requirements for installations and infrastructure in the energy sector at around 1 trillion euros over the next decade (HIS CREA 2011).

Table 4.12 (continued)

	Pump storage	Compressed air reservoir	Hydrogen storage	Synthetic methane
Electricity generation costs	10.3 ct/kWh (6 h storage, charging current: 2 ct)	12.9 ct/kWh (CAES) (6 h storage, charging current: 2 ct) 15.4 ct/kWh (AA-CAES) (6 h storage, charging current: 4 ct)	53 ct/kWh (200 h storage, charging current: 2 ct, 70% efficiency of the electrolyzer, 57% efficiency of the gas and steam pressure power plant)	No data available
Medullary stage	Market maturity	Marketability (CAES) or market preparation (AA-CAES)	Pilot plants, prototypes	Pilot plants, prototypes
Development potential	Underground pumped storage facilities; modernizations; restrictions due to limited locations	Research and development of AA-CAES technology; optimization of efficiency; plants > 1000 MW	Faster reaction processes, higher storage densities, cost reductions and increased efficiency for electrolyzers	Cost reductions and efficiency enhancements

- to create network structures in developing and emerging countries,
- to integrate a high proportion of fluctuating renewable energies, in particular intermittent wind and solar energy, and to overcome the growing distances between decentralized generation and centralized consumption,
- to expand and modernize existing structures in western industrialized countries and to adapt them to new conditions, such as power generation by wind and sun at medium or low voltage levels, including digital control functions.

Today's predominant alternating current grids can only be controlled to a small extent by the grid operator, since physical laws determine the current flows. This can be remedied by so-called FACTS (*Flexible Alternative Current Transmission Systems*), which are available and, in some cases, already installed in transmission networks. Direct current systems are an alternative to alternating current systems. In particular, *high-voltage direct current transmission* (HVDC) is used for long-distance transport, submarine cables and the connection of asynchronous networks. However, the HVDC requires cost- and space-intensive converter installations and is currently mainly used for point-to-point transmissions. Today, direct current transmission is economically viable from distances of 500 km, with submarine cables it can be worthwhile from as little as 40 km. Newly developed systems reduce the *break-even distance* to about 150 km (Akademien der Wissenschaften Schweiz 2012).

Box 2 Definitions (excerpt from IEA 2014)

Micro grids supply local power needs.

The term refers to places where monitoring and control infrastructure are embedded inside distribution networks and use local energy generation resources. They can supply islands, small rural towns or districts. An example would be a combination of solar panels, micro turbines, fuel cells, energy efficiency and information/communication technology to manage loads and make sure the lights stay on.

Smart grids balance demand out over a region.

A "smart" electricity grid connects decentralized renewable energy sources and cogeneration and distributes power highly efficiently. Advanced types of control and management technologies for the electricity grid can also make it run more efficiently overall. An example would be smart electricity meters that show real-time use and costs, allowing big energy users to switch off or down on a signal from the grid operator, and avoid high power prices.

Super grids transport large energy loads between regions.

This refers to a large interconnection—typically based on HVDC technology—between countries or areas with large supply and demand. An example would be the interconnection of all the large renewable based power plants in the North Sea or a connection between Southern Europe and Africa where renewable energy could be exported to bigger cities and towns, from places with large locally available resources.

In the course of new energy strategies with high feed-in shares of solar and wind power plants, whose output can vary greatly,[48] completely new grid architectures are being studied and proposed for implementation. They follow the trend towards decentralized generation, preferably using renewable sources, and optimizing grid utilization. According to Greenpeace (2011) and other sources "*key elements of the new power system architecture are micro grids, smart grids and a number of interconnectors for an effective super grid*", *for* definitions see Box 2. These elements support each other and are interconnected, merging into a (visionary) overall system (see Fig. 4.61). "*Smart grids*", which are the communicative networking and control of electricity generation.

The use, consumption and storage of resources in the transmission and distribution networks are technically feasible and, in some cases, already a reality or the subject of many projects and initiatives. They assume the availability of status information and load flow data from the individual network elements and allow demand to be adapted to fluctuating production with the aid of price signals and *peaks to be* smoothed

---

[48]In solar power plants, the output increases linearly with the solar radiation; in wind power plants, it increases with the third power of the wind speed, from approx. 90 km/h a sudden complete shutdown is to be expected.

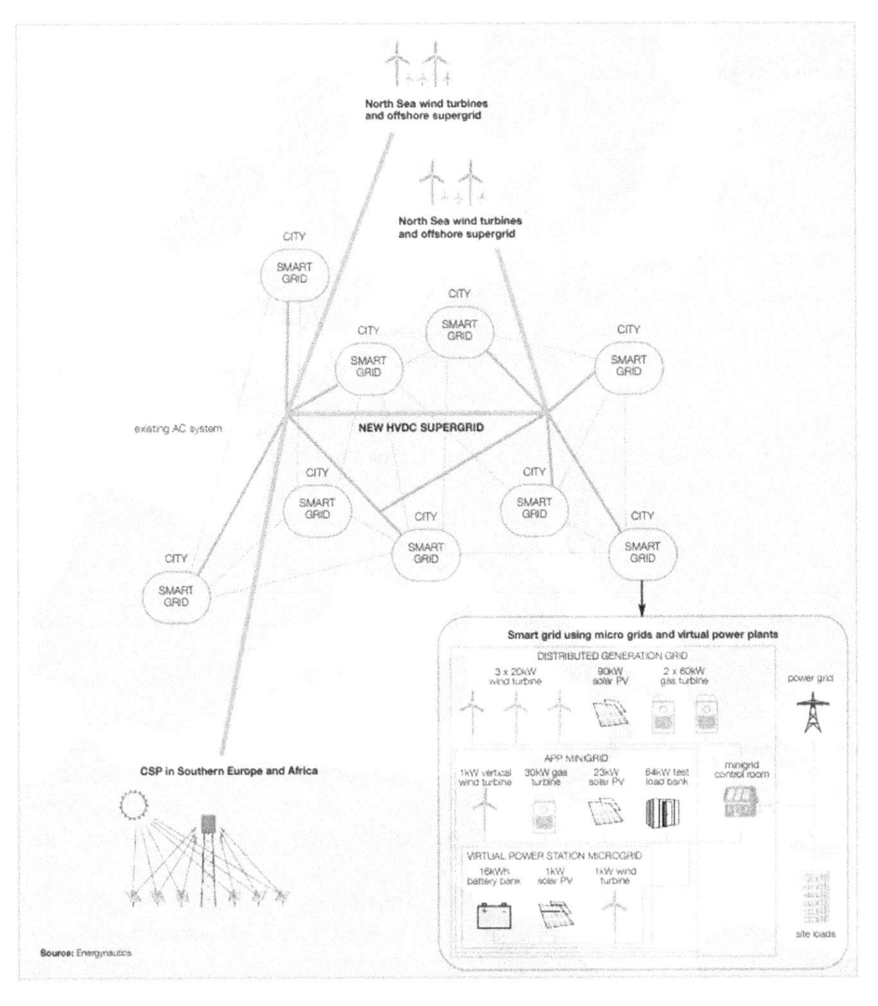

Fig. 4.61 Vision of a European electricity grid with a high share of renewable energies (Greenpeace 2011)

out.[49] This requires support from consumers ("demand side management"), who must be equipped with smart meters in order to be able to adjust their consumption to the current ore production situation. In the course of the German "Energiewende" the development of cellular energy networks was proposed, which means islands in which the production and consumption of electricity and other forms of energy such as heat and gas are balanced ("prosumers") and only in exceptional situations is recourse made to higher-level networks, from which it is also possible to disconnect.[50]

---

[49]On an annual average, networks are often only used to half capacity (in Switzerland only 30–40%); the smoothing potential observed over the day is in the range of a few to 10%.

[50]See also a recent study within the ESYS project in Germany, acatech (2020).

This concept could also play a role in developing countries and in areas far away from extreme densely populated areas ("megacities"). In the exuberance of the advantages of these new concepts, there are now voices that point to the need for sufficient baseload provision.[51]

These concepts usually make even greater use of modern information and communication technologies (ICT), often involving the public Internet, and rely on their further development; for example, so-called *Supervisory Control and Data Acquisition* (SCADA) systems must ensure real-time monitoring of the power grid and generate enormous amounts of data (see Thakur 2018). The adequate protection of enormous amounts of data and the security of the somehow coupled overall system against cyber-attacks, including manipulation, represent major challenges for the future.

The most far-reaching concept for generating "green electricity" at energy-rich locations in the world and transmitting it to consumption regions by means of high-voltage direct current transmission (HVDC) is DESERTEC; a study entitled "*Desert Power 2050*" was presented in mid-2012, the initiative was "redimensioned" in 2014 and is currently on ice.[52]

4.5 Conclusions, Recommendations

The current situation in the energy sector is characterized globally by the dominance of fossil fuels, also in electricity generation, and strongly fluctuating per capita final energy consumption depending on the level of development of the countries. Almost 1.3 billion people today have no access to modern forms of energy supply (electricity), which we regard as one of the greatest social injustices to be addressed.

In the future, decarbonization will be imperative, in addition to conserving resources, primarily for reasons of avoiding anthropogenic greenhouse gas or CO_2 emissions, i.e. climate protection, and other pollutant releases. We assume and recommend that this requirement will be maintained for the period of this study 2030–2050; it will and should determine the energy strategies and drive the development of suitable technologies to market maturity in addition to the desired security of supply and economic viability.

In general, it is assumed—with regional differences—that primary energy demand will increase slightly and electricity demand will increase more strongly; for the latter, the level of use of electricity in non-traditional sectors, notably electromobility is one of the key determinants or uncertainty factors. Future energy technologies must therefore (i) be largely free of CO_2 and pollutant emissions and dependent on limited/rare raw materials only in masses, (ii) be able to replace fossil power plants

[51]On an annual average, networks are often only used to half capacity (in Switzerland only 30–40%); the smoothing potential observed over the day is in the range of a few to 10%.

[52]Cf. https://www.wikiwand.com/en/Desertec (accessed 13-Dec-2019).

and meet increasing demand, and (iii) be accessible and affordable for large parts of the world, including countries that have so far been undersupplied.

Energy technologies must be assessed holistically, i.e. taking into account the entire life cycle. Patterns and methods are available for evaluation, for example in the form of sustainability criteria and their quantification, which can be further developed/adapted and quickly objected to. The results of previous applications show that none of the technologies known today, which can in principle be used, is superior to the others in all areas. Accordingly, weighing processes must be organized taking into account a large number of factors, including decision-making preferences.

In line with prominent roadmaps, we recommend—in the development of energy supply strategies in the medium term (2030)—to focus on technologies whose market maturity can be clearly seen from large-scale experiments and demonstration plants, and in the long term (2050) on those which already exist as ideas and whose *feasibility* can be assumed or which have been partially proven. One should not blindly rely on "miracles" and completely disregard necessary development times, usually about 20 years from invention to market maturity. Accordingly, in our opinion, nuclear fusion and some "exotic variants" of nuclear fission unfortunately fail as bearers of hope. In addition, with the aim of achieving robust security of supply and avoiding surprises and reducing uncertainties, diversity with regard to the technologies used, i.e. a mix of different energy sources, should be relied upon.

Regarding energy sources, a distinction is made between

- constantly renewable energy sources (RES) like wind, sun and practically also water or delayed renewable energy sources like biomass;
- spent fuels which are available in large/practically unlimited quantities and which can be recovered through technical processes such as nuclear energy (uranium/thorium fuel cycle, burn and breed reactors);
- consumption of fuels available in large but finite quantities, such as oil and natural gas or coal, whereby the reserves can be stretched by means of technical processes (gas fracking);
- others like geothermics.

Wind and solar energy could potentially cover the world's energy needs alone. However, it is pointed out, for example by the OECD-IEA in its World Energy Outlook 2016, that this is practically impossible: the technical use of this potential is hampered by site conditions, their intermittent nature and technical, ecological and economic barriers. Hydropower has great potential, but it is already largely exhausted; tidal power plants play only a modest role globally. Geothermal energy also has a large potential, albeit one that varies greatly from region to region, and which can often only be exploited to a greater extent with relatively large risks/costs.

With regard to specific CO_2 emissions, based on the results of life cycle analyses, today's coal-fired power plants rank at the end of the negative scale at about one kilogram per kWh and gas-fired power plants slightly below; more than one order of magnitude below are PV (roof) and biogas plants, while wind power (offshore) and nuclear power plants have CO_2 emissions that are almost 100 times lower than coal-fired power plants; hydroelectric power plants have even lower values. Accordingly, the reduction of CO_2 emissions from the use of renewable energy sources, especially PV and biogas, is one of the goals of technical developments that is considered to be safely achievable for PV systems. Coal-fired and gas-fired power plants can only meet acceptable limits for CO_2 emissions if they are equipped with CO_2 capture and storage (CCS) systems; bioenergy with CCS even permits negative emissions. In principle, the necessary technologies are available and their economic use within the framework of an overall concept can be expected, if at all, at the earliest after 2030.

As already mentioned, research and development work can be identified in the field of renewable energy sources (RES), which, in addition to the "search for something completely new", aims at the further development of today's technologies towards (i) increases in efficiency (PV) and/or higher working temperatures (solar thermal) while at the same time reducing material dependencies, (ii) consideration of environmental aspects and cost reductions (PV cells), (iii) efficiency improvements and more environmentally friendly conversions (wind, especially biogas) and (iv) further evolutionary steps towards commercialization and reliability. We assume that most renewable energy sources will be usable on a large scale during the period under review and that they will be able to generate electricity competitively in further widened siting regions, which is already the case for wind power at favorable locations. However, a similarly positive, reliable picture does not yet emerge for the necessary infrastructures, above all for the storage and transport/distribution systems, the development of which has been going on for a long time but is still waiting for decisive breakthroughs.

Nuclear power plants—technologically and technologically "evolutionary" or "revolutionary" further developed—are already partly under construction today or will be sufficiently mature by 2030 at the latest. In addition to social acceptance, the financing, excessive costs and construction time and related overruns as well as lack of flexibility of large plants are problems that small, modular units promise to overcome. Such concepts are also available in the medium term, especially if they are based on the technology of today's light water reactors. More "exotic" concepts, which promise a further increase in safety, i.e. the total exclusion of serious accidents due to inherent physical mechanisms, but which can also avoid or eliminate extremely long-lived waste products and incubate fissile material, are under development; they can play a significant role in an energy mix only in the long term, i.e. beyond 2030.

A number of energy sources and the corresponding technology are therefore already available, or their development is reasonably certain. They can be used depending on the conditions in the countries of demand and use, whereby each region must and may find its optimum through appropriate processes for the deployment strategy, as long as overarching objectives such as global climate and environmental protection are not jeopardized.

Chapter 5
Aspects of Environmental Compatibility of Energy Systems

5.1 Introduction

The development and preservation of human beings is closely linked to the production of useful energy. Basic considerations must be taken into account here: To satisfy their needs for food, clean air and drinking water, humans and all living beings depend on intact ecosystems today and in the future (sustainably). Energy systems must therefore be required to provide sufficient useful energy at low cost without endangering the natural foundations of life and the environment in the longer term.

Particular attention and protection should be given above all to human health and those parts of the biosphere that are functionally necessary for the survival of humanity and the preservation of the environment with its living organisms:

- The atmosphere (clean air)
- Clean water
- Agricultural land
- Stable climate and vegetation zones (as a prerequisite for the preservation of the environment in its basic substance, biodiversity and the worldwide cultivation of crops).

A key role in keeping the air clean, producing drinking water and ensuring the stability of climate zones is played by the global preservation of large ecosystems, in particular large forests and oceans.

Global biodiversity can be threatened by the expansion of energy systems. The importance of biodiversity in terms of ecosystem stability as well as its role as a potential genetic and pharmaceutical resource speaks in favor of its conservation. Similarly, aesthetic considerations in relation to geological forms speak in favor of these aspects. The sustainable protection of these factors must be given special attention and guaranteed. There is no doubt that changes in these systems and factors are subject to certain natural, evolutionary influences. This happens only slowly over longer periods of time.

C. F. Gethmann et al., *Global Energy Supply and Emissions*,
Ethics of Science and Technology Assessment 47,
https://doi.org/10.1007/978-3-030-55355-5_5

However, energy systems can also pose direct *health risks to* people, and changes in the environment can be caused more quickly by pollutant emissions during normal operation and in the event of major technical accidents. Major accidents, in particular, can cause damage to people and the environment without causing large-scale environmental damage over large areas. Health risks for humans as well as environmental damage should generally be dealt with within the framework of this study not only under local/regional influences but also under the aspect of global environmental compatibility.

Beyond mere survival, people also value nature as a recreational space and among others of its beauty. Against this background, interventions in a landscape should also take aesthetic needs into consideration. In view of the varying severity of the environmental impacts, it is proposed that the environmental impact of an energy system be assessed according to the following priority standards:

1. Interventions in the entire biosphere or in individual ecosystems caused by technical action (in the production and dismantling of the facilities with their materials, consumption of resources, release of pollutants) must not cause any irreversible damage, or irreversible damage to an extremely small extent only, that does not endanger the natural basis for survival of humans and all other living beings today or in the future.
2. Health risks for people caused by energy systems and possible interventions in biodiversity must be in reasonable proportion to the benefits of energy supply.
3. The landscape interventions associated with energy systems should take into account geological structures and aesthetic needs. Possible impairments must be weighed against the benefits of the energy supply in each case.

When pollutants are released, especially into the atmosphere, the global effects must always be taken into account. There is no doubt that the different economic and social circumstances of industrialized, newly industrializing and developing countries must be taken into account in a differentiated manner. These issues will be addressed in detail for the following energy options: Fossil energy sources (hard coal, lignite, crude oil, natural gas), renewable energies (various solar energy options, wind power, hydropower, biomass). tidal and ocean wave energy, geothermal energy), nuclear energy and to a small degree nuclear fusion.

The advantages and disadvantages of these energy options, including their short-term and long-term effects, should be weighed against each other. Finally, an attempt will be made to develop recommendations for the future on this basis.

5.2 The Regulative Idea of Sustainability

Since the Rio de Janeiro Environmental Conference in 1992 at the latest, the concept of sustainability has also dominated the debate on energy policy, having previously dominated environmental and development policy. It therefore seems unavoidable to make at least some preliminary considerations about the concept of the sustainability concept in a paper on energy supply problems.

The Saxon chief mining officer Hans Carl von Carlowitz coined the term "sustainability" as early as 1713 for the management of forests and in 1791 the chief forester in the service of the Prince of Solms-Braunfels, Georg Ludwig Hartig, demanded "sustainability" for the management of state forests in view of the constantly growing demand for timber and firewood:

> No long living forestry can be imagined and expected if the wood levy from the forests is not handled on the basis of sustainability. Every wise forestry directorate must therefore have the state's forests assessed without any loss of time and *try to use* them *'as highly as possible, but in such a way that the descendants can derive at least as much advantage from them as the generation now living can'*.

Hartig mentions the most efficient possible use of natural resources in the present and the consideration of the interests of "descendants", i.e. future generations, as equally important goals between which a reasonable balance must be found. The use of natural resources by present generations, so the maxim could be modernly formulated, finds its limit where future generations are deprived of their chance of an equivalent use.

This understanding of sustainability is meanwhile the basis for the definition of the Brundtland-Commission (Report of the World Commission on Environment and Development) from 1987:

> Humanity has the ability to make development sustainable *to ensure that it meets the needs of the present without compromising the ability of future generations to meet their own needs.*

In its core meaning, sustainability is therefore a moral maxim without concrete content specifications. The way in which it will be ensured that future generations will be able to adequately satisfy their needs remains at least open. The sustainability maxim of the State Forestry Administration has a material content insofar as it specifically aims at the conservation of the natural wood resources in the forests, but it originates from a time when wood had to cover almost the entire energy demand of the country and thus—unlike today—no alternative energy options were available. Sustainability must therefore be determined again and again today, against the background of alternative energy options. Sustainable energy options must be promoted or used to achieve optimum economic growth and environmental protection (including the earth's atmosphere) while conserving resources from a global perspective.

Every energy system is always only a more or less optimal means of achieving the goal formulated in the sustainability concept. In the concretization of the demand for sufficient satisfaction of the needs of today's people worldwide while at the same time taking into account the needs of future generations, the concept of sustainability is characterized above all by a conflict of objectives between these two demands.

In the discussion on the regulatory idea of sustainability, it became clear that its classic formulation contains the maxim of using natural resources in such a way that future generations can still derive an equivalent benefit from them. Ultimately, it is therefore an ethically motivated standard, which, at its core, aims for the *long term* and derives its claim to validity from its responsibility to future generations. Both in everyday practical understanding and from a philosophical-ethical perspective, however, it is quite unclear whether there really is a long-term responsibility towards future generations. The assessment and evaluation of the sustainability of energy technologies is of fundamental importance for the assessment of the environmental compatibility of energy systems (Streffer et al. 2005).

5.3 Environmental Problems Caused by Burning of Fossil Fuels

The release of gaseous pollutants and dust particles from the combustion of fossil fuels causes a number of important environmental impacts. The most important of these are damage to human and animal health caused by air pollutants, immission damage to plants and the anthropogenic greenhouse effect. If the pollutants are chemically inert, such as carbon dioxide (CO_2), methane (CH_4) and other hydrocarbons, they spread in the atmosphere over wide regions to global dimensions. In the case of substances with higher chemical reactivity, such as sulfur dioxide (SO_2), nitrogen oxides (NO_x) and ozone (O_3), the atmospheric dispersion takes place in smaller regions than in the case of inert gases. In addition, dusts with heavy metals (e.g. lead and mercury) including naturally occurring radioactive substances with the fossil fuels, e.g. radium (^{226}Ra), uranium (^{238}U) are released, which sediment in the vicinity of the relevant technical installations. The distance of this range is depending on the size of the particles and the height of the release.

The air pollution with these particles and pollutants is extremely critical for human health and thus a significant risk factor caused by power plants that are run on fossil fuels, especially hard coal and lignite. The particles and pollutants cause in humans mainly cardiovascular diseases, brain infarcts, chronic obstructive pulmonary diseases and lung cancer. There are also increased risks of lung infections.

Air pollution is caused by traffic, industrial plants burning fossil fuels, in particular lignite and hard coal, and burning fossil fuels, in particular shrubbery and other biomass, including dung, in the household. The latter occurs especially in regions where no connection to electrical networks is available. The WHO has estimated that 4.3 million deaths worldwide were caused by this last factor in 2012 (WHO 2014). This high risk occurs to a large extent in Southeast Asian countries (1.69 million), the West Pacific region (1.62 million) and Africa (0.58 million) (Fig. 5.1).

Although these high levels of exposure to these toxic substances and dust particles within the houses during cooking with biomass are decreasing with increasing electrification, they are still widespread in countries such as India, especially in rural regions. In 2013, for example, 240 million people (about 20% of the population) in India had no access to electricity. However, electrification in India varies greatly between the various states, but is on average much lower than the "world average" (Fig. 5.2). In the state of Bihar, more than 60% of households still had no electricity in 2013 (IEA 2016c India).

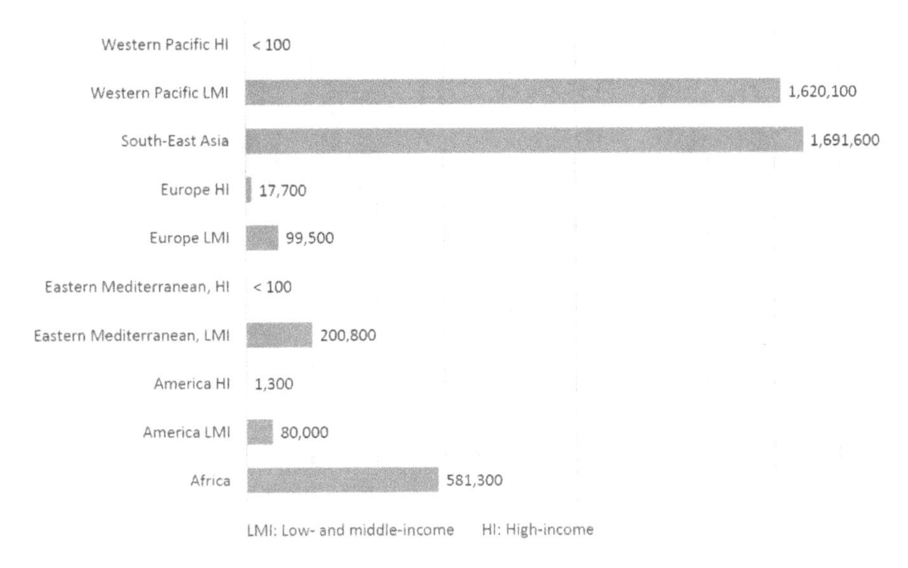

Fig. 5.1 Total deaths attributable to household air pollution in 2012, by region (WHO 2014)

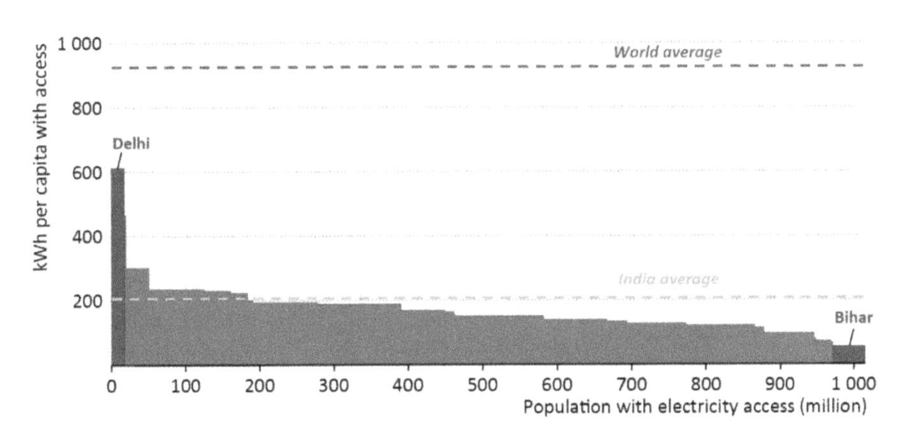

Fig. 5.2 Annual residential electricity consumption per capita by state in India (for those with access), 2013 (IEA 2016c, India)

About 840 million Indians used biomass (shrubs, wood, cow dung, etc.) as energy material for cooking in 2013. On the other hand, increasing electrification leads to an increase in electricity generation from fossil fuels, mainly coal which causes massive release of CO_2. Energy demand in India rose from 441 Mtoe ("million tons of oil equivalent") in 2000 to 775 Mtoe in 2013, while coal consumption rose from 146 to 341 Mtoe, i.e. more than twice as much. Figure 5.3 shows the enormous increase in the use of fossil fuels in India for electricity generation between 1990 and 2013 (IEA 2015).

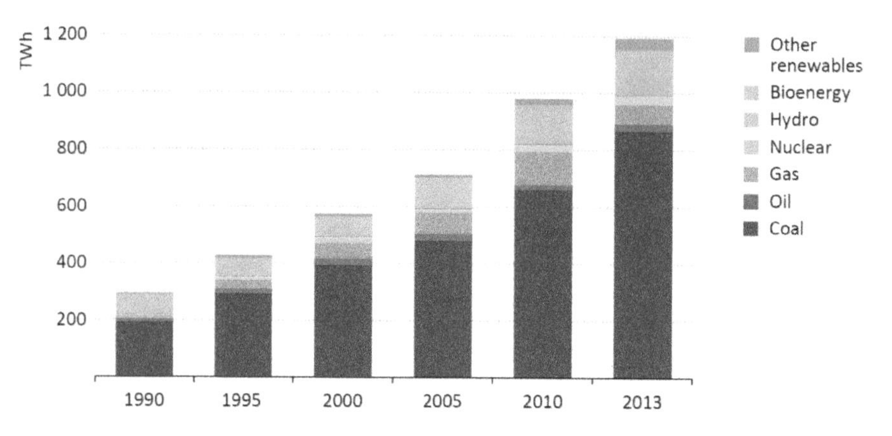

Fig. 5.3 Total electricity generation in India by fuel. *Note* "Other renewables" includes solar PV and wind (IEA 2016c, India)

5.3.1 Release of Green House Gases

Green House Gases (GHG) (CO_2, CH_4 with other gaseous hydrocarbons and, to a certain extent nitrogen oxides) undoubtedly play the most important role in the debate on climate change. Here, CO_2 is in the front line because of its absolute quantity. The CO_2 and the CH_4 with the other gaseous hydrocarbons are distributed globally over wide regions in the atmosphere and by absorption in the water of the oceans. International climate conferences are struggling to limit the release of these gases. Efforts to achieve this are also given high priority in national planning. Nevertheless, emissions of these gases have risen worldwide in recent decades (Fig. 5.4). By far

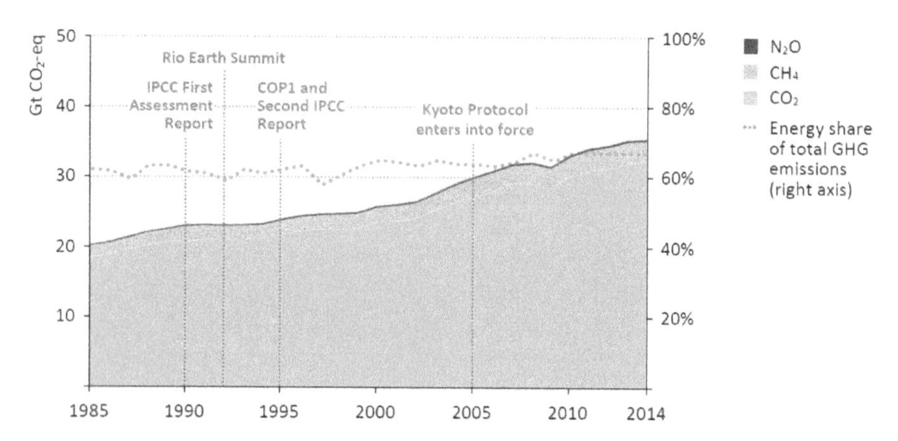

Fig. 5.4 Global anthropogenic energy-related greenhouse-gas emissions by type. *Notes* "CO_2 = carbon dioxide, CH_4 = methane, N_2O = nitrous oxide. CH_4 has a global warming potential of 28–30 times that of CO_2 while the global warming potential of N_2O is 265 times higher than that of CO_2" (IEA 2016a, World)

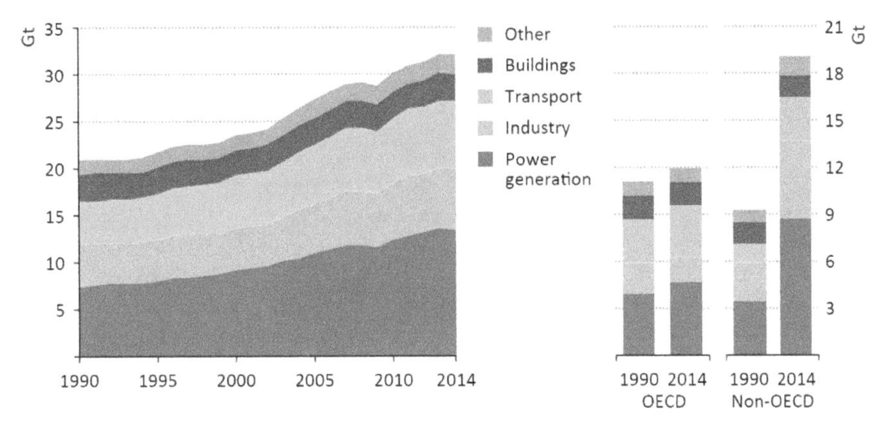

Fig. 5.5 Global energy-related CO_2 emissions by sector and region. *Notes* "'Other' includes agriculture, non-energy use (except petrochemical feedstock), oil and gas extraction and energy transformation. International bunkers are included in the transport sector at the global level but excluded from the regional data" (IEA 2016a, World)

the largest share is in CO_2 emissions. Emissions are reported by the International Energy Agency (IEA 2016d) in Gt CO_2-eq, assuming that CH_4 has 28–30 times and NO_x 265 times the potential for global warming compared to CO_2.

As already mentioned, CO_2 emissions from the combustion of fossil fuel. s for electricity generation make the largest contribution, followed by emissions from transport and industry (Fig. 5.5) (IEA 2016a). The increases between 1990 and 2014 are considerably higher in the non-OECD countries (emerging and developing countries) than in the OECD countries (industrialized countries). This is due to the considerable backlog demand in industrialization to increase the prosperity of the population in the non-OECD countries. This is illustrated by the emissions of China, the United States, the European Union, India and Japan (Fig. 5.6) (IEA 2015). The changes in CO_2 emissions from 2013 to 2014 with a decrease in CO_2 emissions in the European Union, China and Japan are interesting in this context, while emissions in the U.S.A. and especially in India continue to rise (Fig. 5.6) (IEA 2015).

5.3.2 Health Damage Caused by Pollutant Emissions

As already described, the combustion of fossil or organic fuels produces pollutants that can be harmful to human health, animals and plants. The health risks posed by these pollutants have long been ignored or underestimated. The "Great London Smog" of 1952 brought about a change in perception. In December 1952, London sank for several days under a dense haze of smoke and dust. The high level of air pollution led to serious health problems, especially for people with respiratory problems and for the elderly. Tens of thousands of people had to be admitted to

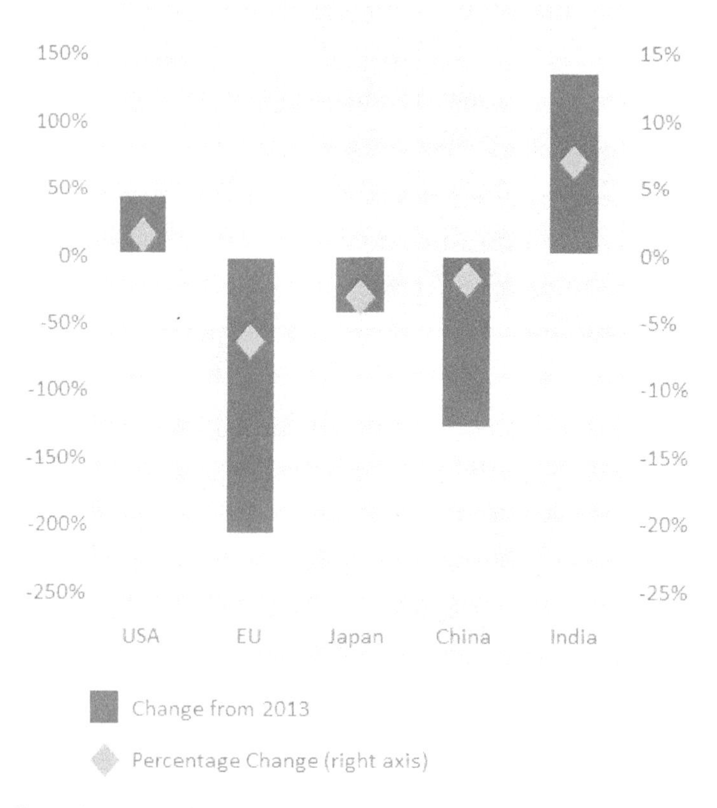

Fig. 5.6 Change in energy-related CO_2 emissions by selected region, 2013–2014 (IEA 2015)

hospitals for respiratory problems, which soon became overcrowded. In the course of the smog period, which lasted several days, about 4000 people died, according to more recent estimates it could even have been up to 12,000. The first epidemiological studies on health hazards from air pollutants were then carried out, on the basis of Western industrialized countries for the first time issued legal limit values for certain pollutants and finally the World Health Organization (WHO) recommended guidelines (WHO 2016). About 72% of the large number of deaths mentioned above are caused by ischaemic heart disease and infarction, about 14% by obstructive pulmonary disease and infection and about 14% by lung cancer (WHO 2016). In 2008, the European Parliament and the EU-Council adopted a directive setting limit values for the pollutants sulfur dioxide, nitrogen dioxide and particulate matter PM_{10} and $PM_{2.5}$.

5.3.2.1 Overview of the Quality of Health Damage

Sulphur dioxide impairs the function of the respiratory tract and lungs, worsens asthma and chronic bronchitis, increases the risk of infections, heart disease and heart attacks. Nitric oxides increase or worsen the risk of asthma, lead to chronic lung diseases, cardiac arrhythmias and infarcts. These oxides promote the formation of ozone in sunlight. Dust particles (PM_{10} and especially $PM_{2.5}$) cause the same health damage and cause lung cancer. Mercury (Hg) and lead (Pb) damage the development of the brain and nervous system in children. These heavy metals also damage the liver and kidneys and can cause prenatal developmental defects (HEAL 2013).

5.3.2.2 Overview of Some Results of Epidemiological Studies

While the first epidemiological studies began with single events, such as short-term smog periods, in the following decades the emphasis shifted increasingly to long-term studies. US-American cohort studies, whose investigation period extended from the late 70'er to the 80'er years. The results of these studies showed that the exposure to dust, which is caused among other things by incomplete combustion processes—as an indicator the initial inhalation of particulate matter with a particle diameter <10 μm PM_{10} (PM_{10} = particulate matter with a diameter of 10 μm and less)—leads to an increase in respiratory diseases, which is reflected in a significant decrease in average life expectancy.

One of the most comprehensive studies carried out on air pollution problems is the European APHEA (Air pollution and health: a European approach) study. The extended APHEA-2 study covered 43 million people in 29 European cities over a period of 5 years in the mid-1990s. Within the framework of the study, an increase in the general mortality rate in the population of 0.6% was determined for an increase in the PM_{10} concentration of 10 mg/m^3 each (Brunekreef and Holgate 2002; Sumjer et al. 2003). The studies thus confirmed the results of an older and similarly extensive study from the USA (NMMAPS 1987–1994). For the same concentration levels of PM_{10}, the latter had determined an increase in the general mortality rate of 0.5%. Overall, the above studies demonstrated the toxicity of PM_{10} and ozone (O_3). However, for other substances such as nitrogen dioxide (NO_2) such evidence has not been provided for the exposures occurring in this study. However, recent studies have revealed various types of damage to health, which will be discussed later (WHO 2014).

5.3.2.3 Risk from Fine Dust (Particle Size <2.5 μm, $PM_{2.5}$)

Among the controversial issues of health hazards from exhaust gases is the extent to which some types of dust are more dangerous than others. Studies that include by the US Environmental Protection Agency (EPA) indicate that particulate matter (PM < 2.5 μm) and ultrafine particles (PM < 0.1 mm), as they occur as a result of incomplete combustion in furnace heating systems, but in particular as combustion products of diesel engines in road traffic as well as when fossil fuels are used in power plants, pose the greatest hazards, since their small diameter during inhalation allows them to reach the deepest and most sensitive tissue areas of the lungs and then also diffuse into the blood US Environmental Protection Agency (EPA 2002). A long-term study presented by the American Cancer Society on the effect of particulate matter has confirmed the suspicion that particulate matter poses a significant risk of cancer. The study of the research group around Pope et al. (2002) compared the individual risk data and causes of death of 500,000 people over a period from 1982 to 1998 with data on air pollution in their respective places of residence in various conurbations in the USA. Through the very precise collection of individual risk factors, such as overweight or smoking, these risks for cancer could be quantified within the framework of the study and compared with the proportion of cancer risk from air pollutants. The researchers succeeded in showing that an increase in the fine dust concentration ($PM_{2.5}$) by 10 mg/m^3 is correlated with an increase in the general mortality rate of 4%. The researchers found an increase in the cause specific mortality rate for cardiovascular diseases and lung cancer of 6% and 8% respectively. These data have been confirmed several times, so that the International Association for Research on Cancer (IARC) and the WHO in general have adopted the conclusions (WHO 2014).

5.3.2.4 Health Damage in People with Respiratory Diseases and Children

As the great London smog, in which many people with respiratory diseases died, has already shown, it is obvious that people with respiratory diseases are pre-damaged by high concentrations of pollutants and therefore react more strongly than healthy people to respirable pollutants. However, studies of the specific pressures of this group led to contradictory results. While the European PEACE (Pollution effects on asthmatic children in Europe) study found no correlation between soot concentration and acute complaints in children suffering from respiratory diseases (Brunekreet and Holgate 2002), the results of a Dutch study point in the opposite direction (Boezen et al. 1999). The group around Boezen et al. (1999), within the framework of a three-year long-term study of 632 Dutch children suffering from respiratory diseases in rural and urban regions, was able to show that an increase in the concentration of pollutants in the air in these children is also correlated with a significant increase in acute pathological respiratory symptoms. Today it is generally assumed that individuals with respiratory diseases react more sensitively to the described pollutants and particles than healthy individuals.

In 2005, the WHO published a summary study on the effects of air pollution on children's health and development (WHO 2005). It has been studied the effect of pollutants during prenatal development in the womb, on the development and function of the child's lungs, respiratory diseases such as asthma, bronchitis and cough, allergies, frequency of infections, cancers and neurophysiological development. For all these health effects, increases in frequency due to air pollutants have been observed. There are certain uncertainties about the cause of cancer. Longer observation times are necessary here because of the latency period until the appearance of these diseases. The studies carried out to date do not allow a clear differentiation to be made as to which pollutants develop specific diseases with regard to these effects. Only in the case of mercury did the WHO working group come to the conclusion that this pollutant specifically causes neurophysiological damage in child development. In all cases, children were found to be much more sensitive than adults (WHO 2005).

5.3.2.5 Do Threshold Values Exist for Health Exposure to Air Pollutants?

Despite extensive long-term studies, a number of important questions in connection with pollutant pollution have not yet been clarified, in particular the question of whether threshold values exist for some pollutants. For example, a PM_{10} threshold of 50 mg/m^3 is discussed for cardiovascular damage. For ozone, such a threshold value could exist below 100 mg/m^3 (Brunekreet and Holgate 2002). The assessment of combined exposures to human health, i.e. the simultaneous effects of different air pollutants, continues to pose difficulties. Combined exposures can both add to and mutually reinforce or weaken their effect on health (Streffer et al. 2003). However, superadditive effects after combined exposure to two or more toxic agents only occur when specific mechanisms are applied, and generally high exposures are present.

Better experimental data on the mechanisms of action are necessary to decide whether threshold doses for the effect of particulate matter exist; the findings to date do not clarify this problem. Therefore, this question remains open. The WHO, however, bases its assessments for PM_{10} particles on threshold doses (WHO 2016). Dose-response relationships have been developed for smaller particles ($PM_{2.5}$) as well as for the other pollutants mentioned in order to determine the number of cases of disease or death caused by the pollutants (WHO 2014). The above-mentioned case numbers have been estimated with the help of these methodical approaches.

5.3.2.6 Health Effects and Government Intervention: Intervention Studies

Despite some uncertainties in detail, the studies mentioned provide sufficient evidence for a damaging influence of the concentration of air pollutants on the number of respiratory and cardiac diseases. Two intervention studies from Ireland and Hong Kong confirm that air pollution control measures have a positive effect on health. In long-term studies, the health effects of legal air pollution control measures were observed:

- Irish researchers were able to show (Clancy et al. 2002) that in Dublin the number of annual deaths from respiratory and heart disease had fallen by 15.5 and 10.3% respectively by 1996, after the burning of coal in domestic stoves and ovens had been banned by law in 1990. The soot concentration of the air in Dublin has dropped by 70% and the sulfur dioxide concentration by 30% over the same period.
- A similar conclusion is reached by an intervention study from Hong Kong, where a limit of 0.5% by weight for the maximum permissible sulfur content in crude oil was set in 1990 (Hedley et al. 2002). As a result of the legislation, the average SO_2 concentration in the air fell by 80% over the next few years. At the same time, the mortality rate in Hong Kong fell by 3.9% for lung and respiratory diseases and by 2% for cardiovascular diseases. Particularly significant, however, was the decline in the traditionally higher seasonal mortality rate in autumn, which fell from 20.3 to 5.3% for respiratory diseases.
- The WHO has repeatedly recommended guideline values for these air pollutants for the last time in 2016 (WHO 2016).

5.3.2.7 Estimates of Current Health Impacts of Air Pollutants in Industrialized Countries

In all western industrialized countries, the health burden from pollutant emissions has decreased due to the setting of guideline and limit values as well as technical environmental standards (WHO 2016; EU 2008). In the Federal Republic of Germany, total dust emissions fell by 84.2% between 1970 and 1996, sulfur dioxide emissions by 76% and nitrogen oxide emissions by 31.1% (Jänecke 2000). Exposure levels in Germany have decreased only slightly since 2000. This decrease is superimposed by weather-related fluctuations.

A study by the Künzli research group (Künzli et al. 2000) commissioned by the WHO and based on current measurements and the dose-effect relationships of suspended and particulate matter determined in earlier studies, determined the number of premature deaths due to air pollution in France, Austria and Switzerland to be at least 40,000 per year. According to information from this working group, more than half of the air pollutants originate from road traffic. The Group estimates that the costs of air pollution for the health system amount to as much as 1.6% of the GDP of the three countries. They are thus higher than the health costs of traffic accidents. Limiting estimates such as those by Künzli et al. (2000), it should be noted that they are based on assumptions that are not shared by all epidemiologists (Brunekreef et al. 2002). The following assumptions are made for the risk assessments:

- The measured noxae should actually be the cause of the observed health damage.
- There is a linear dose-response relationship without a threshold value.

Especially this second assumption seems questionable, as already discussed before, since according to all toxicological findings such a dose-effect relationship can only be assumed for stochastic effects caused by genotoxic agents. It is therefore very unlikely to apply to respiratory diseases and cardiovascular damage. It is more likely to assume dose-effect relationships with a threshold dose. In this case, the previously mentioned risk figures would lead to a systematic overestimation of the health effects of pollutant emissions.

A previously unpublished study by the EPA (USA) indicates emissions of nitrogen oxides (NO_x) in the USA. In the years 1990–2013 NO_x emissions decreased by about 50%. About 51% of these emissions are caused by transport (37% by road, 20% by rail and air) and about 14% by electricity generation (of which about 85% by coal combustion). It should be noted that these emissions occur predominantly very locally in the immediate vicinity of the installations, that they can contribute to exposure of the population and then to acute health effects of the cardiovascular type and respiratory diseases including asthma. It seems that children are particularly vulnerable here again.

In a very comprehensive study "Europe's Cloud—How Coal-Burning Countries are Making Neighbours Sick", prepared by working groups of the "Climate Action Network Europe" (CAN), "HEAL" (Promoting environmental policy contributing to good health) Brussels, "sanbag" and the "WWF" European Policy Office and published in 2016. In 2015, 280 coal-fired power plants generated 24% of the EU's electricity. Emission data (CO_2, SO_2, and NO_x) were available from 257 of the 280 coal-fired power plants. In the case of emission data for mercury, emissions were only available from 156 power plants. On the basis of these data and the population distribution, health damages, primarily "lost life years" and thus "premature deaths" are determined. The already mentioned methodological approaches of the WHO (2014) were used for these assessments. A threshold dose of 0.58 μg/g hair (according to the WHO approach) was used to estimate developmental damage in newborns. In 2013, 1.8 million children were born in the EU with exposures above the WHO threshold dose.

The authors come to the conclusion that 22,900 "premature deaths" are caused each year by the currently operating coal-fired power plants in the EU (with 95% confidence intervals of 14,400 and 31,900). Poland and Germany are at the forefront. Great Britain and Romania follow (Fig. 5.7). Emissions of particulate matter ($PM_{2.5}$) account for 74% of health damage and are thus the main cause, followed by nitrogen oxides with around 14.5%. The emissions of these agents are widespread, so that effects of these emissions occur in neighboring countries. Corresponding risk assessments have been carried out by the authors. They come to the conclusion that the Polish power plants cause about 4700 premature deaths abroad, while the German power plants of these values cause about 2500 cases. France suffers the strongest damage from other countries. The authors estimate this damage at 1200 premature deaths.

Power plants using lignite as power source pose a much higher risk than power plants powered by hard coal. With regard to CO_2 emissions, the lignite-fired power plants predominate among the 10 power plants (7 with lignite, 2 with hard coal and 1 not specified). This is also the case with regard to emissions of pollutants. The diseases that occur, which apply to the effects of the corresponding pollutants, have been previously described. Here, too, among the first 10 power plants with the

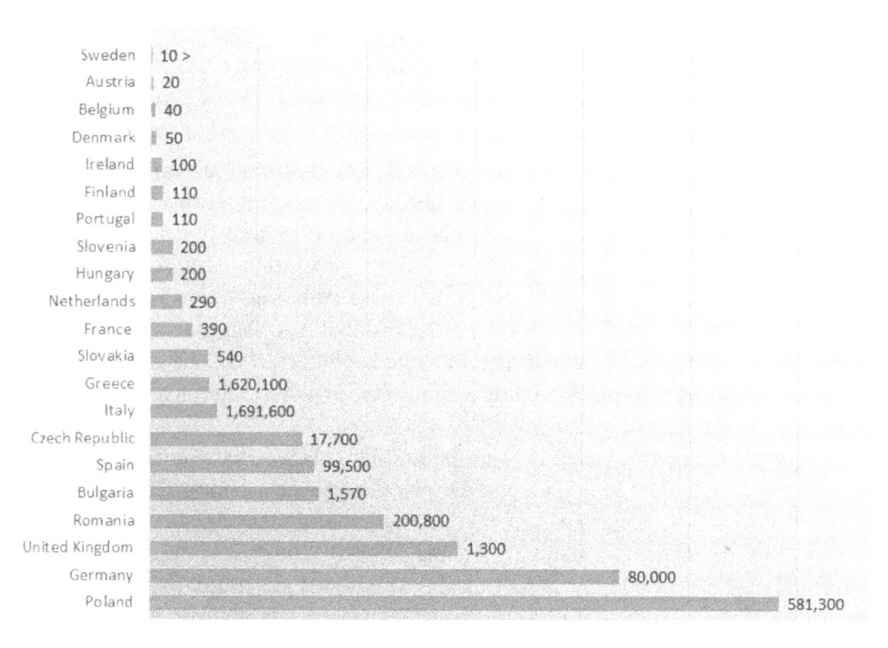

Fig. 5.7 Premature deaths of coal-fired power plants in the respective countries where the power plants are located (Europe's Cloud 2016)

highest emissions, those with lignite-fired power plants generally dominate with SO_2 (7), with NO_x (only 4) and mercury (9). It follows from this that lignite-fired power plants also account for the highest proportion of disease and death frequencies.

5.3.2.8 Air Pollution in Developing Countries

Despite the high pollutant emissions of the European coal-fired power plants described above, it must be borne in mind in the global context that the successes in air pollution control are largely limited to Western industrial countries. In many Third World countries, pollution levels remain extremely high. It is estimated that between 2.7 and 3 million people die every year worldwide as a result of air pollution, 90% of them in developing countries. Approximately 1.1 billion people annually suffer health effects from air pollution outside their homes and 2.5 billion from indoor air pollution. According to conservative estimates, about 780,000 people die each year in China and India as a result of domestic air pollution (UNEP 2002; UNEP 2008). In South Asia, air pollution outside the home has also risen dramatically in recent decades, as already indicated. Calcutta, Delhi, Mumbai, Karachi and Dhaka as well as Beijing and other Chinese mega-cities are among the cities with the highest air pollution levels worldwide. In Delhi, for example, air SO_2 concentrations increased by 109% and nitrogen oxide emissions by 82% between 1989 and 1996. Indian cities also have very high concentrations of suspended and particulate matter. The average annual PM_{10} concentration in Calcutta was measured with 360 mg/m^3 and in New Delhi 270 mg/m^3. In all 23 megacities in India, air pollution exceeded the limits recommended by the WHO (UNEP 2002; UNEP 2008). In China, 98% of the urban population is also exposed to concentrations of air pollutants that exceed WHO recommendations. It was estimated that 340,000 people die each year in India and China due to concentrations of pollutants outside the home (UNEP 2008).

Due to the increasing combustion of traditional biomass and fossil fuels including cow dung in India, a haze veil with an extension of 10 million km^2 has been formed over the emerging countries of South and Southeast Asia over the past few years. This haze veil was examined in detail for the first time within the framework of the Indian Ocean Experiment (INDOEX). The high concentration of aerosols that make up the haze leads to a reduction in direct sunlight and thus to crop losses in rice and sugar cane (Ahmad 2002). Based on the severe health effects of air pollution outlined above, the WHO assumes that an active climate policy, in addition to climate prevention, would also have a direct benefit for the health of the respective regional populations.

5.3.3 Immission Damage to Plants

Plants are particularly exposed to the harmful effects of pollutant emissions in the form of sulfur dioxide (SO_2), ozone (O_3), nitrogen oxides (NO_x) and dusts, as they are site-bound and interact with the ambient air via a very large respiratory surface. In contrast to health impairments, plant damage caused by pollutant immissions was already recognized in the nineteenth century. In particular, the effects of sulfur dioxide were already described and studied as "smoke damage" at that time. However, due to their limited range, the damage remained mostly local in this period.

Large-scale damage to vegetation has occurred in Central Europe, especially since the mid—1950s of the twentieth century, due to the increased use of coal-fired power plants with large chimneys that have carried pollutant emissions into large heights and therefore transported to distant areas. The largest damage ever observed occurred in the Erzgebirge (Ore Mountains), where 60,000 ha of forest were destroyed in the middle of the 80's years and another 200,000 ha were severely damaged. This was mainly due to sulfur dioxide emissions from lignite-fired power plants on the territory of the former Czechoslovakia. In Western Europe, too, there was a growing awareness of the problem of large-scale vegetation damage caused by pollutant emissions. In Germany, the debate about this vegetation damage became known mainly through the term "forest dieback". Although the catchword "forest dieback" (Waldsterben) was used to describe a very complex bundle of causes, the debate on the condition of forests nevertheless led to increased public awareness of the environmental damage associated with pollutant emissions. Due to the sometimes considerable damage to forestry and agriculture, but also due to the adverse health effects outlined above, the emission of pollutants from the combustion of fossil fuels has been dramatically reduced by a series of legal measures since the middle of the 80'er years. One of the most important measures was the installation of desulphurization plants in coal-fired power plants.

Effect of individual pollutant immissions like sulfurdioxide (SO_2, No_x and others) have been described earlier (Streffer et al. 2005). Plant damage has been well investigated in experimental systems with single pollutants as well as mixtures of several pollutants. Generally, it has been found that thresholds exist for these effects. When assessing immission damage, it has proved useful to distinguish between critical levels of airborne pollutants and critical loads of airborne pollutants. Plant damage has been well investigated by experimental exposure to air pollutants both for single pollutants and for mixed immissions. It turned out that plants only show negative reactions to pollutant immissions above a threshold, with the effect increasing exponentially with concentration (Streffer et al. 2005).

5.4 Nuclear Energy

5.4.1 *Environmental Problems Caused by Nuclear Energy, Introduction*

In the case of environmental problems caused by nuclear energy, the effect of ionizing radiation is clearly at the forefront. Various studies have shown that mammals, including humans, are among the most radiation-sensitive organisms. For inhabited regions, it has been assumed until recently that the protection of humans against ionizing radiation also protects the environment in general (ICRP 1991). According to the International Commission on Radiological Protection (ICRP), this continues to apply, but the latest ICRP policy recommendations (ICRP 2007) list a number of special features that must be taken into account, especially in regions that are not inhabited by humans.

The organisms may be exposed from external sources by direct radiation from technical installations, by radioactive substances in the air or from the ground which have been released from the installations (external exposure), or from the inside after incorporation of radioactive substances and their decay in the organism (internal exposure). The radiation or radioactive substances can also come from natural sources. These exposure processes are basically possible at all stages of the fuel cycle (uranium ore extraction, production of fuel elements, nuclear fission and other processes in the nuclear power plant, treatment of spent fuel elements and final disposal).

5.4.2 Fundamental Processes of Radiation Exposure and Epidemiological Data on Causation of Cancer and Genetic Changes After Radiation Exposure

The radiation risks in mammals including humans are relatively well known in a wide dose range due to molecular and cellular investigations in vitro in the laboratory, to animal experiments and to clinical experience with the use of ionizing radiation in medicine and after radiation accidents as well as exposures from technical installations. The following radiation effects may occur: Acute effects including death (days to weeks or months after irradiation), later effects in the tissues, e.g. fibrotic changes, opacity of the eye lens (months to years after irradiation), developmental disorders after irradiation in utero, causation of cancer and genetic defects (years to decades after irradiation).

The level of the radiation effects is strongly dependent on the radiation dose. For risk assessments, the dose-response relationships for the various radiation effects are therefore of decisive importance. In radiation protection two fundamentally different categories of such dose-response relationships are used (Fig. 5.8): There are dose-response relationships that have a threshold dose, such effects only occur after radiation doses that are higher than the threshold dose. These radiation effects include acute effects and tissue late effects (e.g. fibrotic processes in tissues, vascular damage,

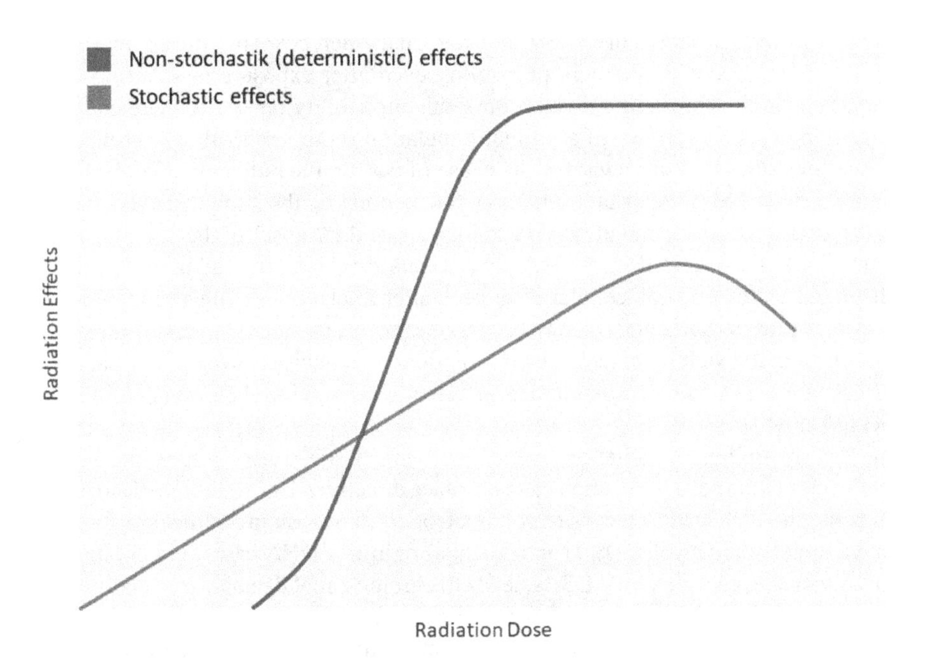

Fig. 5.8 Categories of dose response relationships, **a** dose response relationship with threshold dose, (non-stochastic—deterministic—effects), **b** dose response relationship without threshold dose (stochastic effects such as causation of cancer and genetic effects) (ICRP 1991; Streffer et al. 2004)

cause of malformations, etc., generally referred to as non-stochastic or deterministic radiation effects). The threshold doses for these radiation effects lie in dose ranges that are relatively high (>100 mSv or even several Sv) and by far not reached by nuclear facilities and their effects in the environment during normal operation. Only in the case of serious nuclear accidents (e.g. Chernobyl in 1986, Tokaimura in Japan in 1999, various accidents involving radiation facilities in medicine) do radiation doses cause such effects in a limited region or among personnel in the facility during the "hot" phase of the accident. Even after the after the accident in the Fukushima Daiichi power plant the radiation and no other non-stochastic (deterministic) effects are expected (UNSCEAR 2012). As these events are very rare and have not occurred in the more than 30 years of operation of nuclear power plants in Germany or in EU countries in general, these effects will not be discussed in detail here. There is extensive clinical experience of these effects after high radiation doses, which must also be used in tumor therapy. Also, for radiation effects occurring after exposure of the embryo or fetus in the womb, dose-response relationships with threshold doses are generally observed. There exist only few exceptions which were observed for instance in mice with genetic predispositions (Streffer 2009). Only in exceptional cases and short time periods of early prenatal development are dose effects without threshold dose possible.

For radiation protection to describe the dose dependence of "stochastic radiation effects" in the low dose range, linear dose effect relationships without threshold doses are generally assumed ("linear-no-threshold-", LNT-model, ICRP 2007). This applies to the causation of cancer and hereditary defects. This means that these effects can occur with a certain probability even after exposure to small doses of radiation. With decreasing radiation dose this probability becomes lower and with radiation doses <100 mSv in a normal population (both sexes, all age groups) no longer measurable. With regard to the cause of cancer, the radiation effect becomes smaller than the regional and temporal fluctuations of the "spontaneous" cancer rates, so that a possible radiation-related increase in the "noise" of the "spontaneous" cancer risk disappears (Fig. 5.9). A cancer caused by ionizing radiation cannot be distinguished by clinical, cellular or molecular characteristics from a cancer caused by other causes. Such effects may be possible with low radiation doses; thus they can occur in the vicinity of technical installations, e.g. nuclear power plants but these effects are below the measurable degree. They will therefore be discussed in more detail below.

The absorption of radiation energy in living tissue causes ionizations in the molecules of the cells. Ionization can take place directly in DNA, the genetic material of cells, and then lead to breaks in polynucleotide chains (single-strand breaks, ESB and double-strand breaks, DSB) or to changes or loss of DNA bases (Fig. 5.10). This DNA damage, especially the DSBs, leads to chromosomal damage, especially chromosome breaks as a result of the DSBs. There is strong evidence that such molecular processes are also crucial for the development of cancer (UNDCEAR 2000; Streffer et al. 2004). In contrast to toxic substances, DNA damage does not occur in isolation in individual events, but energy packages are absorbed, so that "clusters" of several, very neighboring damages often occur in the DNA. The number and above all the

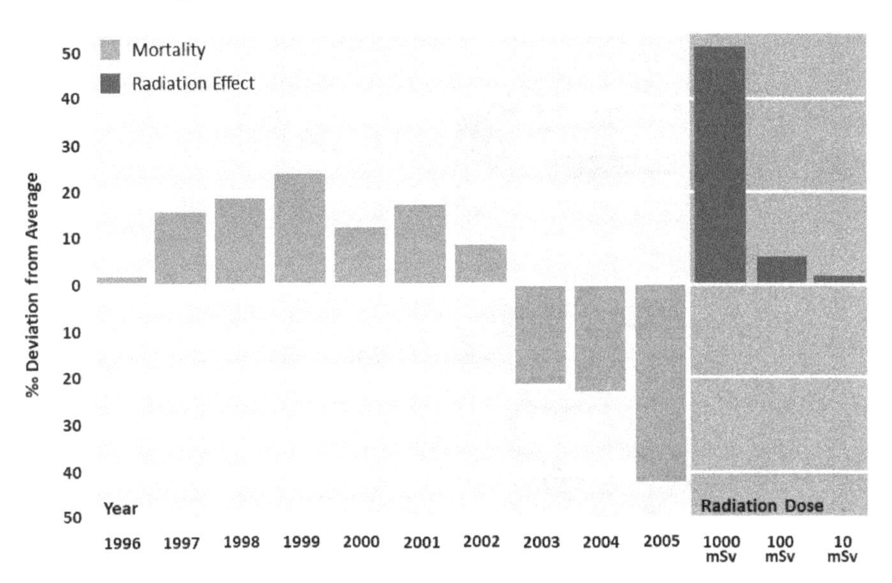

Fig. 5.9 Deviation of cancer mortalities from the average (‰) in 1996–2005 (SEER-USA) and radiation effect (ICRP). The radiation effect <100 mSv is covered within the "noise" of the "spontaneous" cancers (Streffer 2009)

quality of the damage depends not only on the radiation dose but also on the type of radiation and its energy (radiation quality). A biological evaluation of the radiation quality must therefore be carried out (Fig. 5.10).

Some of the damage can be very quickly removed by effective repair systems and the original DNA information restored. However, errors may also occur in this DNA repair and residual damage remains depending on the radiation dose and other exposure factors. DNA damage after exposure to densely ionizing radiation (e.g. neutrons, α rays) is repaired much less and may be with more mistakes (misrepair) than after exposure to loosely ionizing radiation (e.g. X-rays, γ rays) (Fig. 5.11). The efficiency of the repair processes varies considerably from individual to individual (Fig. 5.12). There are thus persons who have a very low repair capacity due to their genetic disposition (ICRP 1999; Streffer 2009) (Examples: AT patients), patient with serious side effects after irradiation "severe-side-effects", Fig. 5.12). Individuals with such a genetic predisposition are much more sensitive to radiation than "normal persons" and also develop more cancer.

For the risk assessment of toxic agents and thus also for the radiation risk, the course of the dose-effect relationships is therefore of decisive importance. Thus, even with very small radiation doses, the probability that the radiation will cause cancer is not zero, although it cannot be measured (ICRP 2007). It must therefore be emphasized that health effects have so far only been measured significantly in a medium and higher dose range (100 mSv and higher). In the low dose range (<100 mSv), which occurs in the environment and generally also in workplaces, no significant data are available for the increase of cancer. The risk can therefore only

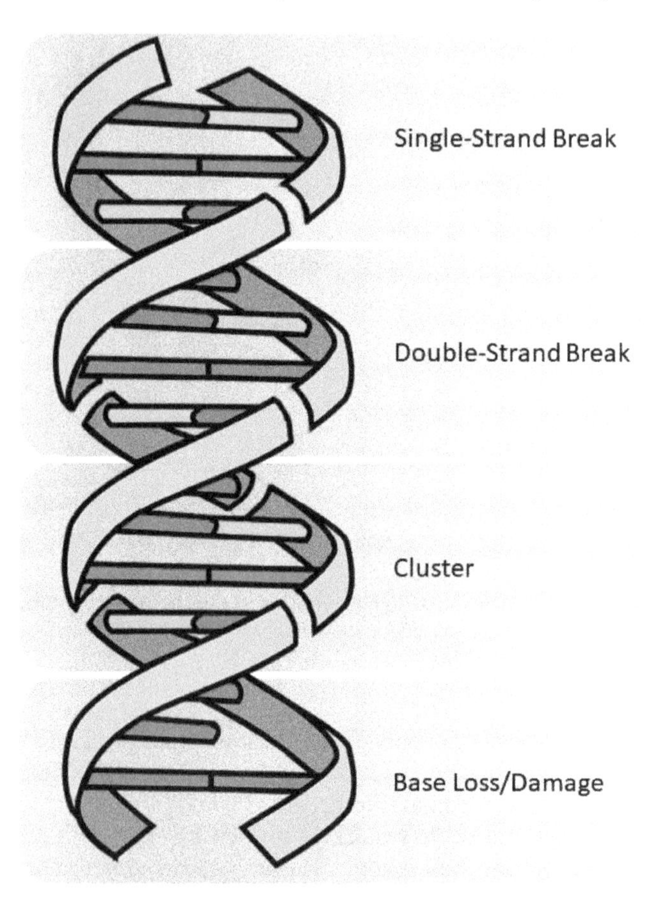

Fig. 5.10 DNA-damage by ionizing radiation: breaks of the polynucleotide strand (Single Strand Break, Double Strand Break, or Cluster of DSB plus SSB) and loss or damage of DNA-bases (Streffer 2009)

be determined by extrapolating effects from the higher dose ranges to the lower dose ranges. The linear dose response relationship without threshold dose is an assumption that requires scientific proof. Uncertainties are associated with this procedure, but for precautionary reasons the adoption of the LNT model for radiological protection is also justified for pragmatic reasons. However, it is intensively discussed that the dose-response relationship could also have a different shape and possibly even a threshold dose. It is also discussed that very low radiation doses lead to a reduction e.g. of chromosome aberrations. Such an effect is also referred to as "hormesis" (UNSCEAR 2000; Streffer et al. 2004) (Fig. 5.13). For health radiation effects, such a shape of the dose effect curve, has not been significantly observed until now for the general cancer causation. However, thresholds may be possible for special cancer entities like osteocarcinoma.

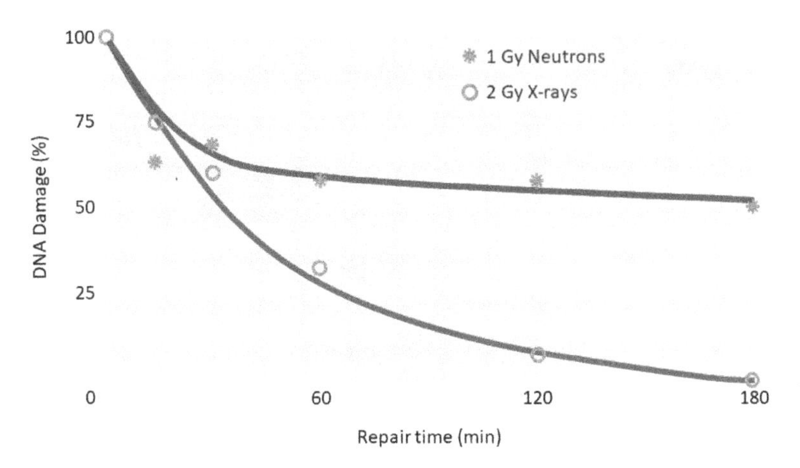

Fig. 5.11 Permanent tumor cell culture MeWo—DNA-repair capacity after neutron and X-irradiation. DNA-repair is much less after high LET than low LET radiation (Streffer 2009)

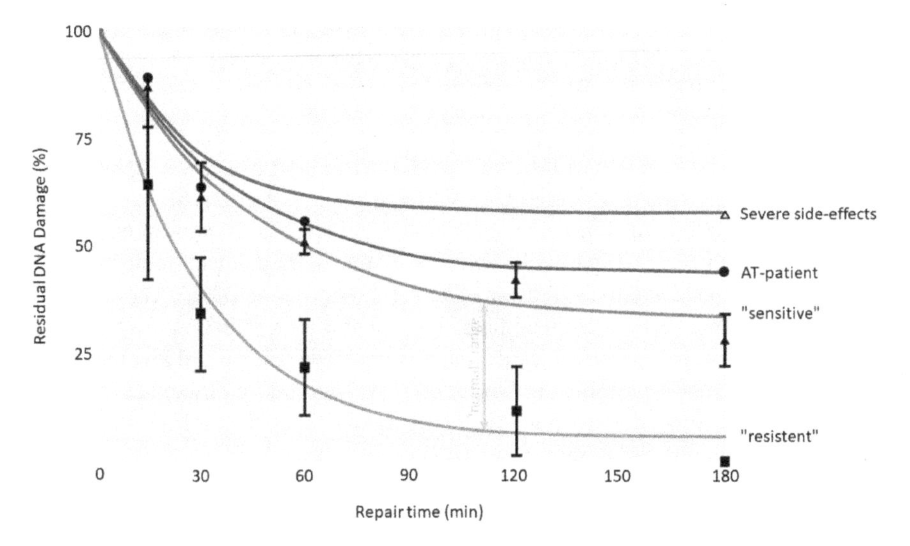

Fig. 5.12 DNA repair kinetics in human lymphocytes after irradiation and subsequent recovery up to 180 min. DNA damage was measured at different times after irradiation. DNA-repair is considerably reduced by genetic disposition in "AT"—patient and in patient with "severe side effects" (Streffer et al. 2004; Streffer 2009)

However, the available data from epidemiological studies on irradiated groups of persons and the experimental data, e.g. on chromosomal changes in cells and molecular radiation damage in the DNA, as well as animal experimental data on cancer causation, can in most cases best be described by a linear dose-response relationship without threshold dose (LNT) (UNSCEAR 2000; UNSCEAR 2008). The most comprehensive data on human cancer causation are obtained from the

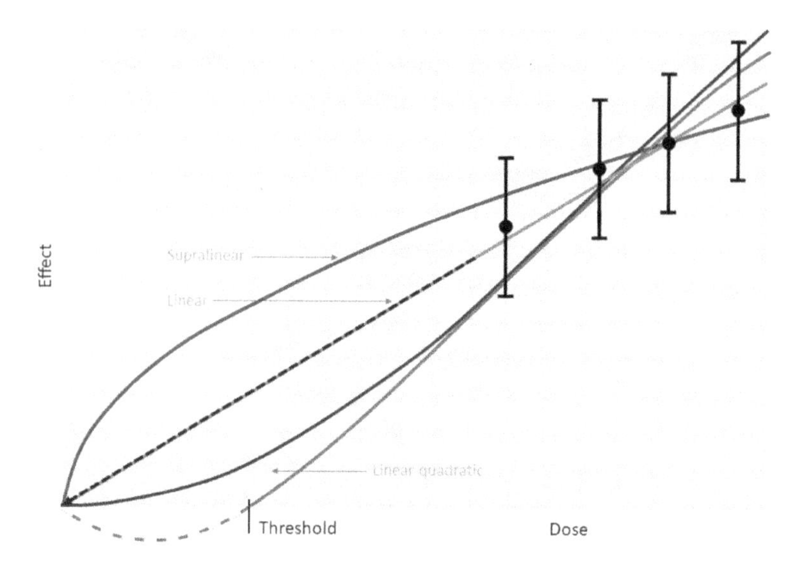

Fig. 5.13 Dose response curves obtained by extrapolation of measured values. Linear, supralinear and linear-square extrapolation and assumption of a biopositive effect at very small radiation doses (hormesis) (Streffer et al. 2004; Streffer 2009)

survivors of the atomic bomb disasters in Hiroshima and Nagasaki. A few years after irradiation, leukaemias were observed to an increased extent in these persons (UNSCEAR 2000). In the following decades, the incidence and deaths of many solid cancers also increased, in particular cancers of the lung, stomach, colon, female breast, thyroid gland and liver (Preston et al. 2007; Grant et al. 2017). The statistical evaluation of these epidemiological data shows that after a radiation dose of about 100 mSv a significant increase in cancer rates is observed (Pierce and Preston 2000). However, in the newest publication it stated: "The lowest dose range that showed a statistically significant dose response using the sexaveraged, linear ERR model was 0–100 mGy ($P = 0.038$)" (Grant et al. 2017).

The values obtained vary around the linear dose-response relationship, but the data can best be described by a linear dose-response relationship. The exposures in Hiroshima and Nagasaki took place within a very short time (less than 1 min). In this case one speaks of an acute irradiation. Many experimental data, especially studies with mice, have shown that chronic irradiation has lower radiation effects than acute irradiation. It has therefore been assumed that chronic irradiations, such as those present in the environment at nuclear installations, present lower risks at the same dose compared to the data from Hiroshima and Nagasaki. On the other hand, data on the causation of cancer after chronic radiation exposure among the population on the Techa River in the southeast Ural region have recently been published, leading to approximately the same risk factors as in Hiroshima and Nagasaki (ERR/Gy, ERR: Excessive Relative Risk) (Table 5.1) (Krestinina et al. 2007; Schonfeld et al. 2013).

Table 5.1 Surveys of the Techa River Population

1950–1999[a]	865,812 person year of exposition 1842 mortality from solid cancer	ERR/Gy 0.92 Male: 0.6 Female: 1.2 Female:male: 1:9
1956–2002[b]	446,588 person year of exposition 1836 incidences of solid cancer	ERR/Gy 1.0 No significant difference female:male
1950–2007[c]	927,743 person year of exposition 2303 mortality from solid cancer	ERR/Gy 0.61 No significant difference female:male

The Excessive Relative Risk of cancer mortality or incidence per radiation dose (ERR/Gy) decreased with the increasing number of cancer deaths. In this case, there was no difference between male and female
([a]Krestinina et al. 2005; [b]Krestinina et al. 2007; [c]Schonfeld et al. 2013)

In addition to the studies in Hiroshima and Nagasaki and on the Techa River, the cause of cancer by ionizing radiation has been investigated in groups of people who have received radiation exposure due to medical indications, at the workplace or in regions with high levels of environmental radiation. Lower risk values have often been obtained than in Hiroshima and Nagasaki (UNSCEAR 2000). This is due to various factors. The exposures in these cases have been fractional or chronic, local irradiations have often occurred and the age groups in the different cohorts are very different. Based on these diverse data, the cause of cancer by radiation depending on gender and age has also been studied. In particular, it has been shown that children are more sensitive to radiation than adults. Very extensive data on radiation risk are therefore available, which give the risk assessment a relatively high level of overall safety despite all the uncertainties identified in the individual studies.

In order to obtain better clarity for the shape of the dose-response relationship in the low dose range, it is necessary to clarify the mechanism of carcinogenesis after radiation exposure. These processes drag on in humans for decades. An increased risk of cancer is still observed today among survivors in Hiroshima and Nagasaki (Preston et al. 2007; Grant et al. 2017).

In recent years, a whole series of biological phenomena after radiation exposure have been investigated primarily in cells and molecular systems, which can contribute to a modification of the course of the dose-response relationship in the lower dose range with a deviation from the linear course (Fig. 5.13). The DNA repair described above, apoptosis (cell death caused by signal transduction within the cells or by the cell surface and apoptosis (the process of cell destruction by hydrolases) are some of the most important factors to be mentioned here. Processes like "adaptive response" (low radiation doses can increase the radiation resistance of cells in the short term under defined conditions), genomic instability (genetic damage can occur to an increased extent after many cell generations), “bystander” effects (radiation damage is also caused in cells that are not themselves affected by the radiation but are in the vicinity of affected cells) (UNSCEAR 2000; Streffer et al. 2004; ICRP 2007; Streffer 2009). It is also discussed that the immune system can remove radiation-damaged cells. All these extraordinarily interesting biological phenomena are intensively investigated, but it is unclear in detail what effects they have on the development or change of cancer in the low dose range and thus also on the dose-effect relationship in this low dose range. Due to these phenomena both an increase and a decrease of the radiation effects appears possible which often show strong differences between individuals.

5.4.3 Exposure to Radiation from Natural Sources and Medicine and Other Sources

Since there has been life on this earth, it has been exposed to ionizing radiation from natural sources. It is therefore possible to investigate the effects of these radiation exposures and to compare them with radiation exposures from civilizational (technical, medical, etc.) sources. Exposures to radiation from external sources (cosmic radiation from the sun, terrestrial radiation from radioactive decay of unstable nuclides in rocks and soils) and internal exposure from intake of natural radioactive substances with food occur. In Germany, these radiation sources lead to an average whole-body radiation level of about 1.1 to 1.2 mSv/year. In addition, the lungs or respiratory tract as a whole receive the highest radiation exposure through inhalation of radon and its radioactive daughter radionuclides. This exposure results in a mean lung dose in Germany of about 10 mSv/year. These data have been described earlier (Streffer et al. 2011).

By weighting the different radiation sensitivity of the individual organs for stochastic effects, one arrives at the effective dose (ICRP 2007). Taking into account the radiation risk from radon exposure, an effective average dose of about 2.1–2.4 mSv/year is then obtained in Germany (BfS). Radiation exposures from natural sources can vary greatly depending on the region and living conditions. The size of the dose from cosmic rays depends on the height at which the human being is present. At an altitude of 1500 m above sea level, this dose is about double that of the North German lowlands. The terrestrial radiation exposure changes greatly with the amount of radioactivity present in the soil. In Germany there are regions such as the Erzgebirge, the Bavarian Forest and regions in the Saarland where these activity concentrations are relatively high. However, the greatest differences are observed in the inhalation of radon and its radioactive daughter radionuclides. This depends decisively on the radon concentration in the rooms of living and working. On average, radon concentrations of about 50 Bq/m^3 air have been measured in Germany. In about 1% of the dwellings, however, the radon concentration is higher than 200 Bq/m^3 and in some regions very high radon concentrations can be measured in the buildings.

As an example, the Schneeberg region in the Erzgebirge can be taken here. In some houses extremely high radon concentrations of more than 15,000 Bq/m^3 have been found. These houses were renovated after the measurements in 1990 in order to reduce the radon concentrations in the rooms.

In addition to radiation exposures from natural sources, people worldwide and thus also in Germany receive radiation exposures through medical and technical facilities. These exposures are summarized in Fig. 5.14. This shows that about 53.8% of the total radiation exposures originate from natural sources, 46.1% of the radiation exposures result from medical indications according to the compilations of the Federal Ministry for the Environment, Nature Conservation and Nuclear Safety (BMU) and the Federal Office for Radiation Protection (BfS) from 2012.

X-ray diagnostics accounts for the largest share of these exposures from the use of ionizing radiation in medicine. Although the individual doses are lower than for radiation therapy, the frequency of examinations per inhabitant is relatively high. This proportion of radiation exposure of the population is likely to continue to increase as CT and other methods of higher radiation exposure (e.g. interventional radiology) increase. Among the occupationally exposed persons, the flying personnel (pilots and cabin crew) receive a dose distribution of up to 8 mSv/a averaged over these

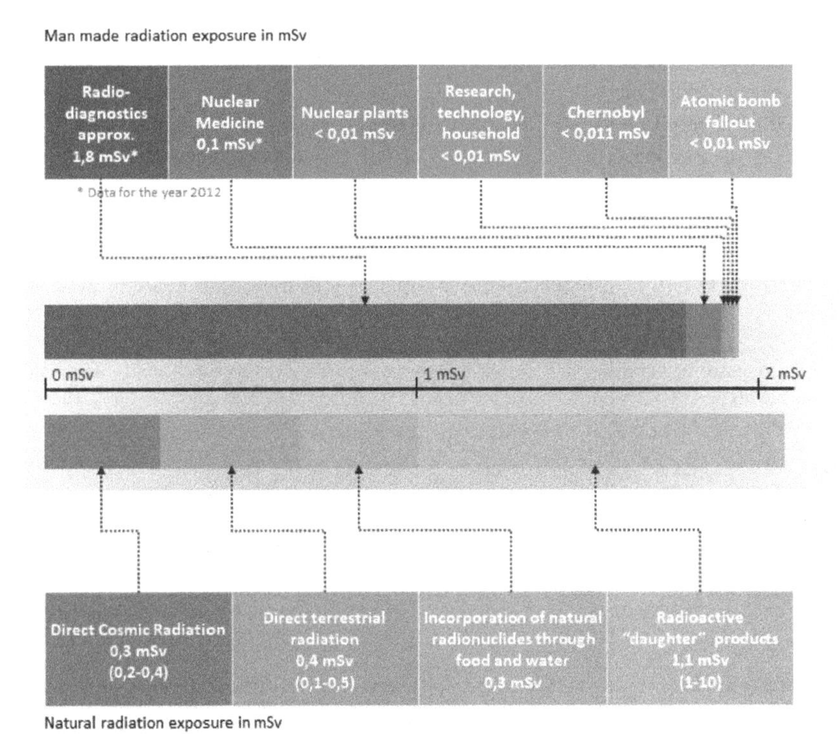

Fig. 5.14 Average radiation exposures from "man-made" (medical and technical) sources as well as from natural sources in Germany (BfS 2016)

occupational groups and is highest with a mean value of about 3 mSv/a (Fig. 5.14). The radiation exposures from nuclear installations (<0.01 mSv/a), the radioactive fallout from the earlier nuclear weapons tests (<0.01 mSv/a) and from the reactor accident at Chernobyl (<0.012 mSv/a) were stated as percentages (BfS 2012).

5.4.4 *Major Reactor Accidents—Release of Radioactive Substances, Radiation Exposure and Damage to Health*

In risk assessments in connection with nuclear energy, the possible reactor accidents, in particular with core meltdowns, are of great importance. There have been three such events since nuclear reactors operate in order to generate electricity:

1. The reactor accident "Three Mile Island" (Harrisburg) in the year 1979 in the USA.
2. The reactor accident "Chernobyl" in April 1986 in Ukraine, then Soviet Union.
3. The reactor accident "Fukushima Daiichi" in March 2011 in Japan.

Table 5.2 Summarizes the releases of radioactive substances and the health damage observed to date for all three major reactor accidents.

The reactor disaster in Chernobyl led to a huge release of radioactive substances into the atmosphere. Due to the high temperatures in the reactor, the radioactivity has been transported to high altitudes and has therefore spread atmospherically over wide regions of Europe. The radionuclides of iodine (especially iodine-131) in the first days and then in the later periods of caesium (caesium-137 and caesium-134) were of considerable importance for human radiation exposure and represented the greatest exposure. In the following years, this radioactivity decreased further, but even today caesium-137 from this reactor catastrophe is still measured in some German

Table 5.2 Releases of radionuclides and the health damage observed to date for all three major reactor accidents

	T1/2 (phys.)	Three Mile Isl.	Tschernobyl	Fukushima
^{131}Iod	8 d	0.39 MBq	1760 PBq	150 PBq
^{134}Cs	2 a	—	~47 PBq	~4 PBq
^{137}Cs	30 a	—	~85 PBq	~8 PBq
^{90}Sr	29 a	—	~10 PBq	Very small
^{239}Pu	24,000 a	—	0.013 PBq	Very small
Acute effect		—	28 death cases	—
Long term effect		—?	Thyroid cancer, cataracts of the eye, leukemias	—?

Acute effects are measured by deaths or damage of organs or tissues within days and weeks after radiation exposure (Streffer 2015)

foodstuffs from regions with the highest contamination with Cs-137 and leads to exposures that accounted for about 0.1% of the total radiation exposure in 2012.

The quantities of radioactivity from the reactor catastrophe in Chernobyl in 1986 were also distributed very differently in Germany. In southern Bavaria (foothills of the Alps) the highest activity levels have been measured. The radiation doses in this region amounted to about 0.65 mSv in the first year after the disaster. South of the Danube the exposure was about 0.4 mSv and north of the Danube about 0.2 mSv in the first year after the accident.

These radiation doses were not high enough to cause health damage in Germany. However, in Belarus (Belarus), Ukraine and the regions of Russia close to Chernobyl, cancer has been caused by radiation from the reactor accident. In particular, an increase in thyroid carcinomas in children in these countries has been observed. There was a clear dependence on the age at the time of exposure. Children who were 0–6 years old during the disaster had a particularly high risk of thyroid cancer (Fig. 5.15). This risk was much lower among older children and adolescents (Kenigsberg 2003). These data have also contributed significantly to the understanding of radiation effects and the assessment of radiation risks. In addition, 28 people died of acute radiation damage as a result of the Chernobyl disaster (Table 5.2). These were people who were used to extinguish the fire in the burning reactor.

No significant health effects were observed during the catastrophe in Harrisburg. Also, in Fukushima Daiichi there were no acute effects. The time is too short to make statements about late effects. However, the radiation doses reported so far are so low that significant increases in cancer are hardly to be expected (UNSCEAR

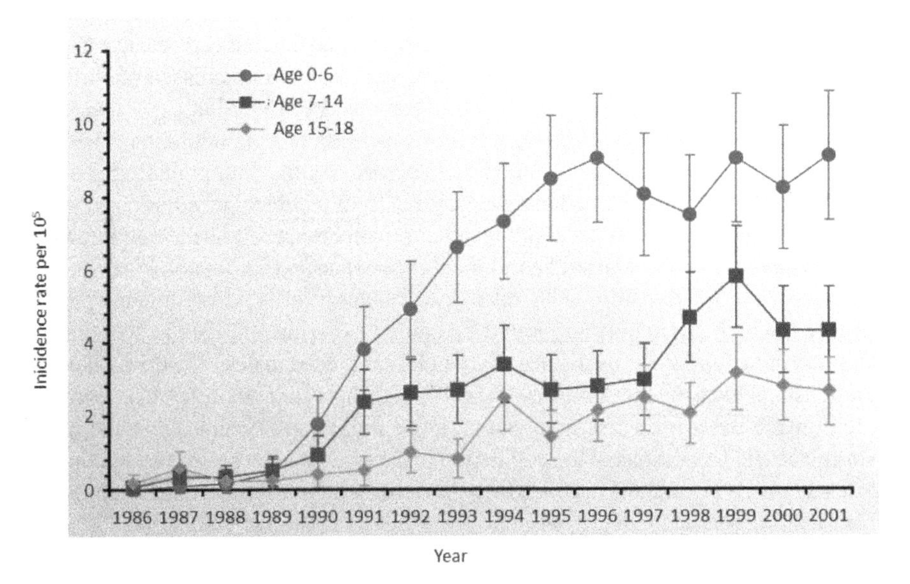

Fig. 5.15 Number of cases of thyroid cancer in children and adolescents (age groups: 0–6, 7–14 and 15–18 years) at the time of exposure in Belarus in the years 1986–2001 (Kenigsberg 2003)

2013). Only in the thyroid gland have a few children been exposed to radiation with the possibility of an increased risk. The main radiation doses of the population are caused by the consumption of radioactively contaminated food. In Japan, a few days after the accident in Fukushima Daiichi, radioactivity in foodstuffs was measured and foodstuffs were withdrawn from circulation in the event of contamination.

5.4.5 Summary of Radiation Exposures in Germany

In a concluding table, the radiation exposures caused by medical exposures, at workplaces exposures and in environmental exposure in Germany have been summarized as an average per person and year. In radiation therapy, high radiation doses must be used to destroy the patient's cancer. In diagnostics, radiation exposure is unavoidable if a good result of the examination is to be achieved. Modern examination procedures such as computed tomography and interventional radiology have increased individual radiation doses in recent years despite improved technologies.

There are dose limits at workplaces that should not be exceeded. They are currently 20 mSv/year in Germany and worldwide. The table shows that radiation doses are reached at individual workplaces that enter this area. After this has been determined by measurements, the radiation exposures at the workplaces concerned have been reduced by appropriate measures and much lower than the dose limits in general. Finally, the individual doses of radiation in the environment are mainly indicated from natural sources. It becomes clear that there are considerable variabilities here that cannot be avoided. In Germany, a standard value of up to 300 Bq/m^3 is currently accepted for radon in newly built houses. This value may be reduced because there is evidence from very intensive epidemiological studies that an increased rate of lung cancer already occurs at such radon concentrations. For radiation exposures from technical sources, there is also a dose limit for persons of the population. The dose shall not exceed 1 mSv/year for individual persons. As the data in Table 5.3 show, these dose limits are far exceeded by exposures from technical installations.

In summary, it can be stated that the radiation effects are relatively well known in humans due to experimental studies, clinical experience and epidemiological studies. Dose-response relationships with respect to radiation effects for humans can be specified, which are mainly supported by animal experimental data and studies on cells and molecular systems in vitro. Extrapolation is used to describe dose-response relationships even in the low dose range, which have uncertainties but represent a reasonable basis for a risk assessment. Dose limits have been established at the workplace and for persons of the population in the environment below those radiation doses at which significant health effects have been measured. Radiation exposures from natural sources make it possible to create comparative standards. The dose limits for persons in the population are below the mean dose derived from natural sources in Germany.

For the radiation dose from nuclear facilities in Germany, the mean value is <0.01 mSv per year for each person of the total population. Although it is very

Table 5.3 Exposures in Germany from medical procedures, at working places and to the population in the environment (BfS: Bundesamt für Strahlenschutz, Germany)

1. Medical use of ionizing radiation (dose per treatment)		
☐	Therapy	Several 10,000 mSv (mGy)
☐	Diagnostics, local, regional	1–50 mSv
2. Exposures at workplaces (effective dose per year)		
☐	Staff in control areas in 2007 (BfS 2008)	average 0.79 mSv
☐	Flying staff in average in 2007 (BfS 2008)	2.2 mSv
☐	— Exclusively flying North Atlantic route	6–8 mSv
☐	Welders (electrodes with ^{232}Th)	6–20 mSv
☐	Workplaces with very high Rn-concentrations (e.g. water industry, Fichtelgebirge, Erzgebirge)	6–20 mSv
3. Environmental exposures (effective dose per year)		
☐	Average of natural exposure in Germany	2.23 mSv
☐	High regional natural exposure in Germany	8–10 mSv
☐	High regional natural exposure in India (Kerala)	15–70 mSv
☐	Nuclear facilities Germany (BfS 2008)	<0.01 mSv
☐	Exposures from the Chernobyl accident in Germany 2007 (BfS 2008)	0.01 mSv

problematic to estimate a health risk for such a radiation dose, as this exposure is far below the dose ranges in which statistically significant data can be collected, a risk assessment will be carried out. Under the conservative assumption of LNT, only a disease or death from cancer can be considered with such a dose. Based on the various epidemiological studies, the risk factor is about 10–4 per mSv, i.e. 10–6 for 0.01 mSv. For a population of 82 million people, 82 cancer illnesses would result from exposure to nuclear facilities in Germany. It should be noted that approximately 500,000 cancers occur in Germany every year. In 2012, there were 477,950 diseases, according to the DKFZ Heidelberg. Furthermore, it should be pointed out once again that the average annual radiation exposure in Germany from natural sources is around 2.4 mSv (effective dose).

5.5 Environmental and Health Risks from Renewable Energies

In this part of the chapter, the possible environmental impacts of electricity production by means of biomass, geothermal energy, solar energy, hydropower and wind energy will be considered together. The section is based, among other things, on the scientific work of the Hirschberg Group (Hirschberg and Burgherr 2015; Hirschberg et al.

2016), Paul Scherrer Institute, Villigen, Switzerland. Hirschberg and his colleagues have developed criteria and indicators for assessing exposures and possible risks for the various technologies for electricity generation in three major areas of influence: Environment including health, economy (economy) and society (social). The main criteria (indicators) defined for the 'environment and health' area, on which the following descriptions will focus, are as follows:

- Consumption of non-renewable resources
- Consumption of energy resources throughout the "life cycle"
- Minerals, the consumption of rare metals
- Possible climate influences
- Possible impacts on the ecosystem
- Possible changes in biodiversity (flora and fauna)
- The consumption and degradation of land.

Thus, the entire chain of different technologies from the production and construction of the technical facilities as well as the normal operation of the facilities and finally also possible serious accidents are considered (Hirschberg et al. 2016).

For the release of greenhouse gases ("Greenhouse" gases), Hirschberg and Burgherr (2015) arrive at comparable, similar values (expressed in kg CO_2 eq/kWh) for nuclear energy and hydropower as well as wind; for photovoltaics, these latter values are about a factor of 8 higher than the first three technologies. When coal is used as a fuel, greenhouse gas emissions are about 100 times higher. With the exception of biomass as an energy source, energy production from renewable energies is free of pollutant emissions during operation of the plants and even the combustion of biomass is largely neutral with regard to the greenhouse gas CO_2. Renewable energies, primarily hydropower, therefore, make a significant contribution to reducing greenhouse gas emissions. However, the high material intensity of the new renewable energies results in considerable indirect emissions from upstream material flows. These are particularly high in the relatively energy-intensive production of photovoltaic cells.

Some renewable energies require massive landscape interventions, not least because of their lower energy density, which can have negative ecological consequences. In addition to conventional large dams, renewable energies, which require large-scale landscaping, also include solar thermal power plants and wind farms. The impact of the construction of dams on the ecosystem is similar to that of nuclear energy in normal operation, according to data from Hirschberg and Burgherr (2015). On the other hand, the estimated values of a "potentially damaged share" per m^2 of land and year in relation to the profit of 1 kWh for photovoltaics and wind energy are about a factor of 10 higher than for nuclear energy in normal operation. This is mainly due to the need for land and the impairment of fauna (e.g. the life of birds). The values for damage or impairment of the ecosystem are considerably higher when coal and other fossil fuels are used. This also applies in general to the production of electricity from biomass, in particular for the monotonous cultivation of certain crops exclusively for these purposes.

With regard to health damage, Hirschberg and Burgherr (2015) assess the use of fossil fuels as having high effects (indicated in lost life years per GWh), especially when using coal, as discussed earlier. This also applies to the use of biogas by these authors. The estimated values for photovoltaics and wind energy are about 2–3 times higher than for nuclear energy, while hydropower is about 3–5 times lower than for nuclear energy. All these values are related to normal operation.

However, the ecological impacts of these interventions depend to a large extent on the ecosystems affected. While large dams represent a massive cut into the river ecosystem, the ecological impact of onshore wind farms is negligible. The ecological side effects of solar thermal power plants are also low, despite an average space requirement of around half a square kilometer per 100 MW. This applies above all if they are found in desert regions or in other regions without agricultural use or on land where there are no harmful effects on ecological values due to a largely lacking vegetation cover. The quite considerable space requirement of around 10,000 m^2 per MW of photovoltaic capacity is of little ecological significance if these systems are installed predominantly on roofs and facades. Possible ecological burdens, however, are caused by energy generation from the sea. Tidal power plants, for example, can adversely affect the resonance characteristics of a coastline, while marine wave power plants and tidal turbines may have a disruptive effect on marine fauna. Due to a lack of commercial use, no empirical data are available on the ecological effects of these energy sources. In general, the ecological burden depends on the extent of the landscape interventions. While a single offshore wind farm does not represent a significant disturbance, a massaging of wind farms off the coast can lead to a significant impairment of marine fauna.

On the other hand, significant landscape changes are unavoidable if renewable energies are advancing into large-scale industrial dimensions of energy production. This is exemplified by the planned updraft power plant in Mildura (Australia) and the planned Butendiek offshore wind farm off the German North Sea coast. Both plants, with a targeted peak output of around 200 MW, will occupy an area of 37 km^2 each when completed. The rapid expansion of wind power in Germany has also led to a major change in the landscape in some regions, which many people perceive as a "saving of energy". The size and space requirements of the Mildura and Butendiek pilot projects also demonstrate by way of example that the often-cited contradiction between "hard" and "soft" energy technologies has become obsolete. The environmental compatibility of new renewable energy systems tends to be high due to a lack of pollutant emissions, but the high land requirements of these energy options lead to conflicting objectives with landscape conservation.

Hirschberg and Burgherr (2015) arrive at the following results through their ratings: The environmental impacts are particularly high for electricity generation through the use of fossil fuels, especially coal and biomass, but also natural gas. The environmental impacts are lowest for hydroelectric and nuclear power plants, while solar power plants and wind power occupy a medium position. With these technologies, however, Hirschberg and Burgherr expect a strong reduction in environmental pollution through improved technologies.

The health risks associated with the use of fossil fuels, in particular hard coal, lignite and nuclear energy, have been discussed in detail beforehand. Hirschberg et al. (2016) have also estimated the health risks associated with the use of renewable energies for electricity generation. The authors estimated the mortality in lost life years per produced electricity in GWh (YOLL/GWh) for the normal operation of the plants. The data were determined with the help of the "Environmental Impact Assessment". The method is based on the "European External Series". The emissions from the plants during operation, but also the indirect emissions of pollutants during the manufacturing processes of the plants, including the extraction of the needed materials, are taken into account. For further details on the methodology see Hirschberg et al. (2016). Hydropower causes the least risk, most other renewable technologies and nuclear energy occupy a middle position and fossil fuels, especially coal, are the technologies with the highest health damage (Hirschberg et al. 2016).

Chapter 6
Knowledge About Climate Change: Significance for Energy Issues

6.1 Overview

In this text, knowledge about climate change is first summarized using the 2014 "key messages" of the UN Intergovernmental Panel on Climate Change (IPCC). Section 6.3 examines the question of "manifestation, detection and attribution"—how climate change presents itself in data, to what extent the diagnosed changes remain within the framework of natural climate variability, and, if not, which external factors allow an explanation of the observed changes (attribution). Section 6.4 deals with the "globally averaged air temperature" and its control and prediction. Section 6.5 discusses the future expectations presented in "scenarios", both the previously used SRES scenarios, which are explicitly based on socio-economic narratives, and the current RCP scenarios outlining "targets" in terms of radiation anomalies due to greenhouse gas emissions at the end of the century. Section 6.6 then deals with the significance of uncertainties before Sect. 6.7 concludes with the politicization of climate science and the de-politicization of climate policy under the heading of "post-normal science".

6.2 The IPCC's Current Assessment of Knowledge

The Intergovernmental Panel on Climate Change (IPCC) Climate Council regularly assesses and summarizes the knowledge about recent and expected climate change for the current century. To date there have been five main reports, in 1990, 1995, 2001, 2007 and 2014. These reports are freely available on the Internet.

This assessment from 2014 is consistent with those from previous years, although of course there were many additional in-depth studies on details; uncertainties in some questions were reduced, but at the same time new uncertainties emerged in details—a process normal for science. The acceptance of the "budget approach",

C. F. Gethmann et al., *Global Energy Supply and Emissions*,
Ethics of Science and Technology Assessment 47,
https://doi.org/10.1007/978-3-030-55355-5_6

according to which the expected changes in the globally averaged temperature can be described as approximately proportional to the sum of the total emissions of greenhouse gases, represents a significant extension. This "budget approach" makes it possible to quantify the share of different economies in man-made climate change—simply as that share of the sum of global emissions. This aspect will be further elaborated below.

When the IPCC speaks of "knowledge", it means scientifically constructed knowledge that is documented in suitable publications, usually after an independent expert review. This knowledge—like any scientifically constructed knowledge—is not associated with an absolute claim to truth, but represents the "best" explanations currently available, consistent with observational data and scientific theory (including models). The IPCC works under the premise of pooling "politically relevant" knowledge, but not "politically prescriptive" knowledge.

The results of the IPCC are highly regarded in climate science circles. In a survey on the Fourth Assessment Report (2007), almost 80% of the climate scientists surveyed responded positively to the question of whether the IPCC reports were of great benefit to climate science.[1] Somewhat less, almost 70%, answered the question whether the IPCC would accurately describe the consensus in the scientific community on temperature changes. In the last 10 years, however, there has also been clear, albeit not majority, criticism that the IPCC would indicate too small expected changes.

All IPCC reports are accompanied by so-called "policy summaries" (SPMs[2]), and the last one even contains a strong summary of "Headline Statements" (IPCC 2014c):

> Headline statements are the overarching highlighted conclusions of the approved Summary for Policymakers [of the 5th, thus last, IPCC report], which, taken together, provide a concise narrative. The four statements in [italic given below] are those summarizing the assessment in the Summary for Policymakers, Sects. 1–4 [of IPCC report 5].

The following non-italic texts are statements of the IPCC office. We bring the complete texts here, without changes or cuts, even if parts of the text are not directly related to the subject of this report.

[1]"The IPCC reports are of great use to the advancement of climate science"; see Bray (2010).

[2]SPMs = Summaries for Policy Makers, for the full report (IPCC 2014a).

We propose to base our assessment of the situation on this 2014 assessment, as the IPCC reports were adopted in the context of a transparent protocol involving a very large number of relevant scientists and following an extensive peer review process.

Box 1: IPCC Climate Change 2014 Synthesis Report

Observed Changes and their Causes

Human influence on the climate system is clear, and recent anthropogenic emissions of greenhouse gases are the highest in history. Recent climate changes have had widespread impacts on human and natural systems.

Warming of the climate system is unequivocal, and since the 1950s, many of the observed changes are unprecedented over decades to millennia. The atmosphere and ocean have warmed, the amounts of snow and ice have diminished, and sea level has risen.

Anthropogenic greenhouse gas emissions have increased since the pre-industrial era, driven largely by economic and population growth, and are now higher than ever. This has led to atmospheric concentrations of carbon dioxide, methane and nitrous oxide that are unprecedented in at least the last 800,000 years. Their effects, together with those of other anthropogenic drivers, have been detected throughout the climate system and are extremely likely to have been the dominant cause of the observed warming since the mid-twentieth century.

In recent decades, changes in climate have caused impacts on natural and human systems on all continents and across the oceans. Impacts are due to observed climate change, irrespective of its cause, indicating the sensitivity of natural and human systems to changing climate.

Changes in many extreme weather and climate events have been observed since about 1950. Some of these changes have been linked to human influences, including a decrease in cold temperature extremes, an increase in warm temperature extremes, an increase in extreme high sea levels and an increase in the number of heavy precipitation events in a number of regions.

Future Climate Changes, Risks and Impacts

Continued emission of greenhouse gases will cause further warming and long-lasting changes in all components of the climate system, increasing the likelihood of severe, pervasive and irreversible impacts for people and ecosystems. Limiting climate change would require substantial and sustained reductions in greenhouse gas emissions which, together with adaptation, can limit climate change risks.

Cumulative emissions of carbon dioxide largely determine global mean surface warming by the late twenty-first century and beyond. Projections of greenhouse gas emissions vary over a wide range, depending on both socio-economic development and climate policy.

Surface temperature is projected to rise over the twenty-first century under all assessed emission scenarios. It is very likely that heat waves will occur more often and last longer, and that extreme precipitation events will become more intense and frequent in many regions. The ocean will continue to warm and acidify, and global mean sea level to rise.

Climate change will amplify existing risks and create new risks for natural and human systems. Risks are unevenly distributed and are generally greater for disadvantaged people and communities in countries at all levels of development.

Many aspects of climate change and associated impacts will continue for centuries, even if anthropogenic emissions of greenhouse gases are stopped. The risks of abrupt or irreversible changes increase as the magnitude of the warming increases.

Future Pathways for Adaptation, Mitigation[3] and Sustainable Development

Adaptation and mitigation are complementary strategies for reducing and managing the risks of climate change. Substantial emissions reductions over the next few decades can reduce climate risks in the twenty-first century and beyond, increase prospects for effective adaptation, reduce the costs and challenges of mitigation in the longer term, and contribute to climate-resilient pathways for sustainable development.

Effective decision making to limit climate change and its effects can be informed by a wide range of analytical approaches for evaluating expected risks and benefits, recognizing the importance of governance, ethical dimensions, equity, value judgments, economic assessments and diverse perceptions and responses to risk and uncertainty.

Without additional mitigation efforts beyond those in place today, and even with adaptation, warming by the end of the twenty-first century will lead to high to very high risk of severe, widespread, and irreversible impacts globally (high confidence).[4] Mitigation involves some level of co-benefits and of risks due to adverse side-effects, but these risks do not involve the same possibility of severe, widespread, and irreversible impacts as risks from climate change, increasing the benefits from near-term mitigation efforts.

Adaptation can reduce the risks of climate change impacts, but there are limits to its effectiveness, especially with greater magnitudes and rates of climate change. Taking a longer-term perspective, in the context of sustainable development, increases the likelihood that more immediate adaptation actions will also enhance future options and preparedness.

There are multiple mitigation pathways that are likely to limit warming to below 2 °C relative to pre-industrial levels. These pathways would require substantial emissions reductions over the next few decades and near zero emissions of carbon dioxide and other long-lived greenhouse gases by the end of the century. Implementing such reductions poses substantial technological, economic, social, and institutional challenges, which increase with delays in additional mitigation and if key technologies are not available. Limiting warming to lower or higher levels involves similar challenges, but on different timescales.

Adaptation and mitigation

Many adaptation and mitigation options can help address climate change, but no single option is sufficient by itself. Effective implementation depends on policies and cooperation at all scales, and can be enhanced through integrated responses that link adaptation and mitigation with other societal objectives.

Adaptation and mitigation responses are underpinned by common enabling factors. These include effective institutions and governance, innovation and investments in environmentally sound technologies and infrastructure, sustainable livelihoods, and behavioral and lifestyle choices.

Adaptation options exist in all sectors, but their context for implementation and potential to reduce climate-related risks differs across sectors and regions. Some adaptation responses involve significant co-benefits, synergies and trade-offs. Increasing climate change will increase challenges for many adaptation options.

Mitigation options are available in every major sector. Mitigation can be more cost-effective if using an integrated approach that combines measures to reduce energy use and the greenhouse gas intensity of end-use sectors, decarbonize energy supply, reduce net emissions and enhance carbon sinks in land-based sectors.

Effective adaptation and mitigation responses will depend on policies and measures across multiple scales: international, regional, national and sub-national. Policies across all scales supporting technology development, diffusion and transfer, as well as finance for responses to climate change, can complement and enhance the effectiveness of policies that directly promote adaptation and mitigation.

Climate change is a threat to sustainable development. Nonetheless, there are many opportunities to link mitigation, adaptation and the pursuit of other societal objectives through integrated responses (high confidence). Successful implementation relies on relevant tools, suitable governance structures and enhanced capacity to respond (medium confidence).

[3]This term refers to the limitation of climate change by limiting the emissions of greenhouse gases.

[4]This statement is made with "great confidence" by the IPCC authors. "The certainty in key assessment findings ... is based on the author teams' evaluations of underlying scientific understanding and is expressed as a qualitative level of confidence (from very low to very high) and, when possible, probabilistically with a quantified likelihood (from exceptionally unlikely to virtually certain). Where appropriate, findings are also formulated as statements of fact with-out using uncertainty

6.3 Manifestation, Detection and Attribution

Now that we have presented the tenor of the last (2014) IPCC report on the relationship between emissions and climate change, and the possible approaches (mitigation and adaptation), we turn to the challenges of determining man-made climate change:

(a) first the documentation of climate change in recent decades ("manifestation"),
(b) then the evaluation of whether this change can be understood as an expression of natural climate fluctuations, or whether the explanation of the change only succeeds if external "drivers" are assumed ("detection"), and finally,
(c) If there is a positive detection, identifying factors which plausibly allow for an explanation (within the framework of current scientific knowledge; "attribution").

In fact, there is a considerable lack of public understanding here; it is often well understood that the reality of man-made climate change is postulated on the basis of model calculations simulating a rise in temperature as a result of increased atmospheric greenhouse gas concentrations. This is only the case to a very limited extent. Rather, in this context the observations of the changing statistics of the weather (i.e. climate) are in the foreground, whereby climate models are used for the interpretation of these observation data.

The two challenges of *manifestation* are lack of, and changing quality, of the data ("inhomogeneity").

Instrumental data collected from thermometers, for example, have only been available for a few centuries, and initially only in very few places (such as cities) which are hardly representative of the state of the global mean air temperature. In addition, there are "indirect" data such as historical reports on noticeable heat or rainy periods, the freezing of rivers and the like, as well as deposits in ice cores or the composition of tree rings. The latter have the problem that they only describe part of the variability. Both data sources must also be treated with reservation because of their "inhomogeneity", namely that the numbers of a measurement or a proxy can reflect different factors at different times and places. For example, a wind measurement may be very accurate, but comparability over time can be severely limited if buildings are erected or removed in the vicinity of the instrument, or if trees grow or are felled. In inhomogeneous data sets, it is often impossible to distinguish between what is a true climatic trend and what has been caused by other changes that are independent of the climate. A good example is the change in the frequency of strong wind events in Hamburg, which decreased considerably from 1950 onwards, not because there were fewer or only weaker storms, but because the measuring station was relocated from the port to the Fuhlsbüttel airfield (see Fig. 6.1).

---

qualifiers." The quantification is given by "virtually certain 99–100% probability, very likely 90–100%, likely 66–100%, about as likely as not 33–66%, unlikely 0–33%, very unlikely 0–10%, exceptionally unlikely 0–1%". These statements are subjective as expert opinions and should not be misunderstood in terms of mathematical probability measures.

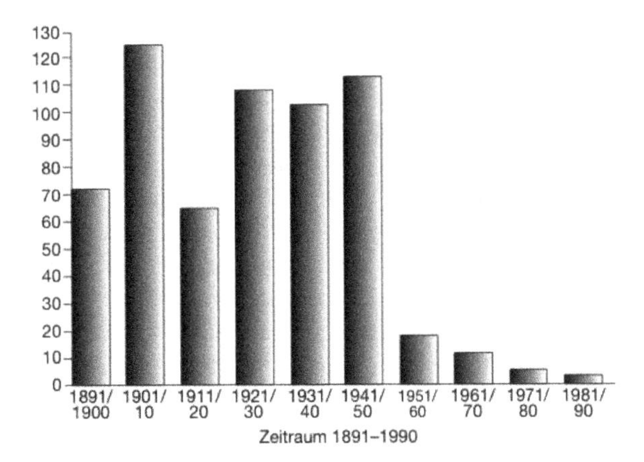

Fig. 6.1 Example of an inhomogeneous climate series: Ten-year frequency of storms (max wind forces greater than/equal to 8 Beaufort) in Hamburg's weather statistics. After Heiner Schmidt (cf. Stehr and von Storch 2010)

This reservation is usually not obvious for laypersons who make use of data offered on the Internet, and sometimes an interesting conclusion proves to be an expression of inhomogeneity.

Data density has also changed considerably over time; in many areas of the world, especially in the Southern Hemisphere and the tropics, there is still little data that is quality-checked and deemed representative. The availability of the satellites has significantly improved the situation, so that since about 1980 quite good data on some meteorological and oceanographic variables are available almost everywhere on the globe.

For the best-known variable, namely the globally averaged temperature of the near-surface air, there is very broad agreement on the development over time—various weather services provided consistent estimates, and the reservations of "skeptics" have largely fallen silent after the BEST working group (Berkeley Earth Surface Temperature (Brillinger et al.), known as critical to the mainstream results, has confirmed the results of the other groups.

After it has been established that the available data series are numbers that actually describe changes in weather statistics (i.e., climate), the *detection* question arises as to whether the changes remain within the framework of "natural variability". This natural variability is composed of the effect of natural drivers, such as changing solar output or the temporary additional presence of volcanic suspended matter in the stratosphere, and of fluctuations resulting from the stochastic character of the climate system. As a system with very many degrees of freedom and many components with different characteristics of memory, the climate system integrates short-term fluctuations, such as those given by daily weather events, so that considerable, slower, stochastic fluctuations ("stochastic climate model" (Hasselmann 1976)) arise. The climate system therefore also generates variability, which cannot be attributed to specific drivers but arises spontaneously from within itself. Smoke without fire, so to speak. This is sometimes conceptually difficult to understand for those who are not experts.

The detection takes the form of a statistical hypothesis test—with the null hypothesis that the trends considered in the recent past originate from the ensemble of natural climate fluctuations. The rejection of this null hypothesis means that these trends can be regarded as very rare in the context of natural climate variability, and it is therefore reasonable to look for explanatory factors other than natural climate variability.

The problem here is above all the determination of the natural climate variability, since the data situation as described above is rather poor—in this respect some reservation remains at this point, which refers to the reliability of the natural variability bandwidths, and will only be reduced over time. Here, the simulation results of quasi-realistic[5] climate models are added,[6] which have been integrated for about hundreds of years with the specification of natural drivers. Here the implicit assumption is made that these models realistically describe natural climate variability. Also, here, some reservations remain.

For the climate-researcher familiar with statistics, it should also be noted that the hypothesis test of the detection approach cannot be performed in a very high-dimensional space, firstly because the determination of a covariance matrix, for example, then becomes impossible, and secondly because the power of the test then becomes low. A reduction of the phase space will therefore be sought, e.g. by considering aggregated quantities such as the globally averaged air temperature or by concentrating on particularly strong changes, as proposed by climate model simulations as a result of increasing greenhouse gases or aerosols.

---

[5]von Storch et al. (1999: 255 pp) explains in German concepts and methods for scientifically educated laymen; although it reflects the state of knowledge at the end of the 1990s, it is topical in its concepts and approaches.

[6]Very different model concepts are used in the various scientific traditions (cf. McGuffie and Henderson-Sellers (1997); von Storch and Flöser (2001), or Müller and von Storch (2004)). The simplest mathematical form are energy balance models, which schematize the influx of short-wave solar radiation and the emission of long-wave and short-wave radiation from the earth's atmosphere and surface. Such models are conceptual descriptions of minimal complexity that describe the fundamental aspects of the thermal engine of the climate system. At the other end of the complexity are those models that are usually simply referred to as "climate models": They have maximum possible complexity (as given by the available computer capacity) and describe as many processes as possible in the climate system. They are considered quasi-realistic and process-based. Since these models are limited by computer capacity, which is constantly increasing, the complexity of these models grows over time. These models are intended to approximate the actual complexity, although they necessarily represent a clear simplification compared to reality. They simulate hourly, sometimes even more frequently stored sequences of weather in the atmosphere, in the ocean and in the cryosphere—for example in the form of temperature, salinity, wind speed, water content in cloud droplets, upwelling, ice thickness, etc. The desired statistics (=climate) are derived from the stored time series. In this respect, working with the output of such climate models is similar to working with observational data. A major advantage with model data is that real experiments are possible (e.g. increasing the atmospheric concentration of greenhouse gases), which is not possible in the real world. On the other hand, these models operate with finite grids, so that the representation of some processes remains a challenge. The interaction of radiation and clouds or the eddy dynamics in the ocean should be mentioned here in particular. Another important, often not understood property is that "the" differential equations of such models do not exist; the considered elements in the differential equations depend on the spatio-temporal resolution and are closed—resolution-specific—by so-called parameterizations.

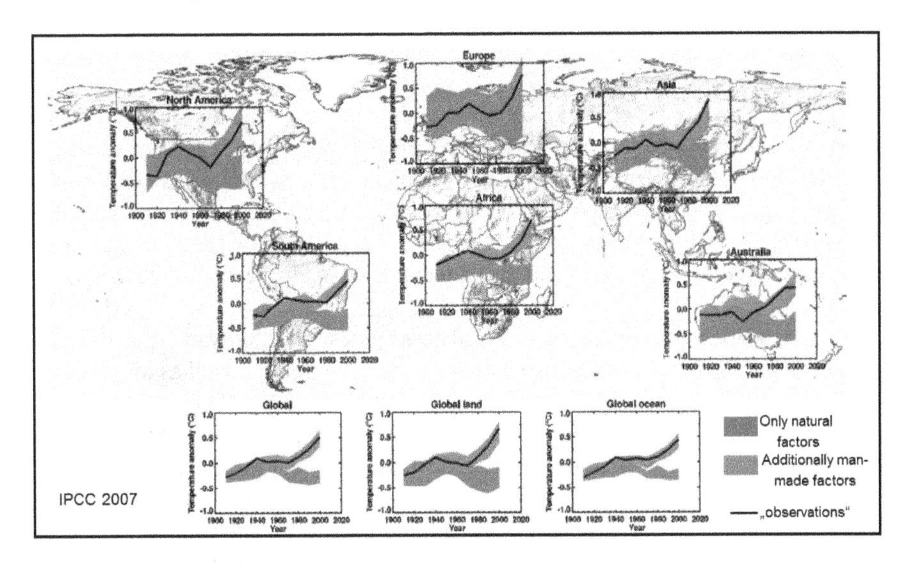

Fig. 6.2 Development of surface temperature in different regions of the world, across all land and sea areas and globally, as proposed by models that take into account the effect of only natural factors (sun, volcanoes; blue) or the effect of anthropogenic factors (greenhouse gases, aerosols; red). The "observations" shown above are shown as a black line. 10-year means are shown to reduce the effect of internal variability. (IPCC, AR4)

With regard to surface air temperature, detection has been successful both on a global scale and at various regional scales such as the Baltic Sea region. For variables such as wind speeds, which could indicate a change in the storm climate, there are no such results for mid-latitudes; changes in tropical cyclones are still the subject of discussion (IPCC 2013). As far as extreme precipitation is concerned, the situation is also somewhat unclear with regard to the changes made so far, as the data situation is very difficult (IPCC 2013).

The third step, *attribution*, is a plausibility argument; it combines proposals from climate model simulations that describe the response of the climate system to, for example, changing concentrations of greenhouse gases in the atmosphere to determine whether a change that has previously been detected as "not within the framework of natural climate variability" can be represented in this way. This has also been successful in the past, particularly for the global and continental development of surface air temperature (IPCC 2013). This is demonstrated by a key Fig. 6.2 from the fourth IPCC report.[7] In numerous simulations with dynamic climate models, the effect of the variable natural drivers (sun, volcanism, internal variability) was first shown (blue band), and in a second series the effect of anthropogenic factors (greenhouse gases and aerosols; red band) was also shown. The observed development is consistent with the presence of only natural factors into the 1950s and 60s

[7]The last (fifth) IPCC report provides a similar but more complex illustration; due to the clarity of the corresponding illustration from the penultimate (fourth) IPCC report, we will use this here.

(blue band); thereafter the development can only be explained by adding the effect of anthropogenic causes (red band).

For regional areas such as the Baltic Sea region, this was only possible to a limited extent (Barkhordarian et al. 2016). Here regional effects of increased or decreased occurrence of anthropogenic aerosols may be necessary for a complete explanation of the change. However, this is still the subject of research. Detection and attribution using satellite data have also been successful since 1982 for the question of the vegetation's response to increased greenhouse gas occurrences and the associated climatic changes.[8]

The reservations that must be expressed here relate to the ability of climate models to realistically represent the "signals" of the various drivers, as well as to the mix of drivers considered potentially effective. At least theoretically, there is a possibility that the expected increase in scientific knowledge will lead to a revision of previous ideas. Moreover, it is not impossible that different explanatory approaches are suitable for deconstructing the observed changes—but so far there are no convincing proposals for such alternative explanations of the ongoing change.

The methodological sequence of manifestation, detection and finally attribution has led to the core statement of the UN Climate Council IPCC that there is climate change and that it can only be explained within the framework of previous knowledge by considering the increased greenhouse gas concentrations in the atmosphere as dominant driver (IPCC 2104a). This statement refers to the changes in the thermal regime in the recent past and is based in particular on the evaluation of observational data.

Simulation models, on the other hand, play a subordinate technical role, namely in improving the data to determine the natural climate variability and deriving the expected effect of external factors. All these assertions, small residual doubts remain, but this knowledge claim is consistent with the previous scientifically constructed knowledge about the dynamics and sensitivity of the climate system.

6.4 Target Value: Globally Averaged Temperature

Highly simplified, the relationship between emissions E, concentration of greenhouse gases C, and global average air temperature T, each as a deviation from pre-industrial levels, can be described by two radically simplified equations[9]:

$$C'(t) = E\beta(t) - C\gamma(t)$$
$$T'(l) = \log \mu C(t) - \alpha T(t)$$

[8]Also the term "greening" is used, see Mao et al. (2016); the authors found an increase in vegetation dynamics in the northern hemispheric land areas in mid-latitudes, such as an increase in (above-ground) biomass and terrestrial carbon fluxes.

[9]Tahvonen et al (1994); this model belongs to the class of highly simplified conceptual models mentioned in a footnote above.

C' and T' denote the time derivatives at a resolution of annual values (t). $\gamma C(t)$ and $\alpha T(t)$ describe the tendency of the system to return to the pre-industrial value in the absence of drivers; the driver for concentration is given by $\beta E(t)$, the driver for temperature by $\log \mu C(t)$. With a simple model of this kind, development paths can basically be estimated (Tahvonen 1994).

Such a simple carbon-climate model invites to consider whether one could not add terms that describe additional sinks in the carbon cycle or that could compete in the temperature equation with the effect of increased carbon dioxide. It is referred to as "geoengineering" because the realization of such terms in the real world would, of course, represent considerable technical interventions in the natural dynamics (Hulme 2014).

In the former case, with additional sinks in the carbon balance, additional storage of atmospheric carbon dioxide in vegetation or in the ocean (by fertilizing oceanic algae, which then sink to ocean depth), or filtering carbon dioxide out of the atmosphere or exhaust air have been proposed.

The option of storage in forests is already being implemented, but does so far not seem to be a really significant contribution to mitigation. For oceanic sinks, there have been only minor trials so far, and effective implementation in the near future appears difficult. In addition, however, there is the fact that it is currently almost impossible to assess the consequences. When filtering, the problem arises where the very large quantities of the filtered material should remain. In this respect, these options may not significantly relieve the situation.

The second possibility, namely to let competing factors act on the development of the temperature, is mostly thought of in the form of obstructions of the incoming short-wave solar radiation by the introduction of radiation-active substances in particular into the stratosphere. In a recent overview article on the effect of this approach, it says:

> ... aerosol injection would generally offset the climate effects of elevated GHG concentrations. However, it is clear that a solar geoengineered climate would be novel in some respects, one example being a notably reduced hydrological cycle intensity.[10]

Here it is an essential caveat that the globally averaged temperature can be regulated in this way, but hardly in such a way that the measures do not manifest themselves very differently from region to region. This means that very considerable risks would have to be taken with regard to the regional distribution of the changes. In the regional, the effects on human and natural systems unfold. In terms of effects, however, the globally averaged air temperature is not a relevant parameter but only a symbolic shorthand that we have become accustomed to using to outline the extent of climate change.

At present, geoengineering is hardly seen as a serious option for dealing with man-made climate change in the near future, although there is a lively debate among scientists and engineers on this issue (e.g. Irvine (2016a) and Keith and Irvine (2016).

[10]Irvine et al. (2016); the "neutralized effects" mainly concern the thermal regime.

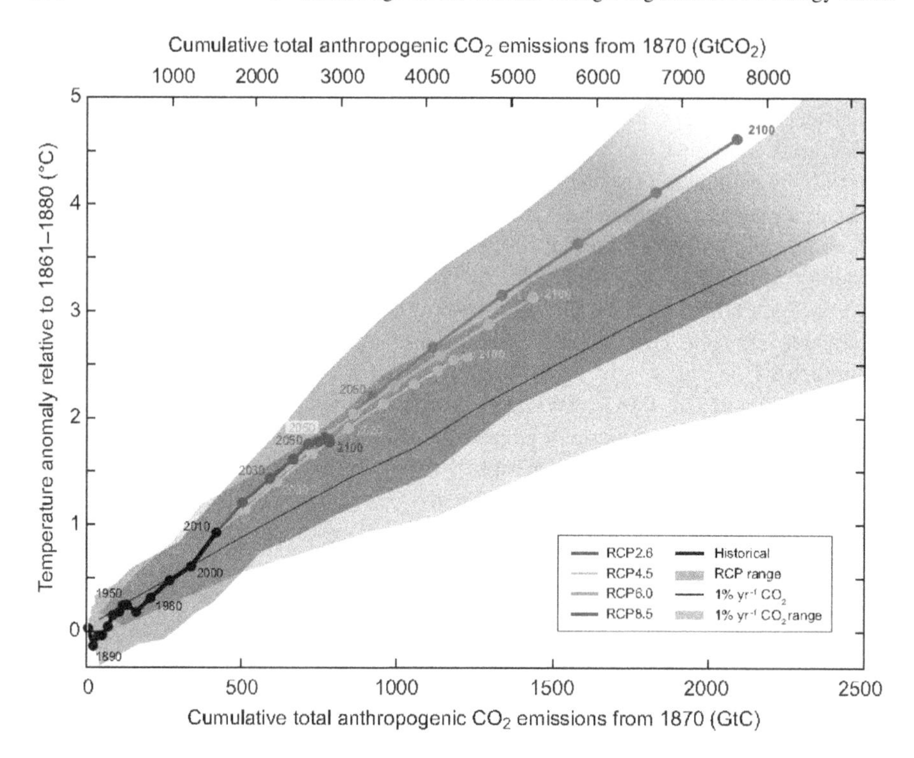

Fig. 6.3 SPM-10 from IPCC (2014a: 28): Global mean air temperature increase at ground level as a function of cumulative global carbon dioxide emissions. Results of ensembles of RCP scenarios (see Sect. 6.5) up to 2100 are marked by colored lines and dots. Results for the past ("historical") period (1860–2010) are marked as a thick black line. The colored tubes represent the range of variation in the RCP scenarios. The ensemble of scenarios representing a 1% increase in CO_2 concentrations is represented by a black line and a grey tube. For further details see (IPCC 2014a)

In the "budget approach",[11] the change in the global temperature at a time t can be estimated well from previous emissions, since a linear relationship can be derived in good approximation between the sum of past emissions and temperature increase in the coming 100 years (see Fig. 6.3).

If we follow this approach and aim for a stabilization of the temperature change of a maximum of 2 °C at the end of the twenty-first century, emissions in the period 2000–2049 must not exceed 600 $GtCO_2$. With the same per capita emission rights, the EU would have consumed its share from now on "in about 10 years" if no transfer of rights were agreed with developing countries within the framework of emissions trading systems.[12]

This is a narrow framework for future emission reductions, as shown by Fig. 6.4, which illustrates how the requirement to reduce global emissions becomes more acute

[11] In terms of modelling concepts, this is a highly reduced, conceptual model, which draws it validity, among others, from simulations with quasi-realistic, process-based models.

[12] Hermann Held, pers. communication; see also Meinshausen et al. (2009).

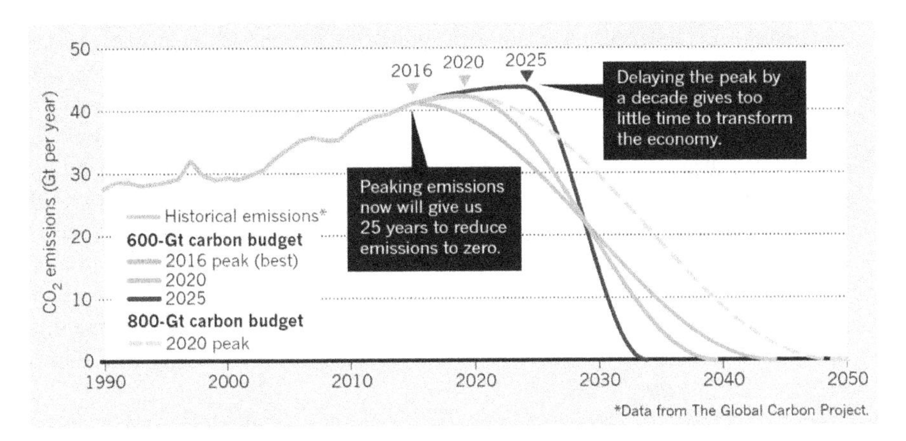

Fig. 6.4 Estimates of possible future emission developments that are compatible with the Paris target of a maximum warming of 1.5–2 °C. The estimate uses the budget approach, whereby 600 Gt of emissions are permitted in one case, and 800 Gt in a second case. Here it is assumed that after a year with maximum emissions in 2016, 2020 and 2025 a kind of "emergency braking" will be triggered (Figueres et al. 2017)

when the start of the reduction is delayed—with peaks in 2016, 2020 or 2025—so that the share of emissions not consumed by 2016 that are compatible with the Paris 1.5 or 2 °C target is between 150 and 1050 $GtCO_2$ (depending on the type of calculation (Figueres et al. 2017)). Currently, about 40 Gt are released annually, so that the target cannot be achieved if they continue for more than four (in the pessimistic case) or twenty years (in the optimistic case). If a mid-range value of 600 Gt is assumed, as shown in the figure, this time is reached after 15 years. If no significant reductions in global emissions have been achieved by then, an abrupt termination of all emissions would be necessary to achieve the Paris target.

Note that the numbers given here originate from different sources and are thus not fully consistent among each other.

6.5 Talking About Futures: Scenarios

An essential consequence of the successful attribution of warming or the increase in thermal energy of the climate system to the increased concentration of greenhouse gases in the atmosphere is that we can estimate future developments by specifying the release of greenhouse gases and the formation of anthropogenic aerosols. Our quasi-realistic, process-based climate models have been successful in the attribution step in deconstructing past changes, so that it is plausible to assume that they can also estimate future changes as a result of such human activities.

In this way, perspective for the future can be constructed. There are two types of perspectives, namely:

- *Predictions*, i.e. the most probable description of the future of the climate, which, starting from an initial state, take into account both the memory of the system (in particular the thermal structure of the oceans and soils) and the drivers. The predictive skill decreases over time because of the stochastic character of the system, which after some time enables a multitude of trajectories, the probabilities of which change over time as these are conditioned by the drivers. However, this does not change the fact that after some time the real trajectory of the climate can no longer be estimated in the sense of a most probable trajectory, and rather an ensemble of possible developments remains. The question of how long into the future such predictions are possible is the subject of current research; the initial optimism seems to have given way to a rather pessimistic assessment of at most two decades of years.
- *Scenarios*, these are ensembles of possible developments conditioned by the prescribed drivers, where no probability can be assigned to the individual members of the ensemble. If all scenario calculations describe a warming of the climate system over time, this can of course be taken as a qualitative prediction, without quantitatively achieving an accuracy that would be desirable for the current answers to many adaptation questions. Scenarios are possible developments that are dynamically consistent, plausible but not necessarily probable. They are suitable for policy and management consulting, in which they allow possible adaptation and avoidance strategies to be examined for their suitability and efficiency (see e.g. Schwartz 1991).
- Constructed emission paths are transferred into scenarios, or projections, of weather streams to, usually, the year 2100. Since the climate models realistically generate internal weather variability, two scenario simulations with the same drivers and the same model each differ if an insignificant disturbance is introduced (e.g. in the initial state) or a different computer is used. Most scenarios are then available as "ensembles" of several, but different simulations for evaluation and further processing.
- Technically, climate change scenarios are thus conditional predictions of distributions—conditioned by the specification of emission pathways.

In the present context, only scenarios are relevant, firstly because the predictions are still in an experimental phase and previous results are only convincing to a limited extent, and secondly because the time horizon for such scenarios is insufficient for many climate change applications.

Technically, the scenarios are constructed in this way: In the phase of the "*SRES*" scenarios (von Storch et al. 2007), which were constructed for the IPCC's 3rd Assessment Report, storyboards on socio-economic developments are first compiled, showing how much greenhouse gases and aerosols are released and when. In the "Representative Concentration Pathways" used in the IPCC's 5th Assessment Report under the abbreviation "*RCP*", scenarios of possible developments of radiative drivers (resulting from human emissions) over the century have been constructed.

The different approaches behind SRES and RCP are outlined in Figs. 6.5 and 6.6.

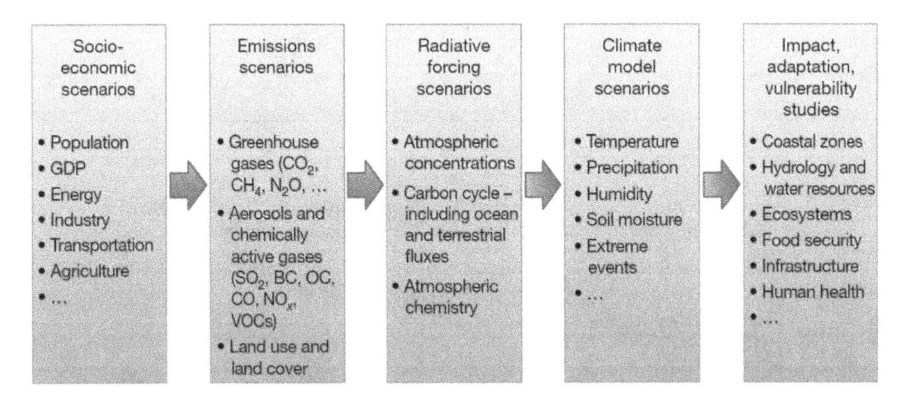

Fig. 6.5 Sequential constructions of SRES scenarios (GDP = gross national product; BC = carbon black; OC = organic carbon; VOCs = volatile hydrocarbons; Moss et al. 2010: 752)

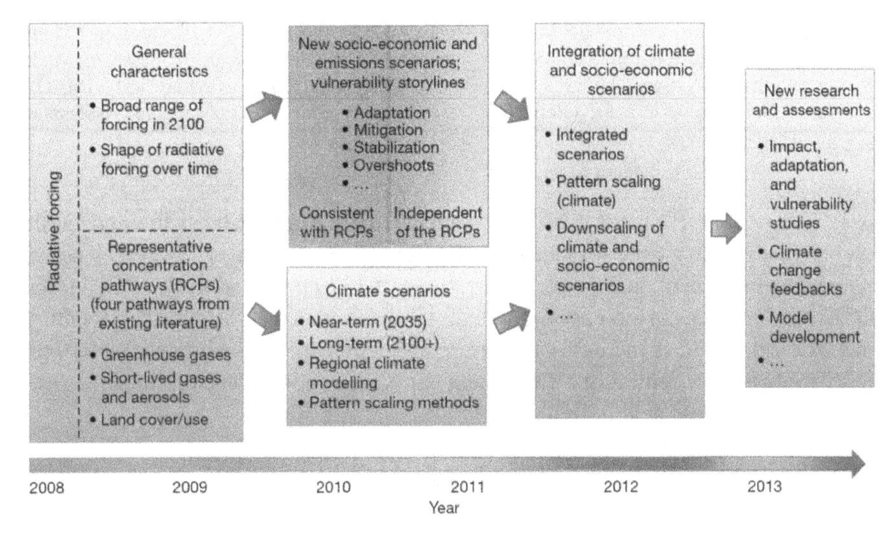

Fig. 6.6 Parallel construction of RCP scenarios (Representative Concentration Pathways; Moss et al. 2010: 752)

A disadvantage of the SRES scenarios was the relatively long lead time before the scenarios could finally be introduced into the investigation of effects of climate changes described in scenarios. Therefore, the sequence of steps in RCP has been changed from SRES. It no longer begins with socio-economic scenarios; instead, construction begins with the selection of radiation anomalies. These scenarios are not initially associated with socio-economic storylines or emission scenarios; in fact, the assumed radiation anomalies can be realized by various technological, demographic, political and institutional developments (Moss et al. 2010: 751). The four main scenarios are outlined in Table 6.1 (Moss et al. 2010: 753). They are referred to

Table 6.1 The four Representative Concentration Pathways (RCPs)

Name	Radiative forcing	Concentration (ppm)	Pathway	Model providing RCP*
RCP8.5	>8.5 Wm-2 in 2100	>1370 CO_2-equiv. in 2100	Rising	MESSAGE
RCP6.0	~6 Wm-2 at stabilization after 2100	~850 CO_2-equiv. (at stabilization after 2100)	Stabilization without overshoot	AIM
RCP4.5	~4.5 Wm-2 at stabilization after 2100	~650 CO_2-equiv. (at stabilization after 2100)	Stabilization without overshoot	GCAM
RCP2.6	Peak at ~3 Wm-2 before 2100 and then declines	Peak at ~490 CO_2-equiv. before 2100 and then declines	Peak and decline	IMAGE

* MESSAGE: Model for Energy Supply Alternatives and their General Environmental Impact (International Institute for Applied Systems Analysis, Austria). AIM: Asia-Pacific Integrated Model (National Institute for Environmental Studies, Japan). GCAM: Global Change Assessment Model (Pacific Northwest National Laboratory, USA, previously referred to as MiniCAM). IMAGE: Integrated Model to Assess the Global Environment (Netherlands Environmental Assessment Agency, The Netherlands)

as RCPX, where X stands for the assumed radiation anomaly at the end of the twenty-first century—the scenario RCP8.6 describes very strong increases as can be expected at greenhouse gas concentrations of 1370 ppm and more, RCP6.0 and RCP4.5 describe mean increases including stabilization in the twenty-second century, and RCP2.6 smallest increases that are consistent with a maximum of 490 ppm greenhouse gas concentrations. Consistent temporal trends of the emissions are given in the work of Moss et al. (2010).

The distinction between "predictions" and "scenarios" as "most probable developments" and "possible developments" is also defined as a language rule within the framework of the IPCC process. Nevertheless, in the scientific community the two terms are often confused or simply misused. A survey of climate scientists found that about a third of respondents used the term scenarios in the sense of predictions, and also about a third speak of predictions when they mean scenarios (Bray and von Storch 2009).

Strictly speaking, scenario calculations cannot be "verified"[13] because they describe developments in an open system that have not yet been observed in this way. To be sure that the models describe realistic perspectives, one would have to wait a few decades. On the other hand, one can take a look at what early model calculations have successfully anticipated the development of the globally averaged

[13]A general discussion of the challenge of speaking about "verification of model simulations" is offered by Oreskes et al. (1994).

air temperature in the last 20–30 years—one of them comes from Hansen and was published in 1988,[14] another one from Allen et al.[15]

6.6 Uncertainties

An essential concept in climate research itself and at the border between policy-making and science is that of uncertainty. Uncertainty is neither unusual for policymaking nor for science. Both social actors can deal with it in principle, both in constructing explanations and in political decision-making.

In the case of the climate change problem, the situation seems to be somewhat different, because we hear and read often of talking about an allegedly scientifically proven necessity. Therefore, the reference to uncertainties in the scientific understanding is a central tool of the opponents of this policy. Science therefore faces the often unspoken demand to provide certain knowledge—although scientific knowledge is always uncertain in character and subject to the current state of knowledge. As a result, some proponents of an effective climate change policy tend to downplay remaining uncertainties, such as the presentation of temperature trends over the last 1000 years, known as the hockey stick.

Sometimes the term "the science is settled" is used to ward off these critical objections. In fact, this assertion is inaccurate in its generality. Rather, there are statements, in particular those on manifestation, detection and attribution, which can indeed be regarded as very largely certain within the framework of current knowledge and empirical testing. In addition, however, there are many questions that are also the subject of controversial scientific debates: one thinks of the expectations of sea-level rise, of changes in tropical cyclones (IPCC 2014a) or of future yields from agriculture or migration movements. It must therefore read "Some science is settled", but this does not stand in the way of implementing a climate protection policy, and in particular a low-emission or emission-free energy policy, because the fundamental mechanism of "fewer emissions = slower change" is very largely safeguarded. What effect such a climate protection policy has on other societal fields is not a question that climate science can answer.

6.7 Politicization

In the 1980s, Silvio Funtowicz and Jerry Ravetz (1985) proposed the concept of "*post-normality*" to describe the conditions and consequences of environmental research when it comes close to policymaking. Accordingly, a post-normal situation exists when the relevant knowledge is unavoidably uncertain, societal decisions

---

[14]See Fig. 1 in Hargreaves (2010).

[15]See Fig. 4 in Allen et al. (2000).

are unavoidable, these are associated with societal values and considerable use of resources. Under these conditions, claims to knowledge are polarized, and produced and used to support societal interests. Methodical quality sometimes takes a back seat to political usefulness.

According to a survey of climate researchers conducted in the mid-1990s (Bray and von Storch 1999), the situation for natural science climate researchers is actually of this sort. The question of the sensitivity of global air temperature to a doubling of greenhouse gas concentrations will not be definitively resolved in the foreseeable future; since the first estimates in the 1970s, the figure has fluctuated between 2 and 4 °C, and even much higher degrees (IPCC 2014a). If the increase in greenhouse gas concentrations and the associated changes in temperatures and other climate variables are to be significantly mitigated, then effective measures must be implemented now; some people now want to assume responsibility for future generations, others refer to the freedom of choice of future generations; if effective measures are taken, then this is associated with considerable efforts—if they are not taken, then all the more significant adaptation measures may be needed later. We observe activism and self-censorship among scientists—"You can't say that, it's exploited by skeptics" was often heard, at least in the past. Consistent with post-normal interpretation, these conditions apparently lead to changes in science itself, namely to dogmatism and a reduction in openness to alternative explanations.

We observe a politicization of climate science, and at the same time a scientification of policymaking, which reacts to societal necessities allegedly discovered by climate science, which manifests itself especially in the 2 °C objective. Science points out that the more one emits, the greater the pressure to adapt; it also estimates which quantities may be emitted in order to make the 2 °C target achievable (see above); science (e.g. in the form of the IPCC), however, does not present the 2 °C target as mandatory, even if prominent individuals do so in public communication. Policymaking, on the other hand, presents itself as in a situation where there is no alternative, where it only has to execute what science pretends there is no alternative. In this way, science is subject to a proviso of purpose, namely to support the right policy, and political decision-making loses its societal negotiating character (see also von Storch and Krauss 2013).

The two societal actors science and policymaking are thus approaching each other and losing their specific strengths—namely openness to explanations and the balancing of societal conflicts—at the same time. More social science research is certainly needed here, since it is clear that these are conditions that are also important for shaping national and international energy policy.

Consistent with the notion of a post-normal climate research situation, there is a market of knowledge claims about the extent, character, and cause of climate change. One of the claims to knowledge is the scientific construction, as mentioned above, which receives very high approval in scientific circles. While in the 1990s only about 40% of climate scientists agreed to the attribution to the increased greenhouse gas concentrations, today this value is clearly above 80% if not 90% (Bray 2010). In competition are other knowledge claims of "skeptics", according to which the seriousness of the problem is exaggerated, or the interpretation of climate change as the incarnation of the loss of a healthy environment and living conditions as a result of overexploitation of resources. In addition, there are other claims to knowledge, such as those rooted in climate determinism. Here, too, further social science research is needed (von Storch 2009).

There is no reason to assume that scientific constructs generally win the knowledge competition in public. This also seems to be the case in climate science: while the acceptance of attribution among climate scientists has risen massively over the years, surveys in the USA since the beginning of the 1990s and in Hamburg since 2008 have shown that the general population is quite worried about climate change, but a systematic increase in the proportion of worried people cannot be ascertained. Rather, the proportion fluctuates around 60%, with shorter variations as known from attention cycles (Ratter et al. 2012).[16]

[16]In the recent two years, a marked increase, with 72% (67%) worried in 2019 (2020), was found in these surveys (Ratter, pers. comm.).

Chapter 7
Economic Problems of Energy Transitions, Resource Scarcity and Climate Change in a Global Perspective

7.1 Introduction: Problems, Solutions, Problems from Solutions

In this chapter we start out in Sect. 7.2 from stating what we consider to be the major global energy problems. Energy problems in poor countries are different from those in rich countries. In particular the problem in poor countries is access to electricity for poor people mainly in the countryside.[1] In contrast, in rich countries the major problem is fossil based energy abundance at subsidized prices causing global climate change. Non-electrical energy has to be included into policy measures in poor and rich countries in order to achieve the two-degree goal of the Paris climate agreement. A policy issue, so far ignored by poor and rich countries, is the international policy coordination for international transport emissions, whose neglect is imposing a higher burden on emission reductions in production and consumption. We then sketch in Sect. 7.3 the basic ideas of the solution to these problems: public investment and policies for renewable energy. These solutions or the deficiencies in the concepts or implementations themselves cause new problems which we discuss in Sect. 7.4. In Sect. 7.5 we study the link between renewable electricity and global trade in electricity.

[1]Energy poverty in rich countries regarding people with access but low and expensive energy consumption is not dealt with in this book, because it would require embedding the analysis and recommendations into poverty analysis of rich countries, which is not the expertise of the current group of authors.

C. F. Gethmann et al., *Global Energy Supply and Emissions*, Ethics of Science and Technology Assessment 47,
https://doi.org/10.1007/978-3-030-55355-5_7

7.2 Global Energy Problems

7.2.1 Problem 1: Electricity Access in Developing Countries

Figure 7.1 shows that for per capita incomes (Gross Domestic Product per capita) up $1,500 per year (in 2005 prices) there is a steep increase of access to electricity in percent of the population. From $1,500 to $4,000 there is a flatter but still positive relation. Interestingly, up to $6,500 there can be a severe lack of access to electricity.

In Sub-Saharan Africa 70% of the population are without electricity connection (Lee et al. 2014). The points with more than 95% of the population having electricity access in spite of incomes below $2,000 are data for Egypt, Paraguay and Vietnam for 2010/11. Below 80% electricity access and income above $1,900 are Angola, Botswana, Rep. Congo, Gabon, Namibia 2010/11, and South Africa 2010.

These results indicate that there will be a strong demand for more electricity when countries get richer and this may come together with much more CO_2 pollution. A similar graph could be made for the density of cars with a similar result. For Africa it may also be the case that electricity is produced mostly without fossil fuel using instead wind, hydro and geothermal power (Lee et al. 2014). However, hydropower has suffered from draughts in recent years.

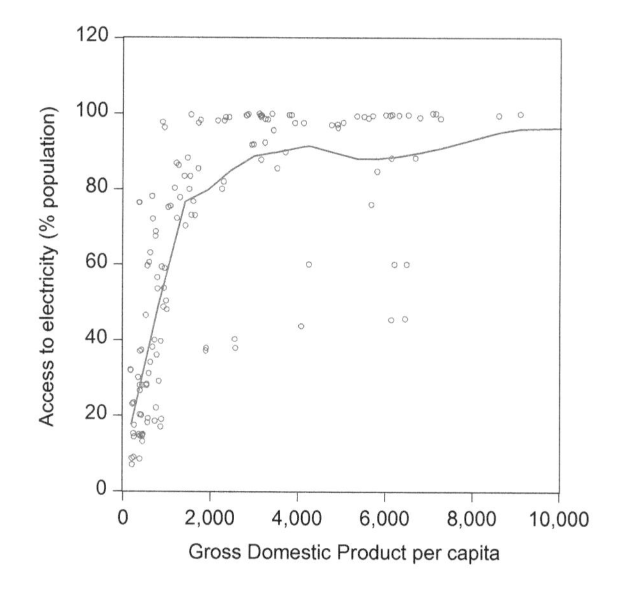

Fig. 7.1 Countries with higher gross domestic product per capita for various years have a higher percentage of the population with access to electricity. *Source* Authors' calculation. Observations of countries with Gross Domestic Product per capita above $10.000 are eliminated; method: nearest neighbor fit

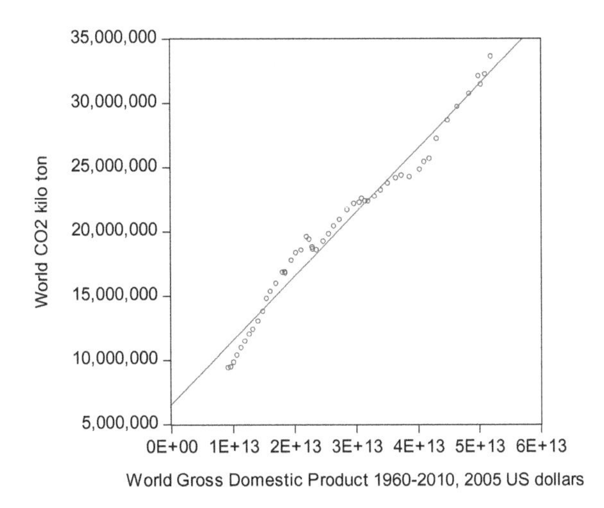

Fig. 7.2 A higher gross product of the world leads to higher emissions. *Source* World Development Indicators, authors' calculation

7.2.2 *Problem 2: Growth Enhances Emissions of CO_2 and Other Greenhouse Gases, the Global Link[2] 1960–2010*

Roughly, a change of the growth rate of the Gross Domestic Product, GDP, of the world, g_{GDP}, leads to an equal change in the growth rate of world CO_2. This can be seen as follows. The best regression is[3] (a simpler version is shown in Fig. 7.2).

Box 1 The impact of world GDP on global CO_2 emissions

$LOG(CO_2WLD) =$

$$\underset{(0.28)}{-0.415} + \underset{(0.1)}{1.1}LOG(CO_2WLD(-1)) - \underset{(0.1)}{0.19}LOG(CO_2WLD(-2)) + \underset{(0.16)}{0.956}LOG(GDPWLD) - \underset{(0.166)}{0.9}\,LOG(GDPWLD(-1))$$

When estimating in first-differences of this equation, the lagged CO_2 values become statistically insignificant and those of world income get a unit coefficient: $g_{CO_2} = -0.009 + 1.0001\, g_{GDP}$. However, for the period 1960–1978, the coefficient is higher than unity, $g_{CO_2} = -0.014 + 1.12\, g_{GDP}$, and for the period 2002–2010 it is

[2]The graph indicates that the simple linear regression would have serial correlation bias.

[3]Econometric note. Standard errors in parentheses. Adjusted R-squared is 0.9976, Durbin-Watson statistic is 1.8. The second lag of CO_2 is statistically significant at the ten per cent level, the constant at the 15.6% level, all other at the 0.0000 level. A time trend has turned out not to be statistically significant. For the log of world income, unit root tests are rejecting both hypotheses 'unit root' and 'no unit root'. The dependent variable has no unit root. Note that vector-autoregressive models, at least in their standard form used in recent literature ignore the current impact of the Gross Domestic Product on CO_2 and include by assumption only the lagged effects. The effects are then much weaker.

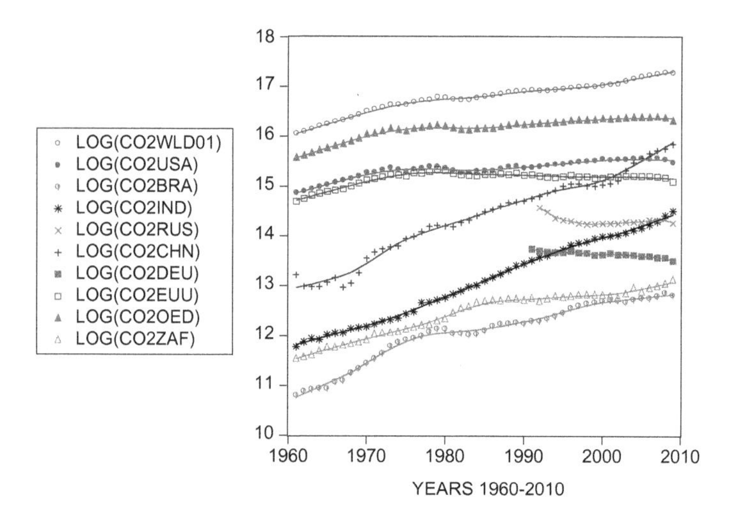

Fig. 7.3 Over time, CO_2 emissions of the world (WLD), USA, Brazil, India, Russia, China, South Africa (ZAF) and the OECD (OED) increase, but not for Germany (DEU), and the European Union (EUU). *Source* World Development Indicators. CO_2 measured in *kt*. Natural logarithm of emissions (not per capita); the slope is the growth rate

lower $g_{CO_2} = 0.01 + 0.75 g_{GDP}$. This shows that growth rate estimates are much less stable.[4] Growth of the Gross Domestic Product (*GDP*) of the World (*WLD*) is here to stay at a rate of 2.7% in the long run as can be calculated from the following regression including a time trend (*LOG* indicates the natural logarithm rounded coefficients)[5]:

Box 2 The growth process of world GDP

LOG(GDPWLD) =

3.26 + 1.137LOG(GDPWLD(−1)) − 0.46LOG(GDPWLD(−2)) + 0.22LOG(GDPWLD(−3)) + 0.0029t
(0.0008) (0.00) (0.019) (0.0765) (0.0043)

7.2.3 World CO_2 Trend Disaggregated

Figure 7.3 shows that China and India clearly have growth rates of emissions above the average. Germany, EU (indicated as EUU) and Russia, are below the

[4]Econometric note. ARMA (auto-regressive moving average) variants do not solve this problem.

[5]Econometric note. Statistical significance levels in parentheses. Adjusted R-squared is 0.999; Durbin-Watson statistic 2.03—it is only roughly indicative of no serious serial correlation. A test for remaining mild serial correlation in spite of using all statistically significant lags requires the Breusch-Godfrey approach.

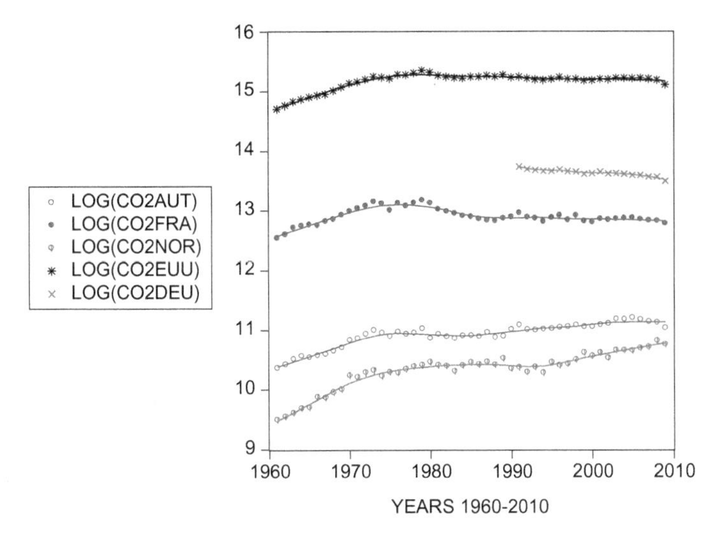

Fig. 7.4 CO_2 development for countries with high share of nuclear or hydro power. *Source* World Development Indicators. CO_2 measured in *kt*. Natural logarithm of emissions (not per capita); the slope is the growth rate

average. Successful growth out of poverty in low income countries implies more CO_2 emissions.

Norway with 94–100% and Austria 54–75% waterpower have growing emissions, to a decreasing extent though in Austria, as shown in Fig. 7.4. We also observe falling CO_2 emissions in Germany and France (since 1980; 1980–1988 fast, then more slowly). Obviously, there are two ways to get out of CO_2 emission growth: nuclear and renewables. Adding alternatives and nuclear, *an*, and energy productivity *GDP/eng* to the above regression[6] we get[7]:

---

[6]Econometric note. Closely related conceptually is the accounting framework where $CO_2 = (CO_2/\text{energy})(\text{energy}/GDP)GDP$. Taking logs results in a linear formula with unit coefficients. But applying it in regression methods leads to some differences. As the share of fossil and alternatives plus nuclear, *AN*, add up to almost exactly 0.9, one can replace CO_2/energy by *AN* when running regressions. As all variables are aggregates across countries and some over energy sources the regression deviates from an accounting relation.

[7]Econometric note. t-values in parentheses. The method is instrumental variable estimation. Least squares and cointegrating regression yield very similar results, but the coefficient of *AN* is less close to negative unity, whereas statistical significance and Durbin-Watson statistic are much better. Exogenous regressors have been found through the Durbin-Wu-Hausman test. The instrument orthogonality C-test has been used to find admissible instruments. If the J-statistic goes too high when adding an instrument and *p*-values are very low, instruments may be endogenous; if the J-statistic does not increase and *p*-values are very high instrumental variables do not remove the endogeneity bias (Roodman 2009). *p*-values which are neither very high nor very low indicate instruments that do not suffer from either of the problems. The resulting instrument specification is: C, LOG(CO_2WLD(−2)) with $p(J) = 0.53$, LOG(CO_2WLD(−3)) with $p(J) = 0.43$, LOG(GDPWLD(−0)), LOG(AN(−1)) with $p(J) = 0.3$, LOG(AN(−2)) with $p(J) = 0.24$,

Box 3 The impact of world GDP, alternative and nuclear electricity, and energy productivity on global CO_2 emissions

$$LOG(CO_2WLD) - LOG(CO_2WLD(-1)) =$$

$$\underset{(-0.16)}{-2.15} - \underset{(-1.94)}{0.94}LOG(CO_2WLD(-1)) + \underset{(3.59)}{0.49}LOG(GDPWLD) - \underset{(-1.7)}{0.95}LOGAN - \underset{(-2.56)}{0.83}LOG(GDP/Eng) + \underset{(1.78)}{0.018}t$$

Observations: 1991-2010; Adjusted R-sq. = 0.46; Durbin-Watson stat = 1.29; p(J) = 0.4.

7.2.4 *Problem 3: Clean Electricity Is not Enough*

Alternative and nuclear energy as a percentage of total world electricity have a statistically significant, negative impact on CO_2 emissions: a one percent increase in these sources reduces emissions also by about one percent. However, this share is falling over time since 2001, because the share of nuclear in world electricity production is falling since 1996 (unlike the corresponding investment levels, which are still increasing; see technology chapter)[8] while that of the alternatives increases (see Figs. 7.5 and 7.6).

Correspondingly, the share of fossil-based energy goes up slightly from 1999 to 2012 and falls in 2013 and 2014. The EU 2020 policy also wants to make progress in regard to energy efficiency. The indicator used is GDP (Gross Domestic Product) in purchasing power parity divided by energy, *eng*, measured as tons of oil equivalent, *GDP/Eng*. The data show a statistically significant but very small growth rate of energy efficiency. Our regression shows that energy efficiency reduces CO_2. Finally, the positive and statistically significant time trend indicates that CO_2 emissions are growing more quickly than the world GDP and the other variables. Once the alternative measures against CO_2 are taken into account, the impact of GDP growth is much smaller than before as the coefficient has fallen from about 0.85 to about 0.5. The World energy Outlook 2015 (IEA 2015) expresses the hope that the fall in preliminary data of CO_2 emissions in 2014 may have brought de-coupling of growth and emissions. Similar drops have happened to occur in the 1980s, 1992 and a bit in 2009 as one can see from the figure above. Conclusions regarding decoupling based

LOG(GDP/Eng(−1)), @TREND. The Jarque-Bera test suggests a normal distribution of the residuals with $p = 58$. The Ramsey RESET test for all specification problems is passed with $p = 0.80$. ADF unit root test with breakpoint selection indicate that there are no unit roots: log(CO_2WLD) has $p = 0.016$ with intercept break and F-test for lag length 7; log(GDPWLD) has $p = 0.14$ for trend break 1974 Akaike information criterion (AIC) for lag length 1, which we interpret as near unit root as the coefficient is 0.565; log(an) has $p < 0.01$ with AIC for lag length 10 and intercept and trend break 1984. Log(GDP/eng) has $p = 0.043$ without break under Schwarz Information Criterion because of observations only for 1992–2011.

[8]Note that a falling share does not exclude growth.

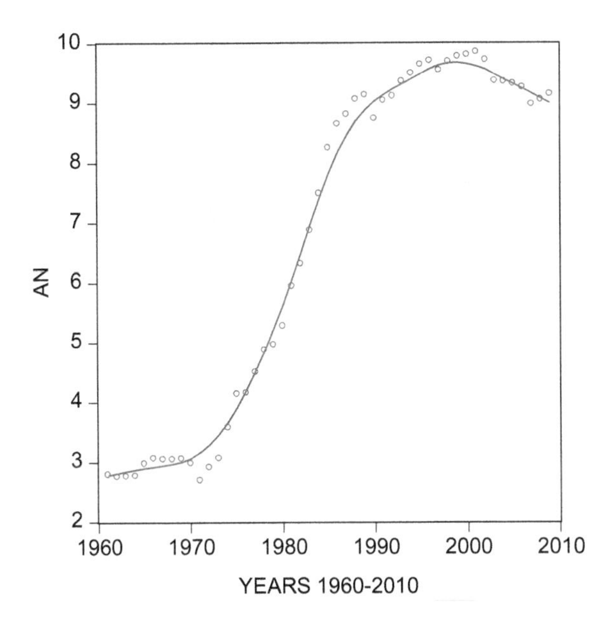

Fig. 7.5 The share of alternatives and nuclear energy, *AN*, as a percent of world total energy use first increases in S-shaped form and then decreases since 2001. *Source* World Development Indicators

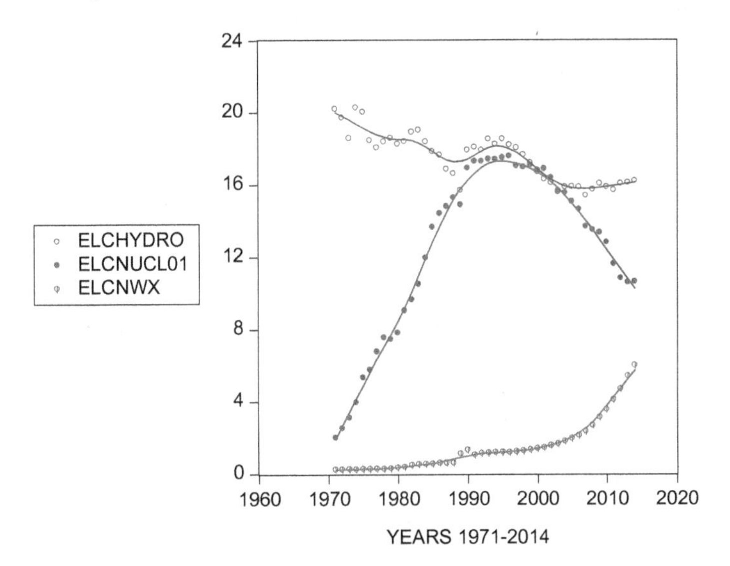

Fig. 7.6 The percentage share of world electricity production from hydro power (ELCHYDRO) falls to almost that of nuclear (ELCNUCL01), which first increases and then decreases; renewables without hydro power (ELCRNWX) is lowest but increasing. *Source* World Development Indicators

on one year only would have been misleading in all three cases. Our first regression shows that the coefficient of GDP is only 0.85 in the first regression and only 0.5 in the last, where more arguments are included. Decoupling is obviously possible, but our first figure indicates how easily one can error when drawing conclusions from most recent data only. It remains to be seen whether a share of 50% of all investment into renewable electricity generating capacity as obtained recently worldwide is lasting and enough to make decoupling sustainable (IEA 2015). So far 66.3% of global electricity generating capacity in 2014 is based on oil, coal and gas (World Development Indicators, December 2017). Given the fact that the remaining carbon budget for emissions is fixed (see chapter on climate change) the positive growth rates of the figure above is worrying (Frankfurt School-UNEP Centre and BNEF 2016, Chap. 2, Fig. 25).[9]

7.2.5 Problem 4: Emissions from International Transport

International specialization of countries in the production of goods and services makes countries richer. Countries specialize in the production of those goods in which they are relatively cheaper than the other countries in the hypothetical situation without trade (autarky). This is based on lower unit costs generating lower prices. Differences in unit costs can stem from differences in productivities (Ricardian theory), differences in factor prices because of abundant or scarce factor supply (Heckscher-Ohlin theory and specific-factor theory), external and internal scale economies[10] in combination with market size allowing being in the lower part on the unit cost curve. If consumers buy each good in the cheapest country, the producers there will produce more than under autarky. Each production expands where it is able to attract foreign consumers and shrinks where consumers move to buy foreign products. Buying at lower prices their income is worth more on average even if factors in shrinking sectors have falling wages. If goods in some countries have monopolistic prices, competition will bring them down. Higher income and demand then trigger higher production. So far, all arguments point to gains from international trade. But higher production leads to more emissions worldwide (Pethig 1976). In addition, transport distances tend to grow and more transport leads to more emissions. Both, national and international emissions reduce gains from trade if they increase with trade as in the case of CO_2 (Soete and Ziesemer 1997). Zhong (2016) uses world input-output tables to show that changes in demand growth cause more additional emissions than changes in international trade for the period 1995–2009 though. Falling energy intensities have reduced emissions. Overall emissions according to our data have increased by 32% in the period 1995–2009; according to Zhong's calculation emissions increased 16% in manufactures and 36% in services. Services

[9]Emissions are predicted to peak in 2026 or later (depending on the origin of the forecast) and be higher than 2015 still in 2040.

[10]The first based on learning effects and the second on fixed costs.

include transport (inland, water, air and others). Emissions from production were shown empirically in the previous regression. In the regression below we show that the volume of world trade as a share of world GDP (Gross Domestic Product) is also enhancing emissions:[11]

Box 4 The impact of global trade, alternative and nuclear electricity and energy productivity on global CO_2 emissions

$LOG(CO_2WLD) =$

$$4.27 + 0.466LOG(CO_2WLD(-1)) + 0.566(EX{+}IM)/GDPWLD - 0.42LOGAN - 0.95LOG(GDP/Eng) + 0.018t$$

(1.85) (4.49) (7.64) (–3.87) (–8.59) (11.02)

Obs.: 1990–2010; Adj. R-sq. = 0.997; DW=2.47; p(J) =0.5485

If the trade volume increases by one percent of GDP world CO_2 emissions increase by 0.566%. This value is slightly larger than that for the world GDP in the previous regression, perhaps because transport is more energy intensive. Both coefficients indicate that emissions can in principle be decoupled from production and trade. The coefficients are now around 0.5 whereas they were around 1.0 when only the GDP variable was included. More renewables and nuclear and more energy efficiency would reduce emissions and the impact of growth.

7.3 Solutions: Public Investment, Markets, Policies, International Agreements

7.3.1 Basic Principles

The basic idea of all economic policy, in particular also competition policy, is to correct for market imperfections: monopoly power, externalities (or more broadly: property rights problems) and a lack of insurance (incomplete market systems not allowing to achieve perfect insurance through private financial investments). The practical principle to ensure this is that all firms should carry the costs they cause and prices should not exceed these costs. Another practical implication is the one of technological neutrality. It says that no technology, product or production factor should get an advantage vis-à-vis its competitors against the benchmark that they

[11] Econometric note. The procedure of footnote 7 is followed again. *P*-values for the instrument orthogonality C-Test are $p = 0.5485$ for both $\log(CO_2WLD(-2))$ and $\log(CO_2WLD(-3))$. The Jarque-Bera test indicates normality with $p = 0.4957$. The Ramsey RESET test is passed with $p = 0.576$. Unfortunately, it is not possible to keep the GDP variable in the regression because it is too strongly correlated with trade variable and the GDP variable then changes sign.

all have to carry the costs that they cause. Similarly, subsidies should follow rules defining the reason and basis for the subsidy and should not prefer for example one trademark over the other. For coal and gas fired power stations or other technologies, which use fossil fuels, this means mainly that they should carry the costs of production and of the emissions and they should not be subsidized as there is no systematic economic or social justification for this. For nuclear power stations this means that they should carry the costs of the waste, risks and scrapping which are currently imposed on others even after the implementation of safety prescriptions of governmental institutions.[12,13]

Public investment in electricity and other infrastructure is related to public goods properties of networks for electricity and other infrastructure such as water, wastewater, roads, and railroads. In the more complicated case of public goods the questions are in how far goods are rival or non-rival and whether or not people could be excluded or not (Russel and Wilkinson 1978: 375). The empirical situation regarding these criteria is often not perfectly clear. As exclusion is often also costly, inefficiencies can get larger if exclusion for privatization is tried. Typically, governments are needed in the starting phase to solve at least the coordination problems regarding electricity networks, roads, and school systems, which can perhaps later be privatized with more or less regulations. In some cases, private and public solutions exist side by side, for example public libraries and book shops, private and public roads, private and public swimming pools. But in case of electricity networks it is efficient to have only one. If that one is not profitable in areas with many poor people, the government needs to set it up as it is also a precondition for running hospitals and avoiding epidemics like Ebola.

7.3.2 Access to Electricity

For the poor developing countries with too little electricity there are three essential steps to solve the problem.

The first step is public investment, which should be higher in poor countries (see Fosu et al. 2016 for Sub-Saharan Africa) and for electricity in particular. The inclination to pay for public investment via taxes is low though and it is not known whether this will automatically get better when countries grow. For Latin American countries, which are much richer than the poor countries of the world, the phase of dominant tax resistance ended only after the 1982 debt crisis in some countries.

The second step is growth, which allows both government and household to invest more money in electricity grids and connections or stand-alone solutions. In densely

[12]Proliferation risks are not caused by the plants alone but mainly by politics, which has to care for it through political decisions.

[13]Most recent information regarding insurance of Suisse and European nuclear power stations can be found at http://www.swissnuclear.ch/de/haftpflicht-_content---1--1028.html, and https://www.bfe.admin.ch/bfe/en/home/supply/nuclear-energy/third-party-liability-in-the-field-of-nuclear-energy.html (accessed 13-Dec-2019).

populated areas like big cities firms and households then connect to the grid. However, without electricity, growth is unlikely to come about. Moreover, grid extension and growth are necessary but not sufficient conditions to get households connected in rural areas with larger distances to the grid (Lee et al. 2014).

The third essential step therefore consists of policies ensuring connections. As early connections reduce costs of further connection, a phenomenon called network externalities, temporary public subsidies to encourage private investment into connections until they are cheap may be important (Lee et al. 2014). Stand-alone solutions with batteries may be more cost-efficient in rural areas. Connecting them bottom-up and sharing batteries could help reducing fluctuations. Some further decreases in costs may be needed to make this possible.

7.3.3 CO_2 Emission Reduction

The social cost of carbon review (Downing et al. 2005) has evaluated the scientific literature and comes up with the estimate that €30/tCO_2 (70 pound sterling/tC) is an adequate initial number—obtained from a range of 35–140 pound sterling/tC—and a yearly increase of 1.4% that includes the cost using a discount rate that allows to achieve the 2-degree goal.[14] This is an interdisciplinary combination of an economic cost calculation and the goal from climate models. In contrast, using a discount rate of four percent, leading to low permit prices in the order of magnitude of 8$/t$CO_2$, is unlikely to reach the 2-degree goal as can be seen from the experience of the EU ETS. When uncertainty is included in more recent studies, estimates go beyond $150/t$CO_2$ in 2050 (Cai et al. 2015). For the reduction of CO_2 emissions there are basically three systems, which try to achieve that the costs of emissions are included in prices and lead to the reduction of demand for products with higher prices: Taxes, tradable permits, and positive incentives such as subsidies for the clean alternatives.

7.3.4 Taxes and Permit Markets

In line with the above-mentioned principle of increasing prices, Switzerland has a tax of €40/tCO_2 and Norway of €50/tCO_2. The EU has introduced a system of tradable CO_2 emission permits (EU-ETS). The systems would be approximately equivalent to the tax policy if the permit price would also be about €40/tCO_2. It is well known that this is not the case, because there are institutional differences between the two systems. Whereas the tax system results in fixing the tax analogous to fixing the price of a permit with variability in the emission quantities, the permit solution in

[14] At that time the price of the pound was about €1.50, and ton carbon $tC = 3.664$ t CO_2. We work here with the assumption that the Euro value remains adequate to express the correct valuation and the pound has lost value through the financial crisis also in terms of US dollars.

fact fixes the supply of the number of permits and the price is flexibly going up and down with the demand for permits. In the current arrangements, supply of the number of permits is determined by national governments (with mild supervision of the EU) who have an incentive to supply many permits when acting in the interest of emission-intensive national firms rather than world welfare. This too generous supply, in connection with the reduced demand as a consequence of the 2007–2014 crises, has led to very low CO_2 prices of €5–7/tCO_2. Since these values are so low the incentive to emit less has led to ignorance of CO_2 as a cost factor. Moreover, the growth dip in 2009 has led to only marginally lower CO_2 emissions in the EU countries—for which we have plotted data above—indicating that the fall in prices is due to excessive supply of emission certificates. One idea in the early debates was that world supply of permits should be fixed in order to get emission growth under control. However, neither for taxes nor for permits can we be sure that policy arranges them in a way that reduces emissions. The governments that have organized a large supply of permits might also have organized a low CO_2 tax.

The emissions from international transport would require integrating international transport into the tax or tradable permit systems.

7.3.5 *Positive Incentives*

The third system besides taxes and permits consists of positive incentives consisting either of subsidies or tax reductions (Diederen et al. 1995; Vermeend and van der Vaart 1997; Ziesemer 2000). The burden then is spread out over the whole community of taxpayers, which leads to less resistance by lobbyists in the political process. Adjustment costs are covered right from the beginning for the recipient. The adjustment costs from raising taxes occurring in the form of reduced demand are spread out over all sectors. Moreover, if time lags are involved and debt financing has a negligible impact on the interest rate because environmental policy has low costs relative to total credit demand, the positive effects may come about earlier than the negative effects of financing the policies. Subsidies may also be justified by learning-by-doing externalities from new technologies (Rezai and van der Ploeg 2014).

Tax deductions have been mainly used for environmentally friendly investments, in Germany for catalyzers against lead, in the Netherlands for solar cells, and also more broadly in the laws EIA and EINP and later arrangements (Steger et al. 2002, 2005). Subsidies have been used in the German system for solar cells and wind energy with expensive bookkeeping regarding the recipients. The traditional rule of the World Trade Organization to keep investment subsidies below 20% of the value of the investment (Esty 1999) is not used here. In the year 2000 the subsidies for electricity from solar cells were told to be a factor 20, meaning 2000%, on the producer cost. Similarly, in the Samsung-Ontario feed-in-tariff case the province pays 20 times the customer rates, which are much higher than the costs of generating electricity. However, World Trade Organization did not come to a ruling against this, which is very controversial (Rubini 2014). One reason why it is difficult to find limits for subsidies for renewable energy is that the Social cost of carbon review (Downing 2005) finds its ideal 2005 value of €30/tCO_2 starting from a range in the literature of zero to €410/tCO_2 (1000 pound sterling costing 1.50 Euro at that time and then divided by the above mentioned tC/tCO_2 = 3.664). Next, they narrow down the range of studies considered to those that find values in the range 14–57€/tCO_2. €30 then is roughly in the middle. This value should then increase every year by €0.40. It is unclear whether or not lawyers at World Trade Organization can use such steps, however plausible they are. They have to justify that they interfere with national sovereignty.

7.3.6 National Policies or International Agreements or Both?

The national policies with positive incentives just described have been invented at times when neither eco taxes nor international agreements seemed to be realistic alternatives. One additional idea behind positive incentives is that subsidizing clean technologies produces a positive distortion in favor of clean technologies and therefore other countries' governments than the one paying the subsidies will get a request from discriminated own firms for paying such subsidies too. This may bring such environmentally oriented subsidies to the whole world (Diederen et al. 1995; Ziesemer 2000). The problem with such an approach is that national energy costs are in the order of magnitude of 8% of GDP for the world.[15] Externalities from them are even much smaller. This may be unable to attract the attention of politicians. If, in addition, countries have budget problems the money may nevertheless be hard to get. By implication, countries try both, positive incentives and international agreements, in addition to taxes and tradable permits.

A special problem arises when trying to arrange taxes or permit arrangements regarding emissions from international transport. In trade agreements they are unlikely to be included because they are seen as an environmental issue and in environmental agreements, they are unlikely to be included because they are seen a as trade issue. A logical way out then may be to have a separate international agreement on emissions from international (and national) transport and the related arrangements regarding the question 'who pays'.

---

[15]https://www.instituteforenergyresearch.org/analysis/a-primer-on-energy-and-the-economy-energys-large-share-of-the-economy-requires-caution-in-determining-policies-that-affect-it/#_edn2.

7.4 Problems with the Solutions

7.4.1 Resistance to Taxation and Public Investment

Tax resistance is one of the fundamental problems in developing countries (Mutén 1985). If there is no money from taxes or government owned resources public investments cannot be financed. This is the reason behind lack of electricity, bad roads, lack of water and wastewater systems, health and education problems (Yepes et al. 2008). If this could be overcome the developing countries would have more growth and emissions.

7.4.2 Scarce Resources Undermining Technical Progress May Limit the Market and Policy Forces: Rare Earths, Silicon, …?

When markets and policies favor renewable energy, investors, among other steps, expand the capacity of solar and wind energy. Solar energy uses silicon. Computers also use silicon. When both markets boom the increasing demand may make silicon and solar energy expensive, inducing a search for alternative materials. Similarly, wind turbines use rare earths,[16] which are also used in cars like the Toyota Prius. Rare earths have become temporarily expensive raising the question whether or not the concentration of supply in China will lead to monopoly prices and cost increases of wind turbines. For the time being the optimistic expectation that other suppliers will open mines have turned out to be realistic and China has relaxed its export restrictions after World Trade Organization intervention. However, this goes together with pollution from mining, which had led to closing of mines in the USA before. These two cases indicate that scarce resources are not only circumvented by technical change as is the presumption in the economics of growth since the discussion on the views of Ricardo and the Club of Rome, but they can also make new technologies too expensive to enter or stay in the market. Technologies that stay in the market may become more expensive and diffusion may slow down. So far technological change did go on but there is no guarantee that this will always be the case for all directions of technical change, here the emission reducing ones. The chapter on technology developments discusses more cases of resource scarcity in relation to specific technologies, product life cycles and the production chains.

[16]This paragraph is based on Barteková (2014).

7.4.3 Policy Interference Weakens CO_2 Reductions

(a) *Low taxes and permit prices.* Costs for buying CO_2 emission permits or paying taxes may increase export prices and reduce export growth or induce re-location of firms (Merrifield 1988). The fear of such reduction of competitiveness has slowed down the speed of implementing policcs against CO_2 emissions globally. The supply of tradable permits by national governments with support from the EU has been very generous and that has led to very low emissions prices. Moreover, exemptions from CO_2 certificate obligations and taxes have been granted generously too. CO_2 certificate prices or equally high taxes should be increased to reasonable levels of about 20–40€/tCO_2 or more if there is agreement about risks, where (a) the first value has been favored by a recent US commission, Australia's first carbon tax and Finland's decision in 2010, and (b) a value of €30—suggested by Downing et al. (2005) and implemented in Switzerland—with a suggested increase of 0.7 pound sterling per ton carbon, which brings us almost to the value of €40/tCO_2, which is also the value of CO_2 taxes in Ireland for some CO_2 emissions, whereas Norway has 50€/tCO_2.[17] Then all energy technologies would carry the costs of the environmental damages they caused and inter-fuel competition for different types of fossil fuels (e.g. coal and gas) would happen to occur. The choice among these values and the implied assumptions on discount rates and uncertainties should be made in line with the objective to limit global warming to two degrees or less and the corresponding carbon budget. Only Switzerland and Norway have reasonably high CO_2 taxes of 40 or 50 €/tCO_2, up from €30. CO_2 costs added on transport costs would have protectionist effects. Gains from free trade may be inefficiently large because of the additional emissions caused by international transport.[18]

(b) The current use of positive incentives like subsidies for renewable energy and voluntary agreements do not reward less polluting coal fired power stations relative to more polluting ones. They also do not provide sufficiently strong incentives to use more gas and less coal. Competition is reduced between different forms of renewables (including all costs along the production chain) if they all receive as much subsidies as they need to be in the market. The current policies of positive incentives are therefore unnecessarily expensive compared to tradable permits for greenhouse gases because they ignore inter-fuel substitution between fossil fuels and also between renewables and do provide a limited incentive to move to more environmentally friendly technologies. Moreover, expectations and long-run orientations become increasingly unclear. For example, the German and Dutch electricity providers find coal and gas fired power stations

[17] Of course, there are also values outside this range like the 150\$/tCO_2 in Sweden. US based literature comes to lower values since several decennia. Calculations are bothered by many details like the tax base, discount rates, exchange rates, (expected) inflation rates and relations with other taxes and the EU ETS. We will not go into the details of these aspects.

[18] See Pethig (1976) and Siebert (1977) for models of perfect competition and Soete and Ziesemer (1997) for monopolistic competition.

unprofitable whereas the German minister of economic affairs states that it is not possible to abandon both, coal and gas. Similarly, the president of the British central bank has warned against the divergence of expectations that may stem from shareholders partly expecting environmental policies as suggested here and other expecting that they can be avoided. If such expectations are disappointed stock exchanges can crash. Expectations between firms, ministers, central banks and shareholders should be aligned requiring a clear announcement of long-term policy plans rather than lobbying to prevent the political majorities. Systems of positive incentives are justified only by the impossibility of getting better systems, but they should be replaced by the latter ones allowing for a socially cost efficient transition to a low-carbon economy. A good guidance for the expectations regarding coal can be obtained from the scenarios in Droste-Franke et al. (2012), Tables 6.5 and 6.6: If the emission reduction should be 90% of those in 1990 for Germany in 2050, then coal has to be reduced to zero. If the goal is only −70% below 1990 emissions, small share of coal may be in the future energy mix. Thus, predictions of an energy mix including coal imply an assumption of limited decarbonization (in addition to assumption on the costs and functioning of carbon-capture and storage).

(c) Tradable permit prices or taxes for CO_2, as well as subsidies increase the share of renewable energy sources. They produce higher fluctuations of electricity supply than the system did without them. The standard recommendation to dampen supply fluctuations used to be the use of gas fired power stations because they can respond quickly, as well as the investment in storage facilities (Droste-Franke et al. 2012). However, electricity cables crossing borders seem to be the much more cost-efficient solution. The larger the network of electricity cables in the sense of covering more regions with different amplitudes, the lower the fluctuations.

7.4.4 *The Impact of Electricity Prices on Foreign Direct Investment in the EU*[19]

The basic problem is that firms may move to other countries when environmental policies are implemented (McGuire 1982; Merrifield 1988). Relocation of firms going outside the EU is not so easy. Where should they go? Countries outside the EU are not very attractive. Alcoa has found a place for aluminum production in Iceland. But this is an exception. The CO_2 intensive sectors have high transport cost in the case of cement or want to be close to their customers or deliverers in the case of aluminum. Customers of the aluminum industry are the automobile industry, and the share of recycling, where less electricity is used, is large for aluminum. Whereas dramatic moves cannot be expected one important question is where the re-investment is taking place. If tradable permits for emissions or environmental taxes

[19]This section is based on Barteková and Ziesemer (2019).

would increase electricity prices trade and foreign direct investment may shift to other countries. Governments, for example in Germany, have avoided this by way of imposing the burden on households. Whether or not this was justified can be figured out by way of investigating whether or not FDI reacts to changes in electricity prices. The question of effects of electricity prices on foreign direct investment has not been analyzed yet for EU countries.[20] We find the following result for 27 EU countries without Luxemburg for the period 2003–2013 (standard errors in parentheses; *, **, *** indicate statistical significance at the 10, 5 and 1 percent level).[21]

Box 5 The impact of electricity prices on net foreign direct investment (%GDP)

FDI/GDP =

0.32(FDI/GDP)$_{-1}$ + 0.322GDPGR − 4.045LOG_EP_SW − 3.294LOG_EP_NE + 0.196DEFL_HI − 11.49LOG_ULC
(0.184)* (0.119)** (1.846)** (1.708)* (0.0748)* (6.060)*

− 0.0532EDU_SEC − 0.296EATR − 1.176PDL0_GFCF + 3.112PDL1_GFCF − 1.006PDL2_GFCF
(0.0217)** (0.0673)*** (0.551)** (1.314)** (0.449)**

Obs.: 188; p(J) =0.374;

The interpretation is as follows. Each percentage point of FDI/GDP (foreign direct investment as share of Gross Domestic Product) in a period leads to another 0.32 % point increase in the next period. Each percentage point of GDP growth, GDPGR, leads to 0.32% points of FDI. Each 10% growth of electricity prices leads 0.4% percentage points less FDI/GDP in south-western EU countries and to 0.33% points less FDI/GDP in north-eastern EU countries. If countries with a GDP deflator on average above the 2-percent target have higher inflation FDI/GDP is increased. Unit labor costs reduce FDI. Secondary school enrolments also reduce FDI because either primary or tertiary schooling is needed by foreign firms. The remaining terms refer to a polynomial distributed lag and imply that current gross fixed capital formation by general government, GFCF, as a proxy for infrastructure investment has a negative impact on FDI/GDP, because they make production factors more expensive whereas earlier values have a positive impact because infrastructure makes business easier. All effects are in the long run a bit higher because the coefficients have then to be divided by $(1 - 0.32) = 0.68$ leading to a factor of roughly 1.5. Our main result is that FDI/GDP reacts to electricity prices to a non-negligible extent. If governments would allow environmental policy to lead to higher electricity prices FDI would

[20] One exception is a master thesis that we have reproduced. Results broke down when adding or dropping some years. An analysis for Turkey by Bilgili et al. (2012) exists.

[21] Econometric note. We use the system GMM method. All estimators and standard errors are robust to heteroscedasticity and autocorrelation. We report one step GMM estimators. $FDI_{i,t-1}$ is treated as endogenous and only its lags $t - 2$ to $t - 5$ are included in the instrument matrix. Unit labor cost, $ULC_{i,t}$, is predetermined and instrumented with lags $t - 1$ to $t - 7$. The collapse option is used. Time dummies are included in the regression.

move out, to a moderate extent though. Environmental policy should therefore not be unnecessarily costly. International coordination and international trade in electric current can help avoiding the price differences and the capital movements of FDI.

7.5 Fluctuations and Global Trade in Electric Currents

Supply of electricity from renewable sources is expected to fluctuate with sun and wind intensity. The early suggestions to deal with this was the idea that gas fired power stations can react quickly and therefore should be used as a complement to wind and solar based electricity supply. When the share of renewables gets larger storage facilities become profitable. In Sect. 4.c.iii above we have suggested that fluctuations could be dealt with in a cost-efficient way by international trade and therefore less of the costly gas fired power stations and storage investment could be used. This would require that there are electricity cables crossing borders and that there are economic incentives to trade. Cables crossing borders exist but at a sub-optimal amount. Therefore, the EU has coordinated plans to extend them. Incentives to trade electric current internationally also exist. Dutch traders buy electricity in France when the sun is intensive there and sell it in the Netherlands or other neighboring countries. Similarly, they buy electricity in the Netherlands or in Germany when the wind is strong there and sell it in other countries. When the Netherlands have a deficit in electricity supply, they buy electric currents from Norway to which there is a cable and a contract. On the other hand, Poland and France have installed technologies to block the supply shocks from northern wind peaks. This type of business press reports does not tell us whether there is a substantial or negligible amount of trade going on and whether or not electricity from renewables has an impact on them.

In this section we look at the data for global bi-lateral international trade in electric currents and some of their determinants, in particular electricity from renewable sources. The data are from the World Integrated Trade Solution (product code 35 in the Standard International Trade Classification version 4 (SITC4) on wits.worldbank.org, freely available). Germany trades exactly with all and only neighbors according to Table 1a, b. Exports go mainly to Switzerland and the Netherlands. Imports come mainly from Austria and France. Germany has a trade deficit (see Table 1c) with France in all years and with the Czech Republic, Sweden and Denmark in some years.

In contrast, the Netherlands trade with 95 countries, Jamaica with 56, and Serbia with 54, to mention only a few. Although the electric currents go through trans-border cables, the trading partner of the two countries on each side of the border may be a third country buying from one country and selling to the other. In short, these are trade data, not technical data.

Our major interest is seeing how electricity from renewable sources affects the trade volume of electric current, because otherwise trade cannot serve as a mechanism absorbing and insuring against fluctuations in electricity production. Therefore, we take all available bi-lateral data for trade in electric current from WITS.[22] Bilateral country pairs are abbreviated as *ij*, indicate reporting country *i* and partner country *j*. There are 737 bilateral pairs reported by exporting countries and 619 pairs reported by importing countries. What country *i* reports as import from (exports to) *j* does not always coincide with what country *j* reports as export to (import from) *j*. Although each exporter has an importing counterpart, these importers seem to report less or less detailed than the exporters. In total 969 different bilateral pairs *ij* are reported by either exporting or importing countries, whereas for about 200 countries this could in principle be roughly 40,000 (Tables 7.1, 7.2 and 7.3).

The statistical explanation of the volume of trade flows in the international economics literature is known under the name 'gravity equation', because the major determinants of the volume of trade flows in regression equations are the GDP of a country (as a product of income per capita and the size of the population) and the distance between the (capitals of) the countries as a proxy for the transport costs. Other variables are added depending on the special research question. The dependent variable is usually the sum of imports and exports called trade volume. As exporters and importers of electric current report very asymmetrically we run regressions for imports and exports separately, denoted as *ecex* and *ecim*.

GDP data are taken from the World Development Indicators. In line with the indices we have two GDP variables, those of the reporting country *i* and those of the partner country *j*. We use the GDP in current US-dollars which has the best data availability and is in line with the fact that trade in electric currents is also measured in nominal US-dollars.

Distance data, denoted as log(*dist*), are taken from DIST_CEPII.XLS (Mayer and Zignagno 2011). As Serbia and Montenegro are one country there, Yugoslavia, we add the information for the two countries separately. Distance from capital to capital has the best coverage among several distance measures.

Electricity from renewable sources is measured as 'Electricity production from renewable sources, excluding hydroelectric (% of total)'. Data are taken from World Development Indicators. They mostly go only until 2014 at the moment of downloading. They are denoted as *RNR* and *RNP* for the reporting and the partner country respectively.

We use dummy variables for country pairs

- ever having been part of the same country, because this may have had an impact on the infrastructure crossing borders (smctry) or other trade inhibiting aspects;
- being neighbor with a common border (contig);
- having the same language (comlang_off) or a common second language (comlang_ethno) spoken by at least 9% of the population.

---

[22]Before having a bilateral panel set of data, we have to eliminate those where the partners are free zones (code FRE), special categories (SPE), 'other Asia' (OAS), unspecified (UNS), bunkers (BUN) or holy see, the latter because there are no GDP data which are needed for the gravity equation. Names and their abbreviations of countries differ per database and we had to harmonize them.

Table 7.1 Germany's exports of electric current 2007–2016 in $1,000

To	2007	2008	2009	2010	2011	2012	2013	2014	2015	2016
Austria	923,944.00	1,371,010.00	1,349,288.00	1,136,945.09	1,266,485.50	1,168,154.47	1,079,294.58	962,452.03	868,568.12	617,891.28
Switzerland	1,133,455.00	1,092,257.00	914,569.00	918,741.35	795,517.91	795,517.91	743,014.57	799,875.88	833,454.80	647,049.67
Czech Republic	64,524.00	133,000.00	109,894.00	44,972.39	161,582.73	207,577.00	180,037.06	237,826.86	277,638.89	242,589.30
Denmark			325,528.00	431,846.77	224,834.28	101,931.62	400,447.35	210,360.93	96,712.96	141,522.42
France	60,690.00	84,312.00	103,074.00	50,389.06	9,953.44	53,961.87	113,865.82	51,853.82	61,866.55	107,757.27
Luxembourg			373,592.00	327,677.94	343,706.56	297,674.68	277,943.03	246,414.44	190,988.71	173,484.36
Netherlands			763,814.00	647,407.10	762,556.74	1,656,615.14	1,776,122.37	1,483,602.91	1,092,408.68	640,643.56
Poland	183,654.00	454,558.00	499,972.00	357,231.11	366,484.10	420,421.90	359,225.55	552,031.95	533,747.19	434,179.63
Special Category	1,657,923.00	2,157,359.00								
Sweden	46,525.00	30,261.00	71,312.00	132,487.09	37,082.24	16,121.46	59,084.87	43,477.95	7,793.95	39,505.65
World	4,070,715.00	3,007,471.00	3,200,447.00	2,657,264.86	3,546,871.48	2,959,725.21	2,409,027.86	2,268,668.16	1,674,668.16	1,107,791.18

Table 7.2 Germany's imports of electric current 2007–2016 in $1,000

From	2007	2008	2009	2010	2011	2012	2013	2014	2015	2016
Austria			548,824.00	485,945.60	431,774.30	490,829.03	508,327.31	349,453.62	310,782.78	225,181.73
Switzerland	202,641.00	239,757.00	205,997.00	172,806.86	187,932.33	208,459.54	219,023.72	296,432.36	176,774.89	141,893.52
Czech Republic			793,656.00	717,351.04	712,735.20	594,690.73	626,309.39	372,121.38	265,224.13	186,752.45
Denmark			569,228.00	185,220.80	390,393.79	587,870.37	277,680.03	204,482.25	252,597.14	117,240.43
France	809,707.00	631,202.00	701,262.00	818,631.33	1,419,173.40	851,545.50	661,961.00	831,282.72	559,763.40	317,571.52
Luxembourg	16									
Netherlands			311,799.00	206,223.82	245,798.35	52,516.85	19,007.23	19,414.82	15,252.96	49,246.63
Poland			11,953.00	13,092.27	35,217.69	11,239.29	34,437.78	3,048.07	962.002	765.21
Special category	1,281,115.00	2,136,512.00								
Sweden			57,728.00	57,993.12	123,846.43	162,563.90	62,281.41	102,720.80	93,310.86	69,139.70
World	2,293,479.00	3,007,471.00	3,200,447.00	2,657,264.85	3,546,871.48	2,959,725.21	2,409,027.86	2,268,956.02	1,674,668.16	1,107,791.18

Table 7.3 Germany's trade surplus of electric current 2007–2016 in $1,000

Exp.-imp.	2007	2008	2009	2010	2011	2012	2013	2014	2015	2016
Austria	923,944.00	1,371,010.00	800,404.00	650,999.49	834,711.21	677,325.44	570,967.27	612,998.41	557,785.34	392,709.55
Switzerland	930,814.00	852,500.00	708,599.00	745,934.49	697,641.86	587,048.38	523,990.85	503,443.52	656,689.91	55,832.85
Czech Republic	64,524.00	133,000.00	−683,762.00	−672,378.66	−551,152.47	−387,113.73	−446,272.33	−134,294.52	12,414.76	55,831.85
Denmark	0.00	0.00	−243,700.00	245,625.96	−165,559.52	−485,938.75	122,767.32	−84,121.31	−155,884.18	24,281.99
France	−749,071.00	−545,890.00	−598.188.00	−768,242.28	−1,409,219.96	−797,583.63	−548,095.18	−779,428.90	−497,896.85	−209,814.25
Luxembourg	−16.00	0.00	373,592.00	327,677.94	343,706.56	297,674.68	277,943.03	246,414.44	190,988.71	173,484.36
Netherlands	0.00	0.00	452,015.00	441,183.28	516,758.39	1,604,098.28	1,757,115.14	1,464,188.09	1,077,155.72	591,396.92
Poland	183,654.00	454,558.00	488,019.00	344,138.83	331,266.42	409,182.61	324,787.78	548,983.89	532,785.19	433,414.42
Special category	376,808.00	20,847.00	0.00	0.00	0.00	0.00	0.00	0.00	0.00	0.00
Sweden	46,525.00	30,261.00	13,584.00	74,493.97	−86,764.19	−146,442.44	−3,196.54	−59,242.85	−85,516.91	−29,634.05
World	1,777,236.00	2,315,286.00	1,310,563.00	1,390,433.02	511,388.30	1,758,250.84	2,580.007.34	2,318,940.76	2,288,521.70	1,936,826.96

The result for a static regression for imports of electric current, not taking into account any lags (with econometric issues put into footnotes), are as follows:

Box 6 The impact of renewable electricity on electricity imports

$$LOG(1+ECIM) =$$

$$1.88 + 0.42LOG(1+RNR) - 0.058LOG(1+RNP) - 0.226LOG(GDPR) - 0.06LOG(GDPP) -$$

(0.51) (0.0004) (0.63) (0.59) (0.87)

$$1.82LOG(DIST) + 0.58\underline{LOG(GDPR)} + 0.36\underline{LOG(GDPP)} - 0.054\underline{LOG(1+RNR)} + 0.025\underline{LOG(1+RNP)} +$$

(0.00) (0.18) (0.35) (0.79) (0.9)

$$1.834CONTIG - 0.02COMLANG\_OFF - 0.385COMLANG\_ETHNO - 0.36SMCTRY$$

(0.00) (0.97) (0.54) (0.35)

Period: 2007-2014; country pairs: 487; obs: 2028

p-values in parentheses indicate the (error) probability (of rejecting the hypothesis) that the coefficients are (statistically insignificantly different from) zero. Underlined variables are the country specific averages over the period 2007–2015, which are all statistically insignificant here.[23] The regression has only three statistically significant coefficients. A larger share of renewables of the reporting country, RNR, leads to more imports of the reporting country. This result in the yearly data indicates that times of low electricity production requires more imports of the reporting country and these are dominating those of peaks, which should lead to fewer imports for the reporting countries. A larger distance between reporting and partner country leads to less imports. Neighboring countries trade more with each other. Renewables in the partner country, GDP variables, language or having been part of the same country seemingly does not matter. However, for renewables of the partner country the result can be cross-checked with the export equation. This also important because the insignificance could come from wind and sun peaks being equally important and therefore causing statistical insignificance.

[23]Mundlak (1978) suggests adding country-specific averages of variables to the regression and use the random effects method for estimation in order to clarify whether there are country specific effects in the intercept (called fixed effects) or in the residuals (called random effects). When they are insignificant as in the import equation, we have no fixed effects but random effects. The percentage share of cross-section random is 74.6% of the total variance of the equation. Details: Swamy and Arora estimator of component variances; cross-section weights (PCSE) standard errors and covariance (d.f. corrected). Adjusted R-squared of 0.16 and S.E. of regression 1.42 are very low and suggest thinking about other regressors. We postpone this to future research.

The export equation is as follows:[24]

Box 7 The impact of renewable electricity on electricity exports

LOG(1+ECEX) =

$$3.926 + 0.454LOG(GDPR) + 0.263LOG(GDPP) - 0.484LOG(1+RNR) + 0.82LOG(1+RNP)$$

(0.12) (0.277) (0.524) (0.0000) (0.0000)

$$- 0.116\underline{LOG(GDPR)} + 0.01275\underline{LOG(GDPP)} + 0.47\underline{LOG(1+RNR)} - 0.46\underline{LOG(1+RNP)} - 1.999LOG(DIST)$$

(0.79) (0.98) (0.0135) (0.012) (0.0000)

$$+ 2.12CONTIG - 0.88SMCTRY + 0.01COMLANG\_OFF + 0.199COMLANG\_ETHNO$$

(0.0000) (0.023) (0.98) (0.68)

Period: 2007-2014; country pairs: 543; obs: 2124

Reporting countries with larger country-specific averages of renewables have higher exports. When partner countries have larger country-specific averages of renewables the reporting countries have lower exports. These effects can be interpreted as capacity effects. In contrast, the time varying effects have the opposite sign: more renewables of reporting countries lead to less exports, and more renewables of the partner country lead to more exports of the reporting country. Both these time varying effects suggest that periods of weak electricity production in reporting and partner country dominate the result. Distance between reporting and partner country again reduces exports. Being neighbors increases exports. Having been part of the same country earlier has a negative effect, indicating that common infrastructure plays a weaker role than other factors hindering trade. GDP and language effects are again statistically insignificant.

Overall, in the period 2007–2014 the percentage of renewables has an effect on international trade in electric current in both dimensions, the time dimension and the cross-section dimension, and more so for exports, where we have more observations, than for imports. In theory, trade then dampens positively but imperfectly correlated price movements, mainly when supply is weak: prices increase less in the country that would have higher price peaks under hypothetical autarky and increase more in countries with lower price peaks under hypothetical autarky. Investments in border-crossing electricity cables could support this effect of dampening supply shocks and equalizing price fluctuations. These seem to be cheaper measures to dampen fluctuations than investments in gas fired power stations and storage capacity. In short, trade has dampening effects under given cable capacity and future investments in cables could increase these dampening effects via trade over time. Of course, neither should trade liberalization endanger the stability of the electricity network nor should stability issues be abused as protectionist measures.

[24] Swamy and Arora estimator of component variances. Cross-section weights (PCSE) standard errors and covariance (d.f. corrected). Cross-section random 0.74. Adjusted R-squared: 0.248; S.E. of regression: 1.39.

Chapter 8
Cooperation in Energy Governance Between China, India, Brazil and the European Union/Germany

The initiation of an internationally binding climate agreement in Paris 2015 serves as a wake-up call for a rapid transformation of the world's energy system (UNFCCC 2015). Reframing the goal of a temperature increase below 2 °C as a dynamic investment signal for renewable energy (RE) paves the way towards a joint undertaking that bridges the North-South divide of global climate policy-making and encourages cooperation with emerging powers. Embedded into the broader framework of a 'great global transformation' (WBGU 2011), the prospect of a wide-ranging decarbonization fosters an energy transition that is not only efficient and comprehensive but also inclusive and fair, insofar as it creates dynamics that ensure fair participation and distribution of RE around the globe (Jänicke 2013).

However, to move this concept beyond the mere buzzword stage, political action is desperately needed. New forms of political cooperation at the climate/clean energy nexus, such as bilateral dialogues between the EU and emerging powers, bear valuable opportunities for implementing the Paris agreement. European States as well as the EU have realized the potential of cooperation with rising powers and established bilateral energy dialogues. Those dialogues are more advanced in the case of dialogues between single EU countries and still underdeveloped at the EU level itself. However, external energy cooperation is a challenging issue that calls for a dialogue about norms, values and interests to find compromises and peaceful solutions in a policy field often driven by merely protectionist security notions.

Unfortunately, though, those bilateral dialogues are carried out separately for climate and energy policy and are embedded differently within the polycentric world order. While the climate regime complex allows embedding bilateral dialogues, the clean energy complex is but, in its nascence, (Andonova and Chelminski 2016) and builds on highly fragmented structures (Baccini et al. 2011). International energy governance is executed in a highly fragmented environment of a highly diversified network of international governmental organizations that deal with the regulation of energy supplies. In a network analysis on IGO development, Bacchini et al. name

C. F. Gethmann et al., *Global Energy Supply and Emissions*,
Ethics of Science and Technology Assessment 47,
https://doi.org/10.1007/978-3-030-55355-5_8

24 different organizations (Baccini et al. 2011: 3), all of them reflecting the political, economic and technological circumstances of their respective founding moment (Fischedick et al. 2011: 2). There does not seem to be a political will to establish a kind of "World Energy Organization". Most existing institutions are limited in either their geographical or their thematic scope, as the examples of the Energy Charter Treaty, the OPEC, or the IAEA and IRENA clearly show (Müller et al. 2015a: 25) (Table 8.1)..

Regarding governance structures, most of these institutions still reflect the political geographies and technological and resource policies of the post-war years, for instance, a belief in growing demands for energy, an emphasis on the security of supply and high hopes for the safe and peaceful use of nuclear energy. Equally controversial issues such as energy poverty, peak oil and indirect land use changing the impacts of biofuels are slowly making their way into these institutions; however they are not yet paralleled by changes of governance structures that would grant more voice to emerging economies (cf. Fischedick et al. 2011: 7; Florini and Sovacool 2011). Thus, to date, multilateral energy governance does not yet properly represent the roles, interests and capabilities of newly emerging powers. In addition, this environment leads to "forum shopping" by the states in energy matters (Müller 2015: 308).

World energy demand is projected to increase significantly by 2035, mainly due to economic development, outsourcing of energy-intense industrial production and changes in consumer habits in the major emerging powers. Thus, knowledge of the markets, policies, bargaining strategies, and decision-making procedures of these countries is vital to constructively shape the external energy relations of the EU and its member states. Through their recent economic development as major powers in the international economy, China and India face similar challenges in energy issues, despite all country-specific differences. On the one hand, energy-intensive economic sectors have risen, and an increasing middle class is consuming more energy than previously. On the other hand, large parts of the (rural) population are still struggling to meet their energy demands. The internal discrepancies and inequalities lead to energy poverty, whilst the expanding economic sectors strive to exploit the energy potential that lies within the countries in their external energy relations. Additionally, the impact of their strategic decisions will affect the EU and its member states' energy policy. On the other hand, Brazil has changed its role from a net energy importer to an important exporter of oil. Especially in the development of the South-South relations and the BRICS network, it plays a major role.

Thus, the main aim here is to assess China's, India's and Brazil's energy situations and policies; their energy relations with the EU/Germany; and finally, their engagement in the international energy system and in South-South Cooperation (SSC).[1]

---

[1]This chapter is based on the outcome of an international interdisciplinary research project. The project was funded by the Volkswagen Foundation, was led by TU Darmstadt, and involved the University of Aarhus, Peking University, Stiftung Wissenschaft und Politik (SWP), University of Cambridge, South African Institute of International Affairs, (SAIIA), Fundación para las Relaciones Internacionales y el Diálogo Exterior (FRIDE), The Energy and Resources Institute (TERI),

Table 8.1 The most important energy organizations (based on Westphal 2015; Lesage et al. 2010, own modifications and further additions)

Scope	Forum	Year of inception	Membership			
			EU MS	Brazil	China	India
Multilateral (membership open for all)	UN Energy	2004	■	■	■	■
	UN SE4All	2012	■	■	■	■
	UN Environment (UNEP)	1972	■	■	■	■
	International Atomic Energy Agency (IAEA)	1957	■	■	■	■
	International Energy Forum (IEF)	1991	■	■	■	■
	UN Framework Convention on Climate Change (UNFCCC)	1994	■	■	■	■
	International Renewable Energy Agency (IRENA)	2009	■		■	■
	International Energy Charter	2015	■		■	
Plurilateral (restricted membership)	Group of Eight/Seven (G8/7)	1975	■			
	G8 + Outreach Five (O5)	2007	■	■	■	■
	G20	2009	■	■	■	■
	Organization of the Petroleum Exporting Countries (OPEC)	1960				
	International Energy Agency (IEA)	1974	◇			
	Gas Exporting Countries Forum (GECF)	2001				
	International Partnership for Energy Efficiency Cooperation (IPEEC)	2009	■	■	■	■
	Clean Energy Ministerial	2009	■	■	■	■
	Energiewende-Club	2013	■		■	■
Europe regional	Energy Charter Treaty (ECT)	1998	■		⊙	

(continued)

Table 8.1 (continued)

Scope	Forum	Year of inception	Membership			
			EU MS	Brazil	China	India
	Energy Community	2006	■			

◇ EU's OECD member countries are members of the IEA, but not all EU Member States are members of the IEA
⊙ China is an observer by invitation of the Energy Charter Conference

8.1 China

In recent years, China's energy consumption has increased substantially. Since 1993, China has become a net energy importer. China's energy consumption sums to over one-fifth of the world's total consumption. Approximately 70% is used by industry, especially by energy-intensive industries such as steel and cement, which have played a key role in China's growth since 1990 (IEA 2016: 22). Lately, the output in some energy-intensive sectors, notably steel and cement has decreased (IEA 2016: 62). According to the International Energy Agency's (IEA) World Energy Outlook in 2013, China not only was the largest coal consumer and importer but also was overtaking the US as the world's largest oil importer in those years (IEA 2013: 55, 61, 62, 67). This change becomes particularly obvious in coal consumption, as China's energy sector displays continuing dependence on coal. The volume of China's coal imports more than tripled in 2009 from a year earlier (China Daily, 10.2.2010), even though China has the third-largest coal reserves worldwide. A huge local availability of coal in China means an abundant, secure and low-cost local supply; however, such heavy reliance on fossil fuels contributes to China's huge greenhouse gas emissions. In fact, China has become the world's largest emitter of CO_2, having overtaken the US in 2007 (The Guardian, 19.6.2007). To fight its air pollution problems in recent years, China eased off its coal consumption and achieved a relative slow growth up to a standstill within the last approximately five years. "Coal demand in China is sputtering as the Chinese economy gradually shifts to one based more on services and less on energy-intensive industries", explained the Annual IEA Coal Report 2015 (IEA 2015a). The IEA stated, "New Chinese hydro, nuclear, wind and solar are also significantly curtailing coal power generation, driven not only by energy security and climate concerns but also by efforts to reduce local pollution" (IEA 2015a). The report concluded that the global coal market is under pressure, in particular because of Chinese economic restructuring (IEA 2015a; Knodt et al. 2017: 155). Related to overall energy consumption, there is a gap between China's energy consumption and production that is continuously widening. This heavy reliance on fossil fuels contributes to China's huge, energy-related greenhouse gas emissions.

Gesellschaft für Internationale Zusammenarbeit (GIZ), Federal University of Rio de Janeiro, and University of Zurich. All data from this project cited as EnergyGov (2014). Parts of this publication are based on Knodt et al. (2019).

Consequently, China's energy consumption and policy are no longer a domestic issue but have worldwide influence (Zha and Lai 2015: 129).

Nevertheless, the trend seems to have passed its peak. "Energy demand growth has slowed considerably in recent years, from an annual average of more than 8% between 2000 and 2010, to less than 3% per year since 2010" (IEA 2017: 471). The IEA in its World Energy Outlook 2016 even calculates that a reorientation of China's economic model towards domestic consumption and services and away from energy-intensive industries will bring down China's industrial coal use. In 2040, the share of coal in the power mix will be less than 45% (IEA 2016: 22). The policy of the Chinese government away from coal and towards renewables is clearly not least driven by air quality problems. IEA estimates that as an impact of rapid economic growth on China's environment, only 2% of the population in China breathes air with a level of fine particulate matter (PM2.5) concentration that complies with the World Health Organization (WHO) guideline (IEA 2017: 496). Overall, the shift away from coal reflects changes in the Chinese economy away from heavy industry and towards domestic consumption and services, its recognition of the environmental problems and a strong focus on energy efficiency policies.

Figure 8.1, which presents China's current energy mix, still shows the mix's reliance on coal. A huge local availability of coal in China means an abundant, secure and low-cost local supply. In contrast, oil and natural gas resources are limited, whilst the development of non-fossil energy is rather recent and involves high costs.

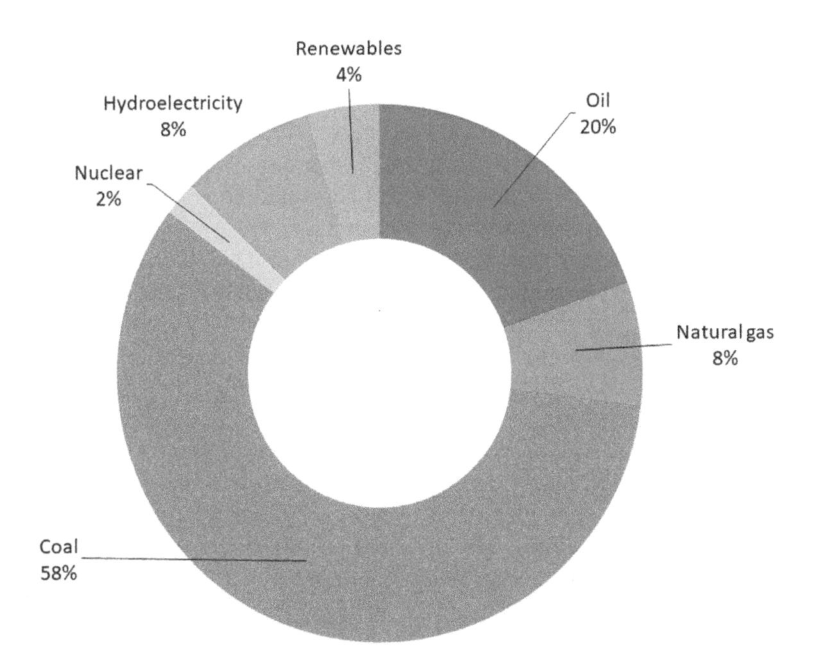

Fig. 8.1 China's energy consumption (2018) (BP Statistical Review of World Energy 2019)

In the last few years, China has taken the lead in expanding renewable-based power generation capacity. For a long time, renewables have been dominated by hydropower, and China ranks first in the world in installed hydropower capacity. If we take hydropower separately, the rest of the renewables, such as solar, biomass, and wind, only constitute up to four percent of the energy mix, but they have developed from approximately 2% in 2016. However, China is the world's largest investor in renewable-based generation (IEA 2017: 471). Solar photovoltaics (PV) in particular have experienced very strong growth. Already, China and India see the largest expansion of PV (IEA 2016: 23). China's installed solar power capacity increased by more than 75 GW between 2010 and 2016, but in May 2018, the Chinese government stopped all subsidies for utility-scale solar projects and reduced feed-in-tariffs in favor of competitive bidding. This change in solar policy will have a negative impact on China's solar capacity and may boost wind power instead. China already stands among the market leaders and added more capacity to its offshore wind capacity than did any other country in 2018 (IEA 2019: 52, 66, 613). The IEA already in 2016 predicted that there will be an expansion of wind capacity to 200 gigawatts (GW) by 2020, "though cuts for hydro (55 GW) and biomass (15 GW) are also under consideration" (IEA 2016: 405). In the Chinese perception, nuclear power also counts as renewable energy, and the country leads in new nuclear power generation. Even if nuclear power has the smallest share in the Chinese energy consumption mix, approximately "half of global nuclear capacity under construction today is in China" (IEA 2017: 477; Knodt et al. 2019: 24) (Fig. 8.2).

Nevertheless, even if the share of coal in the energy mix seems to decrease slowly, for a long time to come, coal will continue to dominate China's energy mix, which poses a growing challenge for the country. Given China's huge population, the availability of energy resources per capital is indeed scarce. In addition, in most of its policy papers and in the five-year plans, China has always highlighted its will to further urbanization, industrialization and economic growth. To cope with this problem, the Chinese government is promoting energy efficiency measurements. Moreover, the environmental and social burdens generated by energy-related carbon emissions are a major concern of the Chinese government (Zha and Lai 2015: 130).

In recent years, China has expanded its natural gas share of its energy consumption to a large extent, from two up to eight percent, and the IEA predicts up to 13% in 2040. From only 2017 to 2018, China's natural gas consumption increased by 33% (IEA 2019: 175). Part of this rise seems to come from the expansion of liquid natural gas (LNG). China has developed into the world's second-largest LNG importer. "In addition to the large national oil companies, the gradual opening of China's gas markets is allowing a growing cast of midsize utilities and private enterprises to contract for new LNG supplies, develop transport infrastructure and facilitate trade, turning China into a focal point for innovation in the use of LNG" (IEA 2019: 65).

Thus, confronting the twin challenges of energy security and climate change requires strong governance structures in China (Kong 2011). The role of net importer of energy has created China's self-perception of high energy-insecurity based on energy resources in the Middle East, Russia, Central Asia and Africa. With this perception, China's energy foreign policy expanded during the 1990s, and the country

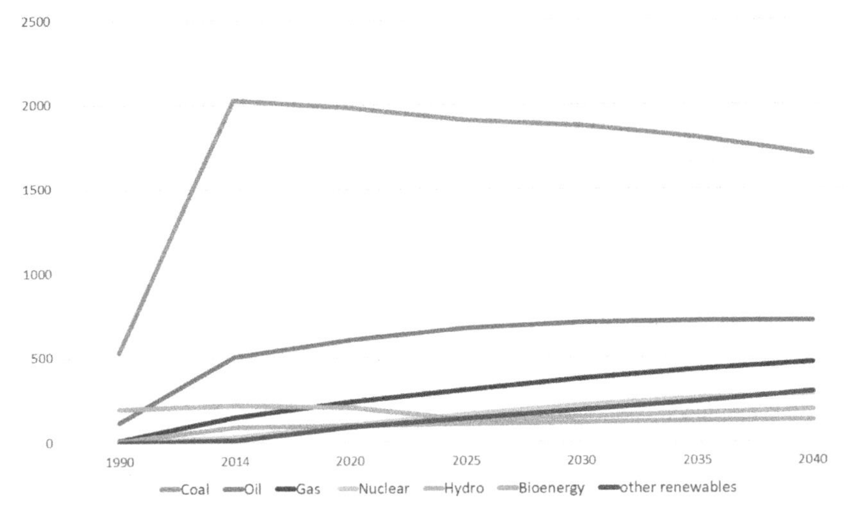

Fig. 8.2 Chinese Energy Demand in million tons of oil equivalent (Mtoe, total primary energy demand) according to the New Policy Scenario of the IEA (2016, 2017), (total primary energy demand (TPED) is equivalent to power generation plus other energy sectors excluding electricity and heat). (The *New Policies Scenario* (NPS) represents the newest Scenario by IEA. It "is the central scenario of this *Outlook*. In addition to incorporating the policies and measures that affect energy markets and that had been adopted as of mid-2016, it also takes into account, in full or in part, the aims, targets and intentions that have been announced, even if these have yet to be legislated or fully implemented. The scenario includes the greenhouse-gas (GHG) and energy-related targets of the Nationally Determined Contributions (NDCs) pledged under the Paris Agreement" (IEA 2016: 627)) (2017)

established new relations with countries that China had no or only few relations with previously. However, energy still does not seem to be a priority issue for the Chinese government (Zha 2013). The Chinese government even had decided to abolish its Ministry of Energy, which was formed in 1988, at the same moment China turned into a net importer of crude oil. Despite having energy matters bundled within one ministry at a national level, energy governance in China is complex and fragmented, with overlapping competences (Downs 2008; Vaclav 1981; Leung 2011).

In the period between 1993 and 2003, the national energy administration was characterized by high fragmentation, which has increased since the responsibilities from the National Coal Industry Bureau and the National Petrochemical Industry Bureau were redistributed to other ministries (Knodt et al. 2019, p. 27). In 2003, Hu Jintao and Wen Jiabao took over power in the government and brought a new vision and policy approach to China's governance (Meidan et al. 2009: 592; Knodt et al. 2010, 2019: 28). They introduced the National Development and Reform Commission (NDRC).[2] This agency reports directly to the State Council in the areas of energy, economic and social affairs. Within the NDRC, the Energy Bureau is responsible for

[2]NDRC was the predecessor of the state Planning Commission (SPC), which was founded in 1952 and changed into the State Development Planning Commission (SDPC) in 1988.

energy planning and control. To overcome the Energy Bureau's lack of authority, China set up the National Energy Leading Group (NELG) in 2005, composed of 13 members, from the NDRC and other key ministries including consultation links with think tanks.[3] The NELG is in charge of energy strategies and major policies, the development and conservation of energy resources, energy security and emergency responses, and energy cooperation with foreign organizations (Yu 2010: 2163). Administration support is provided by the subordinate ministerial State Energy Office (SEO), while the NDRC Energy Bureau still retains its functions on policy implementation. Nevertheless, it seems that the NELG is not able to operate independently of the influence of the national oil companies (NOCs) or ministries (and especially the NDRC) (Knodt et al. 2010).

The failure to establish an Energy Ministry in March 2008, which in the future could have enhanced leverage, is one proof of the strength especially of the NDRC (Meidan et al. 2009: 595). With the 2003 changes, the sustainable use of energy was introduced as a new key priority for China's energy policy (Yu 2010: 2161; Knodt et al. 2010). Additionally, since 2003, the electricity issue was regulated by the State Electricity Regulatory Commission (SERC). SERC was responsible for the united electricity regulatory system and electricity-trading mechanism. SERC was merged with the National Energy Administration (NEA, vice-ministerial level), which was established in 2008 under the jurisdiction of the NDRC. Following this recent reform, NEA's mandate covers formulation and implementation of energy development plans and industrial policies, regulation of energy sectors, promotion of energy system reform, and leading China's participation in international energy cooperation. Noteworthy is that former leaders in SERC are those who formed the leadership of the current NEA after the SERC-NEA merger (Zha and Lai 2015: 132; Knodt et al. 2019: 26).

As energy governance in China remained fragmented notwithstanding the reform efforts at the time, the government created the National Energy Commission (NEC) in January 2010 to formulate a comprehensive and coordinated national energy strategy. The NEC is headed by the Premier, and its vice-head is assigned to the head of the NEA to intertwine both institutions. Other members of the commission are recruited from different Chinese ministers tasked with energy matters. The NEC is responsible for the political strategic guidelines, only meets in a one- or two-year rhythm, and is rather crisis-driven; however, the day-to-day work is still executed by the NEA (Zha and Lai 2015: 132; Knodt et al. 2019: 27).

The fragmentation within China's energy governance is amplified by other ministries and national agencies holding power in certain energy-related matters. One important one is the Ministry of Science and Technology (MOST), which is in charge of energy aspects within technology, innovation and research and development. All trade-related energy issues are under the jurisdiction of the Ministry of

[3]For the energy policy field, in particular, the Development Research Centre of the State Council and the NDRC Energy Research Institute are of relevance. They produced the two most authoritative reports on the main priorities for China's energy policy in 2004 at the request of Wen Jiabao: the National Energy Strategy and Policy Report (Beijing: Development Research Centre 2004); and the Medium and Long Term Energy Conservation Plan (Beijing: NDRC 2004).

Commerce (MOFCOM). The State Administration of Coal Mine Safety regulates and monitors the safety of coal mining. The distribution of specialized competences among ministries results in a considerable overlap of competence with NEA and NDRC in many areas. This overlap has been acknowledged with the 12th Five-Year Plan on Energy Development published by the State Council in January 2013, which created a 'division of labor' table. Therein, besides financial issues, the NDRC remains the main leader of the responsible ministries within one issue (Zha and Lai 2015: 133–135; Knodt et al. 2019: 27).

Energy firms play a central role in Chinese energy governance, as they are state-owned. They are powerful players, as they are supported with public money and government backup. The most relevant ones are the State Grid Corporation of China (State Grid), the three Oil Giants—China National Petroleum Corporation (CNPC), China Petrochemical Corporation (Sinopec) and China National Offshore Oil Corporation (CNOOC)—and the major power-generators—China Huaneng, China Datang Corporation (CDT), China Guodian Corporation, China Huadian Corporation and China Power Investment Corporation (CPI) (Zha and Lai 2015: 133; Knodt et al. 2019: 28). However, the solar panel and wind turbine industry is also gaining power among Chinese business firms. As an example, "China Longyuan Power Group ranks as the largest producer of wind power across Asia, while the China Three Gorges Corporation (CTG)—previously known for its hydroelectric projects—is one of the world's largest energy companies and has become actively involved in the offshore wind industry" (IEA 2019: 616). State-owned enterprises including energy-related firms are under the jurisdiction of the State-owned Assets Supervision and Administration Commission (SASAC). SASAC is subordinated under China's State Council (IEA 2017: 499; Knodt et al. 2019: 28).

The Chinese governance system not only appears to be highly fragmented in a horizontal dimension but also stretches in the vertical dimension over central, provincial and municipal levels. Energy issues regarding central goals and regulations are allocated at the central level, whereas the implementation is located at the provincial or municipal level (Zha and Lai 2015: 133). Within the setting-up of the main instrument of China's planned economy, the Five-Year Plans, all levels are involved. In March 2016, the National Development and Reform Commission (NDRC) and the National Energy Administration (NEA) released their 13th Five-Year Plan (2016–2020) on Energy Development (IEA 2017: 494). Together with the market reforms in China and "especially after 2013 when the government launched a far-reaching modernization of administrative structures that streamlined procedures, delegated authority, strengthened regulation and regulatory bodies and limited the scope for central micromanagement of economic affairs" (IEA 2017: 501), NDRC and NEA led a shift from central planning to policy guidance and regulation. They delegated authority towards the provincial level concerning decision making on energy infrastructure and production capacities and abolished national guided approval procedures (Hu and Guan 2017: 122; Knodt et al. 2019: 27).

Both the 13th Five-Year Plan on Energy Development and the 2017 Energy Production and Consumption Revolution Strategy are guided by the president's call for an 'energy revolution', which should serve as a long-term goal of China's energy policy "to build a more secure, sustainable, diverse and efficient energy future" (IEA 2017: 471). The goal focusses on energy consumption and supply, technology and institutions. The 13th Five-Year Plan on Energy Development in particular points out that international cooperation in energy policy should be enhanced beyond bilateral cooperation with the US and should focus on issues of new energy development, establishing international R&D funds and technology transfer (Hu and Guan 2017: 120–123).

8.2 India

India's energy consumption has almost doubled since 2000 due to its rapid economic rise and GDP growth rates of up to over eight percent through 2016. Only recently did GDP begin to fall towards five to 6%, but it is expected to rise again (IEA 2019: 2). Thus, the potential for rapid growth in energy consumption is vast (IEA 2015b: 11). While China is the top world energy consumer, "India is the largest source of demand growth". Together with China and the U.S., India "accounted for 70% of the total energy demand growth" (IEA 2019: 34, 38) and needs substantial growth in its primary energy and electricity supply. For this demand, India relies on its vast coal reserves, which make up 64% of primary commercial energy (MoSPI 2018); 60% of power generation (CEA 2018) and 56% of energy consumption is coal-based. Overall, fossil fuels will continue to be a crucial part of India's energy mix; coal accounts for 56% of its energy consumption (Fig. 8.3), and 70% of power generation

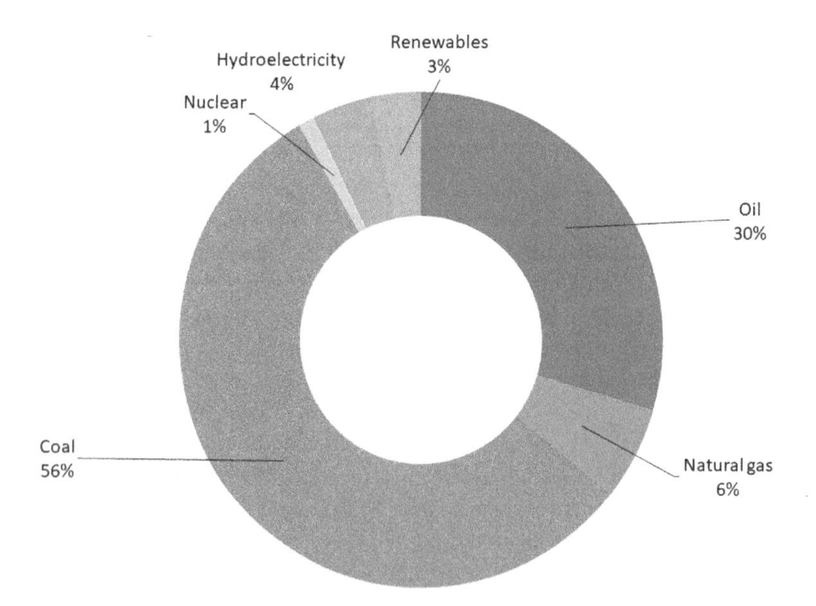

Fig. 8.3 India's energy consumption (2018) (BP Statistical Review of World Energy 2019)

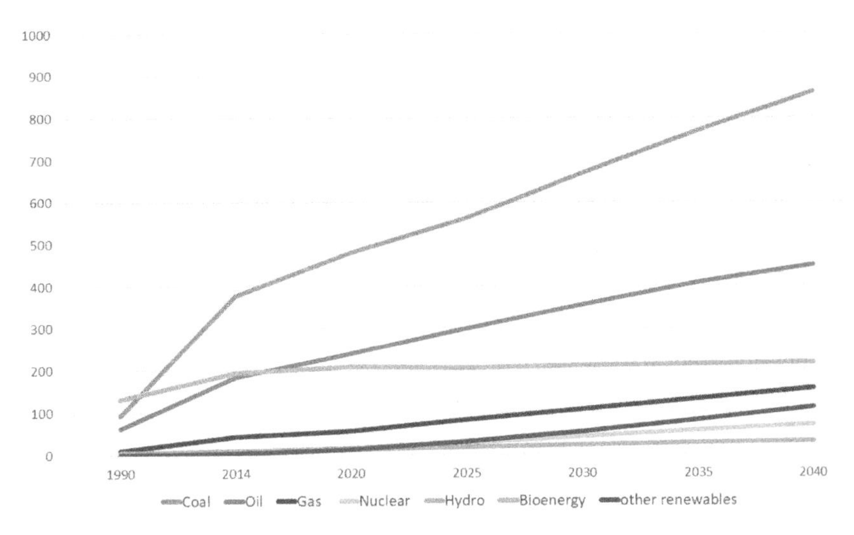

Fig. 8.4 Indian Energy Demand in million tons of oil equivalent (Mtoe, total primary energy demand) according to 2040 Scenario IEA (2016. 2017), (total primary energy demand (TPED) is equivalent to power generation plus other energy sectors excluding electricity and heat)

is coal-based (IEA 2013). According to the IEA, a "large expansion of coal output makes India the second-largest coal producer in the world, but rising demand also means that India becomes, before 2020, the world's largest coal importer, overtaking Japan, the European Union and China" (2015: 13). Thus, the country has developed into a net importer of energy (Knodt 2019: 16).

This trend is likely to continue and is essentially due to abundantly available domestic coal resources with better security and competitive prices (Joshi and Khosla 2016). India's energy strategy first and foremost aims at utilizing all available domestic resources for the optimal output, including also hydro power, oil and gas, nuclear energy, biofuels and other renewable energies (Knodt et al. 2010) (Fig. 8.4).

The most obvious characteristic of India's energy governance is the close link between energy access and development, especially in the light of India's economic growth. The increasing domestic demand and the limits of domestic supply result in discussions on energy security and energy poverty. In the Indian context, energy poverty is identified as a lack of access to modern energy services, especially concerning electricity and clean cooking facilities, which do not cause indoor air pollution (Jaeger and Michaelowa 2015: 236). The discussions are dominant because of approximately 25% of the population not having access to electricity and 66% having to rely on traditional biomass for cooking (Jaeger and Michaelowa 2015: 237). Thus, Indian energy strategies always include social development instruments to guaranty a secure, affordable, and sustainable supply of energy. With its Integrated Energy Policy from 2006, the Indian Planning Commission delineated India's energy security objectives for the next few decades. Clearly, the objectives are supplying energy to all Indian citizens and producing safe and convenient energy at competitive prices at all times (Planning Commission 2006; Joshi and Gansehsan 2015: 151). However, significant progress was made in the last two years with respect to energy access. Between October 2017 and March 2019, the government provided access to electrification for 26 million households (99% through the grid) (IEA 2019: 97).

In addition to energy access, the norm of energy security is also prominent in India's energy governance (Betz and Hanif 2010: 12–14). Particularly in the preference for exploring domestic oil, gas and especially coal and for expanding domestic energy production through new nuclear plants, India's striving for self-reliance is obvious. Regardless, India's striving for energy security collides with India becoming more vulnerable and dependent on foreign energy sources through its strong economic growth (Knodt et al. 2010). One of the outcomes of this discussion is India's strategy for clean energy to open institutional opportunities (Dubash 2011: 66) and strong interest in renewable energy sources. The latter also stems from the motivation to obtain good energy relations with industrialized countries for several reasons: first, the transfer of technologies, which could enhance energy security; second, the continuation of projects within the Clean Development Mechanism (CDM); and third, strengthening its international role as an equal partner in climate negotiations (Dubash 2009).

India's latest energy policy shows the trend of shifting towards the use of renewable energies in response to the COP 21 in Paris. Even before the COP Summit, India's climate vision outlined its intention to follow "a cleaner path than the one followed hitherto by others at a corresponding level of economic development" (cited in IEA 2015b: 3). The outcome of supportive policies of the government and the introduction of competitive bidding combined with falling costs increased the share of renewables in energy consumption up to 3%. Over the past five years, India's renewable power investment has doubled, reaching approximately 20 billion dollars in 2018 (IEA 2019: 114).

The Indian governance system is a multilevel system, as India is a federal state. As in the case of China, India's energy governance is highly fragmented. The main actors with regard to energy issues are the Planning Commission (set up in 1950) and now transformed into the National Institution for Transforming India (NITI Aayog), the Department of Atomic Energy, the Bureau of Energy Efficiency and the four energy ministries—the Ministry of Power (MoP), Ministry of Coal (MoC), Ministry of New and Renewable Energies (MNRE), and Ministry of Petroleum and Natural Gas (MoPNG). With the stronger focus on climate change in recent years, the Ministry of Environment, Forest and Climate Change (MEFCC) starts to play a more important role. Furthermore, independent regulatory authorities for electricity, oil and gas and coal have been established within the last decade (Dubash 2011: 68). Within India's federalism, the central government takes over exclusive competences on mineral and oil resources and nuclear energy, whereas the states have responsibilities on natural gas infrastructure, water issues and taxation on mineral rights. In the case of electricity, both levels share power. In the latter case, the national level is responsible for policy direction, whereas State electricity Boards (SEBs) at the provincial state level control generation and distribution at the very end (Chatterjee 2017: 4). Chatterjee called them, 'vertically integrated monopolies under the politicized control of State governments, which controlled almost three-quarters of generation and virtually all distribution by 1991' (2017: 4). Horizontal coordination between institutions at the federal level is insufficient, even though the Planning Commission strives to assert control over the others. Furthermore, state-owned enterprises (SOEs) are important in Indian energy policies with Coal India Limited, the national Oil and Natural Gas Commission and the State Electricity Boards being the most important ones. State-owned Coal India Limited is the largest coal producing company in the world and the largest employer in the country (Coal India Homepage 2012) (Knodt et al. 2012). Since 2014, the Indian administration has promoted a model of cooperative federalism—also used in the energy system of Germany's federalism—to coordinate the different levels. As an example, the model promotes a higher regional share of hydrocarbon revenues in some cases to enable the regional level to participate financially in energy transformation processes. With these changes, it also introduced a wider set of regional responsibilities to speed up implementation and approval of the state-level clearances required for investment projects, which could be a reason for the increased investment, especially in the renewables described above (IEA 2015b: 42; Knodt et al. 2017: 197, 2019: 18).

To create and execute a more coherent energy policy has been one of the objectives of India's government for decades. Nevertheless, new elements of the system were for a long time imposed onto the old institutions while eliminating pre-existing structures to a reasonable extent, a process called institutional layering (Chatterjee 2017: 5; van der Heijden 2011). One major step to overcome the institutional layering was a Cabinet Resolution on January 1, 2015, replaced the long-lasting Planning Commission with the National Institution for Transforming India (NITI Aayog). As one of the first outputs, it presented India's Intended Nationally Determined Contribution (INDC) in October 2015. The INDC aims at advancing India's energy and environmental policies towards sustainable energy policy. It commits India to increase the share of non-fossil fuel power sources to 40% and to reduce emissions' intensity by 33–35% (measured against the 2005 baseline) by 2030 (Knodt et al. 2017: 196f, 2019: 18).

8.3 Brazil's Energy Policy

Since the 1970s, Brazil has changed its energy policy to a great extent. Approximately 50 years ago, the country was an energy-independent country with a high rate of imported oil (Ribas and Schaeffer 2015: 173f). Nevertheless, the country always had immense renewable resources that were exploited early to reduce this high dependency on energy imports such as hydroelectric and sugarcane-based power plants, ethanol, and biodiesel. Despite the current economic recession and the political crisis since 2015, Brazil as the largest South American country has been showing impressive economic growth rates in recent decades, which caused an increasing in Brazil's energy demand. Brazil's energy consumption rose 40% in 2005–2015, making Brazil the seventh-largest energy consumer in the world (BP 2015, 2016;

Enerdata 2016). Remarkably and given its fortuitous environmental circumstances, it has turned for the most part into an independent energy actor due to its own energy production (particularly renewable energy). Hydroelectric power plants and ethanol supply energy to Brazil's electric and transport sector to a great extent, and newly discovered pre-salt oil reserves add to Brazil's energy independence (Ribas and Schaeffer 2015: 173). Brazil is close to attaining its goal of being energy independent. Brazil also shows its strong engagement in fighting climate change and achieving an international climate agreement. It is responsible for less than 1.3% of the world's total CO_2 emissions from fuel combustion and from energy use (Ribas and Schaeffer 2015: 174). In Paris COP 21, Brazil played a key role in the negotiations and formulation of the 2015 agreement. It affirmed its commitment to high reductions of greenhouse gas emissions (WBG 2016a).

Since the oil crisis of the 1970s, Brazil has undergone immense changes concerning its energy mix and production approach from its large dependency upon oil imports at that time (approximately 80% of its oil consumption). To limit Brazil's vulnerability on the oil price, the Brazilian government launched the National Ethanol Fuel Program (Proalcool) in 1975. The program subsidized the national gasoline supply with Brazilian ethanol production (Hira and de Oliveira 2008) and developed its ethanol fuel-production sector. At the same time, ethanol production offered an alternative for Brazil's sugar industry as a main export commodity to cope with falling sugar prices. The program and other accompanying instruments such as tax exemptions on ethanol-fueled cars established a solid market for ethanol and increased ethanol production approximately 6% per year on average from 1980. By 2004, Brazilian ethanol became fully competitive with gasoline on the international markets (Goldemberg 2007; Goldemberg et al. 2008; Ribas and Schaeffer 2015: 174f). The increase in sugarcane-based ethanol also enhanced the biomass energy production, as Brazilian sugar mills combined their production with biomass facilities (based on bagasse—a by-product of sugarcane processing), producing their own electricity and—with elevated production and power sector deregulation in Brazil—fed it into the grid (Knodt et al. 2017: 120, 2019: 8).

Even more than for its ethanol, Brazil is known for is hydropower plants. As home to one of the largest dams in the world, Brazil is also a forerunner in hydropower, with most installed capacity coming from hydro sources. Hydropower currently accounts for 30% of Brazil's energy consumption (Fig. 8.5) and 80% of electricity production. The hydropower plants all over the country are coordinated over large interconnected transmission networks, especially those connecting the dams with the demand in the (south)east. The networks thus provide the energy supply a certain degree of flexibility (IEA 2013: 309f). However, the discussion on the construction of 'Belo Monte', a hydropower dam affecting indigenous areas in the Amazonian rainforest, has led to social uprisings and strong international criticism. Thus, the construction of hydropower dams remains a socially and environmentally controversial issue. The National Energy Plan 2030 still contains projects for new hydroelectric power plans. However, they tend to be smaller and hence more susceptible to the effects of rainfall (Knodt et al. 2017: 120, 2019: 9).

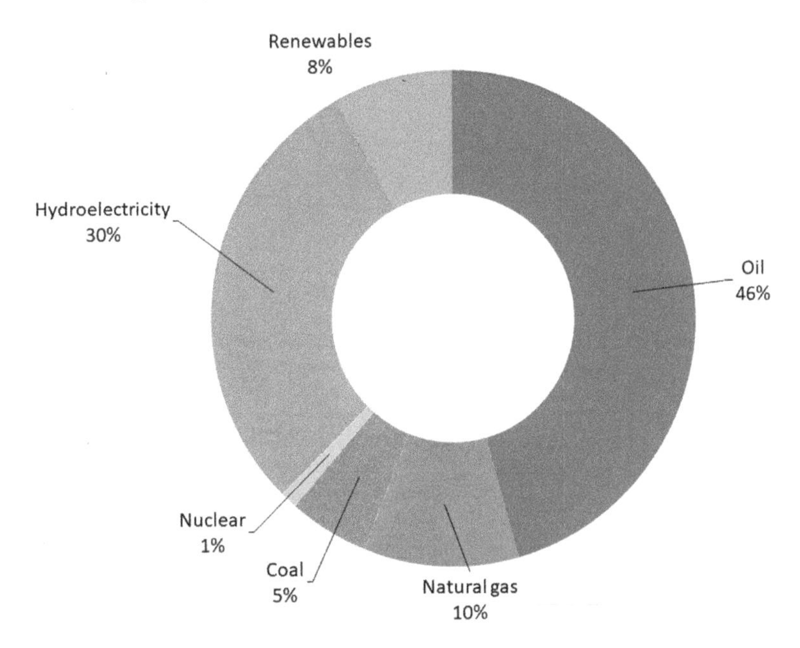

Fig. 8.5 Brazil's energy consumption (2018) (BP Statistical Review of World Energy 2019)

In the latter sense, Brazil refers to small hydropower plants, and to wind, solar and biomass, as "new" or "alternative" renewables compared to old large hydropower from the beginning of the 2000s on, when Brazil enlarged its capacities for renewables (Hochstetter 2017: 11; Knodt 2019: 8). The background of this change was the heavy blackouts in 2001 due to the over-reliance of the system on large hydro power systems and extreme dry seasons. In response to the situation, the Brazilian government launched the PROINFA Program (Incentive Program for Alternative Electric Generation Sources) in 2002. It consists of a first phase mainly of a feed-in tariff system. In a second phase, the program set a 10% rate of alternative renewable energy by 2022. In addition, a bidding system was introduced in 2003 (Lucas et al. 2013). To curtail foreign investment, Brazil auctions included local content requirements such as locally manufactured inputs, enforced as a condition for connection to the grid. For the area of power generation specifically, Brazil is investing primarily in these small hydro, wind, biomass and solar facilities. Among those technologies, wind power constitutes the fastest-growing source of electrical power in Brazil as laid down in the most recent Ten-Year Energy Expansion Plan of the Brazilian Ministry of Mines and Energy (published December 2015) (MME/SPE 2015; Knodt et al. 2017: 121–122; Knodt 2019: 8).

In addition to these renewable energies, Brazil recently discovered major oil resources in its offshore territories, in which huge 'pre-salt' oil fields have been located deep under a thick layer of salt. These reserves are anticipated to almost triple the country's oil production by 2021, and Brazil is expected to become an important oil-exporting nation (Ribas and Schaeffer 2015: 177f). Currently, oil accounts for

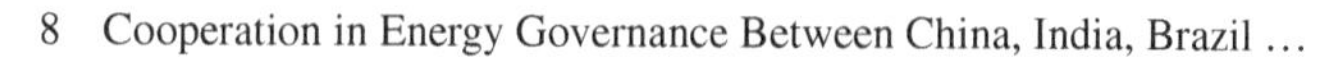

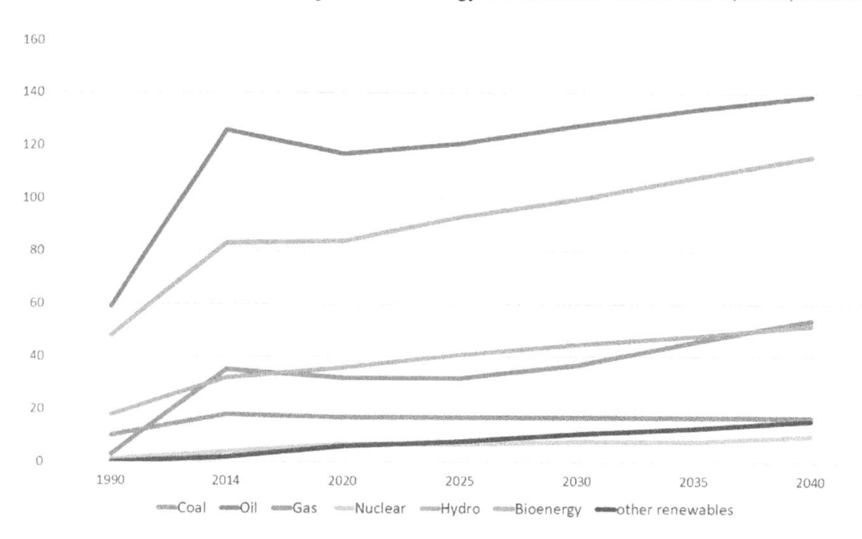

Fig. 8.6 Brazilian Energy Demand in million tons of oil equivalent (Mtoe, total primary energy demand) according to 2040 Scenario IEA (2016, 2017), (total primary energy demand (TPED) is equivalent to power generation plus other energy sectors excluding electricity and heat)

46% of Brazil's energy consumption (see Fig. 8.5, Knodt et al. 2017: 119). In addition, the oil extracted from the pre-salt fields is accompanied by a certain amount of gas. This associated gas will also play a role in the future energy mix of Brazil. Thus, the highest overall energy investment in Brazil in the next ten years will be in oil and gas and might convert Brazil into an "important oil-exporting nation with one of the cleanest energy matrices" (Ribas and Schaeffer 2015: 178, see Fig. 8.6).

Similarly, to India, Brazil is a federal state, and its energy governance is also fragmented to a certain extent. The Minister of Mines and Energy (MME) is the key ministry responsible for energy policies. It formulates the guidelines and policies for the national energy sector and is supported by two regulatory federal agencies: ANEEL (Agency for Electric Energy) for the electricity sector and ANP (Agency for Oil, Gas and Biofuels) in the oil and gas sector. ANEEL was created in 1996 and addresses all electricity topics delegated to it by the MME and the federal government. The Agency is financially and politically independent (ANEEL 2016). The tasks of ANP (created in 1997) are related to the assurance of oil, natural gas and biofuels supply and establishing technical standards (ANP 2016). Furthermore, since 2004, the Power Research Company (EPE) has had the mission of developing an integrated long-term plan for the energy sector in Brazil. Moreover, the Ministry of the Environment (MMA) is addressing renewable energy issues. In energy foreign policy, such as the energy dialogues with the EU, the Ministry of External Relations (MRE) is an important actor. To reduce the fragmentation, the National Council for Energy Policy (CNPE) was introduced in 1997 as an advisory body for the President on energy issues. The Council is chaired by the MME and consists of eight ministers and three representatives appointed by the President of Brazil (Knodt et al. 2017: 122).

Within the Brazilian energy sector, private actors also play an important role. The two large, state-owned enterprises, Petrobras for oil and gas and Eletrobras for the power sector in particular, are closely linked to the MME (Ribas and Schaeffer 2015: 181). In 2016, the Brazilian electric sector was facing a deep crisis not least due to the problems challenging Eletrobras and Petrobras in their role in Brazilian energy governance.[4] Another key actor is the Brazilian Development Bank (BNDES). It channels and provides the right incentives for private investments in portfolio investment, including energy, especially in the auctions. Brazil's medium-term renewable energy targets are laid down its ten-year energy expansion plans (PDEE). The PDEEs are updated annually. The National Energy Plan 2030 (PNE) from 2006/7 has been running out, and the National Energy Plan 2050 is currently being developed. Overall, the Brazilian energy market is growing and has the potential to thrive if it overcomes its bottlenecks. The obstacles include a lack of long-term strategic policy planning, a lack of infrastructure investment, and other issues discussed above. However, Brazil's energy policy also aims at strengthening and reforming its energy governance and the development of long-term plans to provide the proper incentives to all energy stakeholders. Brazil is challenged by a need to balance exploiting new energy resources and pursuing and sustaining a strong energy sector attractive to investors while considering the socioeconomic effects of new hydro dam projects, securing its green energy mix and ensuring its current low carbon-emissions status (Knodt et al. 2017: 123). Beside the new oil boom, wind has become the cheapest of Brazilian electricity sources. "Brazilians hope to emulate their success in wind industry development in the solar sector as well" (Hochstetter 2017: 47). Due to the recent change in government to the right-wing president Jair Bolsenaro, the future of renewable energies in Brazil is unclear, but nuclear power seems to witness support within the new PNE.

8.4 EU Energy Governance

Common energy policies seem to be one of the most truly 'European' issues. Looking back at the times of the European Coal & Steel Community (ECSC) and the European Atomic Energy Community (EURATOM), one would clearly regard this policy field as an example of early supranational cooperation, based on the idea of peaceful and reliable energy supply for a growing community. Instead, the EU for a long time only exhibited weak competences in the treaties, as European energy policy has

[4]Since 2014, Petrobras has been involved in a corruption scandal, the so-called Operação Lavo Jato (Operation Car Wash). Construction companies are claimed to have been overcharging orders from Petrobras and transferring the surpluses to Petrobras executives and politicians. In the aftermath, the involvement of Eletrobras is also being investigated (http://newsroom.ucla.edu/stories/what-s-behind-brazil-s-economic-and-political-crises; http://www.bloomberg.com/news/articles/2015-12-22/in-petrobras-s-shadow-one-company-races-to-contain-graft-probe (accessed 13-Dec-2020).

been regarded as a strictly national issue for most of the time and was only recently regulated at the European level as a shared competence. With gradual integration over time, the European Commission could acquire a stronger energy profile (Piefer et al. 2015a). Thus, the Commission could extend its energy competences (1) within the new integrated environmental competences in the 1980s, (2) the development of a single European Market also for electricity, petroleum and natural gas and (3) through appending cross-border construction of energy infrastructure such as 'trans-European networks' (see Fischer 2009).

For the first time, European Energy Governance was brought to a contractual basis by the Lisbon Treaty. Article 194,1 TFEU regulates that "In the context of the establishment and functioning of the internal market and with regard for the need to preserve and improve the environment, Union policy on energy shall aim, in a spirit of solidarity between Member States, to (a) ensure the functioning of the energy market; (b) ensure security of energy supply in the Union; (c) promote energy efficiency and energy saving and the development of new and renewable forms of energy; and (d) promote the interconnection of energy networks". Nevertheless, the EU is not provided with competences to intervene into the national energy mix (Art. 194,2 TFEU).

Those norms have already been developed in the 2006 Green Paper of the Commission on 'A European Strategy for Sustainable, Competitive and Secure Energy' (European Commission 2006). It states that European energy policies should strive to fulfil three main objectives (Knodt et al. 2015a: 64):

> Sustainability: (i) developing competitive renewable sources of energy and other low carbon energy sources and carriers, particularly alternative transport fuels, (ii) curbing energy demand within Europe, and (iii) leading global efforts to halt climate change and improve local air quality.
>
> Competitiveness: (i) ensuring that energy market opening brings benefits to consumers and to the economy as a whole, while stimulating investment in clean energy production and energy efficiency, (ii) mitigating the impact of higher international energy prices on the EU economy and its citizens and (iii) keeping Europe at the cutting edge of energy technologies.
>
> Security of supply: addressing the EU's rising dependence on imported energy through (i) an integrated approach – reducing demand, diversifying the EU's energy mix with greater use of competitive indigenous and renewable energy, and diversifying sources and routes of supply of imported energy, (ii) creating a framework that will stimulate adequate investments to meet growing energy demand, (iii) better equipping the EU to cope with emergencies, (iv) improving the conditions for European companies seeking access to global resources, and (v) ensuring that all citizens and businesses have access to energy (European Commission 2006, pp. 17–18).

Article 194, 2 states, "Without prejudice to the application of other provisions of the Treaties, the European Parliament and the Council, acting in accordance with the ordinary legislative procedure, shall establish the measures necessary to achieve the objectives in paragraph 1. [...] Such measures shall not affect a member state's right to determine the conditions for exploiting its energy resources, its choice between different energy sources and the general structure of its energy supply". Thus, the EU and the Member States share energy competences within the European Union.

Article 192,2 regulates that the Council within a "special legislative procedure and after consulting the European Parliament, the Economic and Social Committee and the Committee of the Regions, shall adopt [...] measures significantly affecting a Member State's choice between different energy sources and the general structure of its energy supply" (Art. 192,2 TFEU).

In addition to this lack of competences, the EU developed its energy policy since the Lisbon treaty and has lately intensified its endeavors. In January 2014, the Commission proposed the EU 2030 framework for its energy policy, which contained an emission reduction target of 40% compared to 1990 levels by 2030 and a share of renewable energy on the order of 27% by 2030, thus developing its former 20-20-20 goal. Some months later, in July 2014, the Commission also proposed a new energy efficiency target on the order of 30% by 2030 (European Commission 2014). This communication began a highly controversial discussion on the EU's climate and energy policy and especially the sustainability frame. The debate was held in the European Council, which is responsible for decisions on such long-lasting strategic issues, as prescribed in the former 2020 goals. Two discourse coalitions, which had already begun to emerge in 2008, can be identified in the discussion. On the one hand, member states such as Germany and Denmark led a group of environmentalist and climate-friendly governments within the EU and pushed for the package proposed by the Commission. On the other hand, the Visegrád states (Poland, Slovakia, the Czech Republic and Hungary), along with Bulgaria and Rumania and under the leadership of the Polish government, opposed the new targets and insisted on national sovereignty over decisions on their national energy mix and a limited role for the EU. The latter camp pleaded for greater emphasis on the security frame rather than pushing for sustainability targets (Fischer 2014: 2f). To settle this compromise, the Central and Eastern European states made a highly strategic move. The Polish government, which had been forced onto the defensive because of its obstructionist attitude, developed a new concept. Donald Tusk, then Polish Minister President and the current president of the European Council, launched the concept of an energy union.[5] The concept pushed the energy security frame to the center of the European discussion. The Central and Eastern European member states united in a discourse coalition held together by the narrative of how Central and Eastern European households would be left without heat in the winter due to Russian gas policy. They linked the discussion with the European solidarity frame, and Tusk asked for a new policy under the umbrella of a European Energy Union. The Central and Eastern European member states' problem definition quickly attracted sympathizers due to the Ukraine crisis, which accompanied the discourse even without being mentioned explicitly (Fischer and Geden 2015a). The cleavage deepened between the Northern and Western sustainability-oriented member states in one discourse coalition and the Central and Eastern European member states asking for solidarity in energy supply in another. A final agreement could only be secured because of the substantial ambiguity in the formulation of the

[5]Tusk revisited an old idea of Jacques Delors and then-president of the European Parliament Jerzy Buzek from 2010, who pleaded unsuccessfully for a European Energy Community to integrate the Central and Eastern European member states into a system of common energy security.

conclusions. The targets for 2030 that were agreed upon were an 'at least' 40% cut in greenhouse gas emissions compared to 1990 levels, as proposed by the Commission, an 'at least' 27% share of renewable energy consumption and 'at least' 27% energy savings compared with the business-as-usual scenario—thus failing to satisfy the Commission's plea for 30%. Moreover, the extent to which the decision could be revised was framed ambiguously. The European Council's conclusions included a review of the framework after the climate conference in Paris in December 2015. The two discourse coalitions interpreted this review differently. Whereas the Northern and Western member states hoped for greater greenhouse gas reductions, the Central and Eastern member states expected an unsuccessful outcome of the international negotiations. Thus, both sides were willing to agree. In addition, the European Council, at the direction of the Eastern camp, stated that 'these targets will be achieved while fully respecting the Member States' freedom to determine their energy mix. Targets will not be translated into nationally binding targets' (European Council 2014: 5). To achieve consensus, extensive concessions were also made to the governments of Central and Eastern Europe in the form of financial compensation and exemptions from the regulations (Fischer 2014: 3–5; Knodt 2017).

The former president of the European Commission, Jean-Claude Juncker, adopted the idea of the Energy Union after his election in 2014 and asked Vice-Commissioner Maroš Šefčovič (responsible for the Energy Union) and the Commissioner for Climate Action and Energy,[6] Miguel Arias Cañete, to draft a framework for the Energy Union (Fischer and Geden 2015). Obviously, the Commission and Juncker are pursuing the project to work towards developing broader mutual consent on all three frames within the Energy Union. Already in February 2015, the Commission composed a communication called the 'Energy Union Package. A Framework Strategy for a Resilient Energy Union with a Forward-Looking Climate Change Policy' (European Commission 2015). In the communication, the Commission presents its vision of a European energy system, which unites all three of the existing frames—being secure, sustainable, and competitive, while also producing affordable energy. The Commission makes explicit that 'achieving this goal will require a fundamental transformation of Europe's energy system' (European Commission 2015: 2) away from the 28 different national regulatory frameworks and towards one common European framework. The Commission's strategy contains 5 dimensions that were adopted by the European Council in March 2015: (1) energy security, solidarity and trust; (2) a fully integrated European energy market; (3) energy efficiency contributing to moderation of demand; (4) decarbonizing the economy; and (5) research, innovation and competitiveness (European Council 2015a: 1; Knodt 2018 forthcoming).

To bridge the gap between the member state camps and to gain greater acceptance of the Energy Union, Vice-President Šefcovic toured all of the member states from May 2015 until January 2016 and sold the Energy Union as a concept that comprises all three frames. In addition, the Commission had launched several legislative acts

[6] The new directorate general on climate action and energy underlines the close relationship between the climate and energy topics and, thus, is committed to the sustainability frame.

within its 'summer' and 'winter package' to implement aspects of all five dimensions of the Energy Union The European Council's decision in March 2015 also entailed a reporting duty for the European Commission, which has already delivered its first report, in November 2015. In this report, the Commission described the first nine months of the Energy Union as a success story. A close examination reveals that many of the listed projects and implemented instruments had already been agreed upon before the Energy Union and were not affected by the dispute among the different frames. The recently launched and future legislative acts and further strategies will reveal the lines of conflict between the two discourse coalitions within the Parliament and the Council, and within the European Council (Knodt and Ringel 2017; Ringel and Knodt 2018). As mentioned above, according to article 194 TFEU of the Lisbon Treaty, the EU lacks energy competences, especially with respect to national policy mixes. Nevertheless, the European Commission tries to bridge the cleavage with its newest legislation act, the "Clean and Secure Energy for All Europeans" or so-called "winter package" of November 2016. The package is designed to set the goals for the next decades and find a governance mode to nevertheless push Member States in the direction of more ambitious and better coordinated climate and energy policies. It codifies the politically agreed energy and climate targets of the EU and proposes a set of both regulatory and nonregulatory measures to reach the overall Energy Union objectives. The procedural aspects of the transformation within the Energy Union are conceptualized within the Regulation on "Governance of the Energy Union" (European Commission 2006e) proposed by the Commission as part of its winter package and entered into force after the trialogue with the co-legislators on 24 December 2018. With the Governance Regulation, the Commission reverted to soft governance mechanisms known from the "open method of coordination" (OMC) but attributed these mechanisms to harder elements that allow a deeper engagement into national energy sovereignty. In addition, the governance provisions are linked with the international level, especially the Paris Agreement on climate change, to legitimize the mechanisms (European Commission 2016f; Knodt and Ringel 2018: 209f).

This proposal of the Governance of the Energy Union is relying on monitoring tools. There will be an assessment of the national aims, strategies and measures according to their ambition by the Commission, and 'ambition gaps' are to be detected on the Member States side. In addition, national measures will be assessed according to their goal achievement/delivery by the Commission to draw attention to possible 'implementation/delivery gaps'. If ambition or implementation/delivery gaps are found, the Commission will recommend to the Member States. They will have to take due account of the recommendations. The burden of proof will be on the Member States side. If Member States do not follow the Recommendation by the Commission, the Commission will be able to take potential corrective actions. The proposal lists delegated acts on the one hand or a voluntary financial contribution to a fund to be set up on the other hand. In addition, the proposal introduces bilateral horizontal coordination parallel to the joint horizontal coordination, which should take the form of regional cooperation. The reporting of Member States to the Commission also includes Member State activities relevant to the external dimension of the Energy Union.

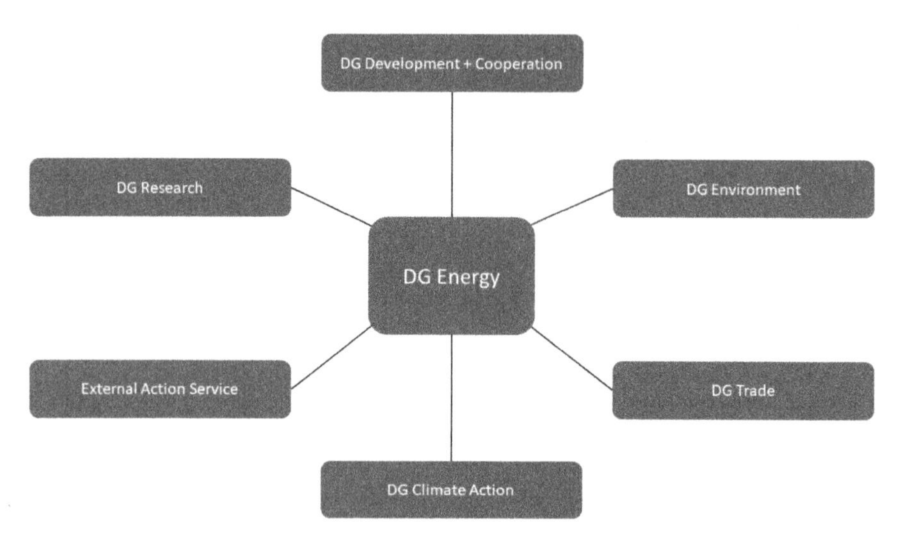

Fig. 8.7 European Actors involved in the bilateral Energy dialogue with BICS (EnergyGov 2014)

As a cross-cutting issue, energy involves the policy fields of external relations, trade, energy, development, research, environment and climate policies. Thus, more than one EU Commission Directorate General (DG) and the European External Action Service (EEAS) are part of EU external energy governance (Knodt et al. 2015a: 61–64, 2019: 6). Figure 8.7 provides an overview of the different actors at the European level. The arrangement of the actors already gives an idea of their position in the bilateral energy dialogues with the BIC countries (Brazil/India/China) concerning their importance. As DG Energy is the most important one, the distance of the others in relation to DG Energy expresses their importance. The further away, the less important they are. The exact position and importance will be shown in each dialogue below.

Within the Commission, 'DG Energy' has the main responsibility for all energy-related issues, and it is clearly the leading DG of European Union energy policies. DG Energy has established several supranational activities, with the objective of strengthening relations with third countries and transnational energy companies. Internal and external EU energy strategies and policies are drafted, and DG Energy takes the lead in energy dialogues with emerging powers. The work of DG Energy is challenged by its limited budget lines for cooperation programs with third countries, and it calls for close collaboration with other DGs with more funding for energy projects. In particular, DG DevCo (see below) in recent years was one of the main cooperation partners of DG Energy in this respect (Knodt et al. 2015a: 61, 2019: 6).

With regard to energy issues, the actions of 'DG Climate Action' and 'DG Environment' show considerable overlaps with DG Energy. The discussions on the Commission Green Paper "A 2030 framework for energy and climate policies" (European Commission 2013) and the 20-20-20 goals of the Energy 2020 Strategy prove the close linkages between both issues. Indeed, these DGs are the two who are closest to

DG Energy in the dialogues. However, our survey has interestingly shown that the DG Environment is not very knowledgeable within the EU-China energy dialogue, although our respondents attribute high importance to it. DG Climate is more prominently visible, which could be related to the link between climate and energy issues (Knodt et al. 2015a: 61, 2019: 6).

For 'DG Trade', energy issues are one of the few topics that are not yet part of a consistent liberalization strategy, and according to our interviews with DG Trade officials, their activities concentrate on bringing this topic back into the liberalization discourse, e.g., consistent with the demands of the Singapore issues[7] (Interviews Brussels, May 2012). 'DG Research' shares interest in issues such as intellectual property rights and technology transfer with DG Trade. Generally, a shift in R&D relations with emerging powers can be observed intensely with China (Knodt et al. 2015a: 62), but both DGs are more marginal in the discourse (Knodt et al. 2019: 6).

'DG Development and Cooperation' (DevCo) is a necessary partner for DG Energy concerning the implementation of energy projects, which are funded under the development heading in emerging powers. Within the financial framework (2014–2019), development cooperation with emerging powers was phased out, which will be of great significance for BIC and especially EU-China energy relations in the future (Knodt et al. 2019: 6). As measures and programs in the energy cooperation between the EU and China have been predominantly financed through DG DevCo, new financial strategies have to be identified. In its 'Agenda for Change' (European Commission 2011), the EU has set out a new development cooperation strategy, also with emerging powers. The broader aim behind this strategy is the need for a change of the relationship between the EU and emerging powers from a development (recipient) relationship to more-equal partners. The new Partnership Instrument and the Industrialized Countries+ Instrument are now pooling funding for cooperation projects beyond development cooperation. After the phasing out, all new activities are funded by the new Partnership Instrument and the Industrial Cooperation+ Instrument (ICI+) from 2014 on. China, India and Brazil are countries that are eligible to receive funding through these two instruments (Piefer et al. 2015a: 36f; Knodt et al. 2019: 11f). Certain budgets were allocated to energy projects but are overall lower than before. With the phasing out of development cooperation comes a reduction of staff in the EU Delegations abroad, so that many energy contacts will be lost. As other staff in the Delegations—e.g., from DG Trade as in the case of China—have to be in charge of energy in addition to their other responsibilities, the importance of energy policy in the relationship is impaired (Knodt et al. 2015a: 62f). Nevertheless, it does not seem to have an active part in the dialogue and is not as central as DG Energy or DG Climate Action and DG Environment.

From 2014 onwards, all new activities will be funded by the new Partnership Instrument and the Industrial Cooperation+ Instrument (ICI+). Brazil will be one country among many from the group of emerging economies and industrial countries

[7]The Singapore Issues are a package of strategies for deep trade integration taking place beyond borders, namely competition in energy services should be introduced, with liberalization of government procurement and investment conditions.

in the European neighborhood and other parts of the world that will be eligible to receive funding through these two instruments (Piefer et al. 2015a: 36f).

Created under the Lisbon Treaty, the EEAS plays a very interesting role in EU energy governance and is characterized by several distinctive features; it aims at being the overall coordinator of EU energy governance by framing energy relations as 'energy diplomacy'. Thus, it tries to streamline energy into general external relations. The EEAS had reported this strategy in an internal non-paper on energy and EU foreign policy (EEAS 2012) and acknowledged changing dynamics and power relations in energy governance. It sees itself as bridging political and business interests. Thus far, the EEAS lacks issue-specific competencies due to a clear role in the EU's institutional system. The EEAS' understanding of energy governance as 'energy diplomacy' is not yet accepted as a positive advantage by the other institutions that compete on governing energy issues. 'Energy diplomacy' would imply a more flexible way of connecting public and non-state energy actors both in the EU and in the respective country, which might thwart the competencies of other EU actors involved, as long as roles and competencies are not clearly defined (Knodt et al. 2015a: 63).

Already the reference towards the limited power of the EU concerning the national energy mixes above hints that the coordination of EU and Member State energy activities should exhibit an important role in the EU's external energy governance. Indeed, research showed (Knodt et al. 2015) that agreements and partnerships of the EU and Member States towards the BIC in the energy sector have at times rivalled each other, and there is a considerable lack of information on Member States' activities in DG Energy. The perception of the European public and non-state actors is clearly towards great importance of the coordination of the bilateral MS-EP and EU-EP dialogues. The European Commission has reacted to this perception and the obvious problematic non-coordination within the external energy policy. In 2012, the focus has also been placed on an information exchange mechanism with the EU Member States (Decision 994/2012/EU in Official Journal of the European Union 2012). According to the new mechanism, Member States have to inform the Commission of all their new and existing international agreements (IGAs) with third countries in the field of energy; otherwise, they risk sanctions. The mechanism should cover only intergovernmental agreements having an impact on the internal energy market or on the security of energy supplies in the Union. Regarding future intergovernmental agreements to be negotiated between Member States and third countries, the Commission should be informed. The Member States have the possibility to invite the Commission to participate in the negotiations as an observer or an assistant. Actually, the decision also covers the possibility of Member States opposing the participation of the Commission in the negotiations. It is stated that "The Commission should also have the possibility to participate as an observer at its own request, subject to the approval of the Member State concerned". In addition, the Commission should facilitate and encourage coordination among Member States. With this measure, the EU aims for more transparency, which should allow the Commission to take coordinated action for the EU and thus promote a common voice in energy issues. However, this approach could also lead to higher tensions between the Member States, as the

Commission could induce pressure on them (Interview Brussels, November 2012). Most EU Member States oppose this regulation, as they see their confidentiality and sovereignty in setting up energy supply agreements endangered (EURactiv 2012). Nevertheless, they comply and inform the Commission about their IGAs. In addition, a database has been created with more than 114 IGAs (European Commission 2013a; Knodt et al. 2015a: 59).

The Commission has taken additional initiatives to enhance the exchange of information between the Member States on external energy relations. It created discussion forums, for instance, monthly discussions in the Council's Energy Working Party, where the Commission updates the Member States on the EU activities and/or prepares common EU positions. Formal meetings of the Energy Council continued to include international energy relations on the agenda, where discussions at a Ministerial level are taking place. The establishment of a Strategic Group for International Energy Cooperation (Commission, EEAS and Member States' Energy and Foreign Affairs Ministries) allows identifying and discussing common priorities, which could lead to the development of joint initiatives and positions vis-à-vis third countries and regions; however the Strategic Group has hitherto only discussed cooperation with China but not with any other BIC country (European Commission 2013a). Thus, the establishment of such a forum under the umbrella of the EU-BIC energy dialogues bringing together the working level of EU and Member States would be advisable (Knodt et al. 2015a: 60).

Despite overall positive ratings for bilateral cooperation, stated by the participants of the respective energy dialogues (see below), relations and tensions between the EU and the Member State level are issues of concern, as our data on coordination reveal. There does not seem to be an institutionalized coordination between the EU and the Member State level. In some cases—India in particular is a good example—informal coordination activities have been developed, but they do not involve all participants of the energy dialogues. In the country cases—most strikingly EU-Chinese cooperation—a significant number of EU actors state that there was no coordination between the supranational and the Member State level. While interpreting the data, it has to be considered that actors from the Member State level were not part of our survey. Including those actors might have changed the outcome.[8] At the same time, an overwhelming majority (86.1%) of the EU actors state that coordination between both levels would be an important issue. This discrepancy suggests that external energy relations tend to run on parallel paths, whereas possible synergies between the supranational and the Member State level are not yet highly recognized. It also seems that for the characteristic challenges of 'realpolitik'—for instance, the wish to negotiate over concrete technology transfers—the Member State level is preferred over the supranational level. Coordination between the different negotiation arenas thus seems to be an issue that needs to be optimized, as enhanced coordination would be beneficial for energy cooperation (Knodt et al. 2015a: 60f).

---

[8]Please note that the data this contribution is based on date back to 2012/2013; this mechanism was either not yet in place or newly established.

This institutional mapping shows that the governance network of European institutions resembles a polyphonous structure. Obviously, the EU does not 'speak with one voice', and while the cross-cutting nature of energy issues underlines that there is not a need to do so, this institutional structure bears both potentials and risks. Ideally, these tailor-made modes of governance allow creating flexible relationships with the EU's strategic partners and allow releasing synergies deriving from close-knit coordination between the different DGs that are involved in energy relations. Knowledge management may be facilitated through close informal connections and through regular personnel exchange, which is one of the bureaucratic principles within the EU administration. At the same time, the institutional fragmentation and inadequately defined roles and responsibilities (e.g., the EEAS' positioning) may foster concurrence between different EU actors involved, possibly resulting in highly fragile forms of coordination. Additionally, the absence of a clear leading position may hamper the cooperation with the EU's strategic partners. This point is where the cross-cutting nature of energy issues may also weaken the role of DG Energy, as we can think of constellations where other actors such as DG DevCo or the EEAS may be in a more apt position to meet the interests of both the EU and its partners (Knodt et al. 2015a: 63).

With the Energy Union and the European ambitions under the Paris agreement, the EU tries to partly overcome the horizontal fragmentation of the EU (Knodt and Ringel 2020).

8.5 EU-China Energy Dialogue

As one of the most advanced dialogues, the energy dialogue between China and the EU started almost 35 years ago with cooperation in basic energy science following the EU-China Trade and Economic Cooperation Agreement (1985). The EU and China initiated biannual energy conferences between the European Commission and the Chinese Ministry of Science and Technology in 1994. The energy dialogue in its current form started in 2005 between the European Commission and the National Energy Administration (NEA). Since then, six priority areas have since been identified: renewable energies, smart grids, energy efficiency in the building sector, clean coal, nuclear energy and energy law. Furthermore, energy issues are discussed during the annual high-level summits (DG Energy Homepage 2013; Knodt et al. 2015a: 39, 2019: 23).

The year 2012 marked interesting developments and a clear uplifting of China-EU energy cooperation. The first China-EU High Level Meeting on Energy took place in May, and for the first time, the EU Commission plus the 27 energy ministers of the EU Member States met with Chinese representatives in this format to enhance cooperation on energy issues. The EU-China Partnership on Urbanization and the Joint Declaration on Energy Security are milestones in the cooperation, as they were the first documents, which were agreed by all 27 Member States plus the European Commission with the Chinese side (Knodt et al. 2019: 30). It is remarkable that a dialogue on energy security has been initiated, giving evidence that for the EU, questions of the geopolitics of energy with China gained in importance. The EU discusses issues of Central Asian resources and the role of Russia in the region with its Chinese partners (see Zha et al. 2011). Additionally, the question of how to engage China in international energy governance reached the political agenda (for example IEA, ECT), as Commissioner Oettinger mentioned in a speech in the European Parliament (May 2012). China was also keen on talking about global energy markets. The EU-China energy dialogue resulted in the joint implementation of projects, such as the EU-China Clean Energy Centre (EC2). The European Commission and the National Energy Administration (NEA) and the Ministry of Commerce of China (MOFCOM) initiated the EC2 in 2010 and aimed at supporting the Chinese government officials in the clean energy sector through capacity building, policy advice and the provision of services in technologies in the focus areas of range of issue. Among them were clean coal technology, sustainable biofuels, renewable energy sources, energy efficiency in energy consumption, and sustainable and efficient distribution systems. For each of the above-mentioned activities, there is an EC2 project team of Chinese-European tandems who jointly agree on final results (EC2 2014). EC2 acts as facilitator between the EU and China. At the 2013 EU-China Summit, the 'EU-China Agenda 2020' was agreed including the six points on energy in the chapter on 'Sustainable Development' (EEAS 2013). Most importantly, the EU and China want to reinforce cooperation on energy issues, with a special emphasis on global energy security within the framework of the Energy Dialogue (Knodt et al. 2015a: 39f, 2019: 23).

The 2015 EU-China Summit celebrated the 40th anniversary of bilateral relations between the two partners. The Summit was marked by breakthroughs on several previously contested issues. Willingness to cooperate was displayed on highly disputed Intellectual Property Rights and climate change (European Commission 2015). Importantly for our research, the two sides agreed on the Statement on Climate Change, an area of cooperation closely linked to energy matters in the dialogue between China and the EU. Both parties demonstrated their willingness to take steps towards development of low carbon emissions and cooperation on emissions trading (facilitated by China's plans to introduce a nationwide carbon emissions trading system by 2020). China and the EU stressed that they „commit to work together to reach an ambitious and legally binding agreement" (European Council 2015b: 2) with regard to the Paris Agreement, which both signed subsequently. However, for the EU, such a common commitment would be of importance, particularly at a time when the EU's significance and influence in international politics is perceived to be decreasing. Notwithstanding this imperative for closer collaboration on energy and climate change, Lee argued that „EU-Chinese relations have ebbed and flowed much since 2007" (Lee et al. 2015: 6). Instead, China seeks closer ties with other actors, such as the US, with whom China agreed on a „Joint Announcement on Climate Change and Clean Energy Cooperation" (IEA 2015: 50). New momentum was given when the EU and China agreed on the EU-China Energy Roadmap (2016–2020), which focusses on common energy and climate challenges, including the security of the energy supply, energy infrastructure, and market transparency, and resulted in a

Work Plan in 2017. Moreover, it has to be noted, that official Chinese policy papers on China-EU cooperation focus much more on economic and financial cooperation than on energy issues (Lai and Shi 2016). Overall, it looks as though, compared to EU Member States, the EU is gaining less attention from China, which is due not only to the Eurozone sovereign debt crisis and changes in Chinese foreign policy priorities but also to the attractiveness of cooperation partners and China's attitude of choosing to focus less on multipolar multilateralism and more on bilateral national ties (Lee et al. 2015: 3; Piefer et al. 2015a: 42–43; Knodt et al. 2017: 141f, 2019: 26).

Based on the following data on the interaction structure within the EU-China energy dialogue, such as within the other energy dialogues described here, there has been a network analysis carried out. The respective relational network data were analyzed by social network analysis (SNA). SNA has by now become a standard form of analysis, especially in inter-organizational social scientist research. The basis of the network question was an organization list of all actors that participated regularly in the dialogues—35 (India), 41 (Brazil), or 49 (China). The list was then presented to all those actors that form part of the list. Attention was limited to only two types of information, that is, first, the reputation of domestic and supranational public and private organizations within the energy policy domain and, second, the exchange of strategically important information occurring between these organizations. For each bilateral dialogue, this process produced networks containing confirmed and unconfirmed, in the sense of reciprocal and unilateral, ties of reputation and contact. The responses to these questions were then stored in rectangular data matrices that, after first elaborations, contained entries ranging between 1 and 3 depending on whether actor "I" chooses actor "j" (1), actor "j" chooses actor "I" (2), or both choose each other, meaning a reciprocal relationship of, in our case, mutual reputation and mutual exchange. With the help of the appropriate software, both these types of matrices—on reputation and information exchange—were then submitted to a procedure of visual representation, thus producing network graphs whose basic information can now easily be grasped by the reader. Thus, the maps can be used as a heuristic tool to draw insights on the form of EU-Emerging Powers energy cooperation, such as which actors occupy central positions in the respective structure, prominent gatekeepers or brokers. With regard to the graphic representations of our network, the reader can easily distinguish between unilateral (thin ties) and reciprocal relations (bold ties). We have considered all ties in our networks independent of whether they are unilateral or reciprocal, when we proceed with information on individual institutions and organizations, e.g., indegree and betweenness centrality, which we show in the respective tables or in an aggregated way in the block modelling as a representation of the relational performance across different types or categories of organizations. For the block modelling, we regrouped the data matrices according to the affiliation of our network members to one of four categories: domestic public, domestic non-state, supranational public, and supranational non-state institutions and organizations. Matrices were then partitioned, and the density values were calculated for each of the resulting four sub-matrices. Partitions obtaining a density value equal to or greater than the density of the respective overall network were attributed a "1" entry. We thus obtained 4 × 4 image matrices, which we converted into graphic representations (Müller et al. 2015a: 27–29).

The network structure in the EU-China dialogue shows that the Chinese actors are considered the most important, i.e., the Ministry of Commerce (MOFCOM), the National Development and Reform Council (NDRC), the National Energy Administration (NEA), and the Ministry of Foreign Affairs (MFA). The NEA has recently undergone a re-structuration process and has been merged with the State Electricity Regulatory Commission (SERC) (see above). As this change occurred after the beginning of our survey, we still analyze both actors, keeping these changes in mind. The NEA is one of the main partners in the energy dialogues. The European Commission calls NEA its "natural partner in energy cooperation" (European Commission 2014b). However, one would have assumed that the Ministry of Science and Technology (MOST) as another key partner ranks higher in importance (Knodt et al. 2015a: 40). MOST is the oldest Chinese partner in energy cooperation (European Commission 2014b), but it seems that MOST has lately lost importance. The position of the Ministry of Housing and Urban Rural Development (MoHURD) is quite surprising. MoHURD is the partner in the Urbanization Partnership, so that we also expected it to be considered a very central and important actor (Piefer et al. 2015a: 41f; Knodt et al. 2019: 26) (Fig. 8.8).

Table 8.2 shows the most important five actors in the EU-China dialogue. The table lists the actors with highest indegree centrality values, in both the importance and exchange network. In the third column, the data of the betweenness centrality

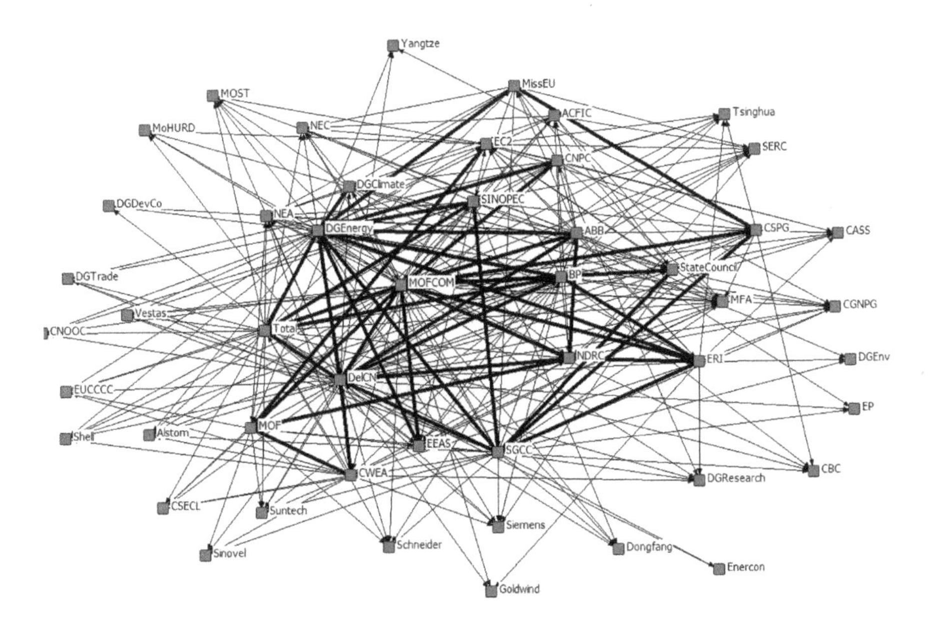

Fig. 8.8 Importance network for China-EU energy cooperation; importance data: unilateral (thin ties) and reciprocal relations (bold ties). (EnergyGov 2014; Piefer et al. 2015a: 41; for actors, please see Table B.1 in Appendix B)

Table 8.2 Top five actors in China-EU energy dialogue (EnergyGov 2014; Piefer et al. 2015a: 41)

Importance network Five most important actors (indegree centrality)	Exchange network Five most important actors (indegree centrality)	Broker function (exchange) Five most important actors (betweenness centrality)
NEA 33.3	DG Energy 18.8	MOFCOM 9.0
MOFCOM 29.2	Tsinghua 16.7	DelCH 8.8
NDRC 29.2	CSPG 14.6	CSPG 4.5
MFA 29.2	ACFIC 14.6	DG Energy 2.9
DG Energy 22.9	MFA 14.6	CNPC 2.5

within the exchange networks are added for the five most important actors to create data for the broker function.

Again, it shows that the Chinese actors are dominating the network. For the EU, we can only find DG Energy and the Delegation of the EU in China. DG Energy appears in each category and shows up as the most important actor in the exchange network. The Delegation of the European Union to China (DelCH) could be called a key actor. It holds a broker function, which seems quite reasonable as it serves as the first contact point with the EU (Piefer et al. 2015a; Knodt et al. 2019: 28).

Due to the economic orientation of China-EU energy cooperation, the private sector[9] plays an important role. Thus, business actors are represented in the network, with state-owned companies (SOCs) in the Chinese case. SOCs standards and policies are being harmonized by the government. Among the most central actors in the network are the China State Power Grid (CSPG) and China National Petroleum Group (CNPG). For the EU side, the EU Chamber of Commerce in China (EUCCC) in Beijing has a quite central place within the network and serves as a gatekeeper for European companies. Its main function is to inform European companies of developments concerning market access, and in its energy working group, provide advice on best strategies to engage with the Chinese authorities (Interviews Brussels, February 2012; Beijing, March 2012; Knodt et al. 2019: 28). Furthermore, business summits are organized parallel to the summit meetings (which are held without non-state actor participation) to benefit from Müller (2015: 42).

A glimpse at the network block matrices on importance and exchange indicates that EU public actors show strong interest in cooperating with China. The EU's interest in Chinese public actors is not reciprocal. While EU public actors are actively contacting their Chinese counterparts, Chinese public actors do not consider the EU to be important or want to reach out towards them. Unlike EU public actors, Chinese public actors attribute greater importance to the Chinese non-state actors and exchange information mainly with them. They also have active contact with European non-state actors, because of their great interest in technology transfer and export of renewable energy technologies, showing the Chinese government's

[9]For the purpose of this analysis, Chinese state-owned companies are also regarded as part of the "private sector" due to their business logic of functioning versus that of a public sector institution.

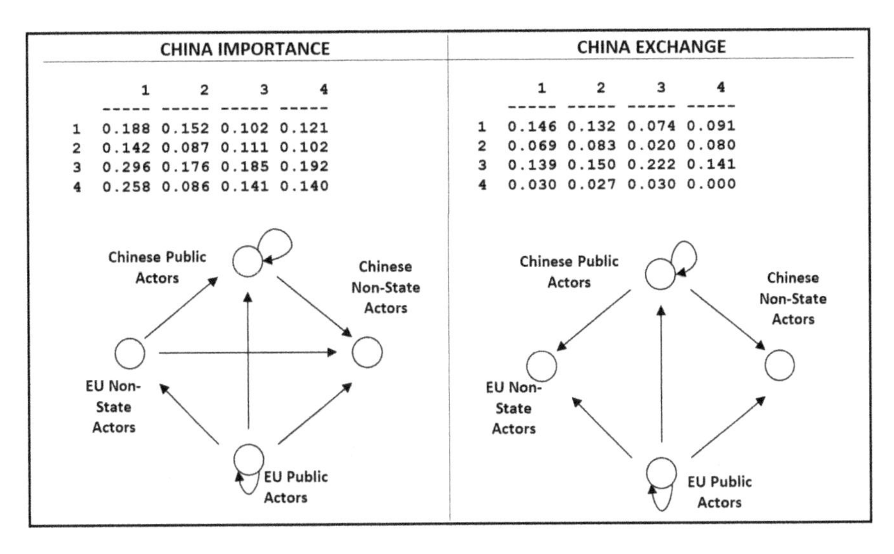

CHINA IMPORTANCE

	1	2	3	4
1	0.188	0.152	0.102	0.121
2	0.142	0.087	0.111	0.102
3	0.296	0.176	0.185	0.192
4	0.258	0.086	0.141	0.140

CHINA EXCHANGE

	1	2	3	4
1	0.146	0.132	0.074	0.091
2	0.069	0.083	0.020	0.080
3	0.139	0.150	0.222	0.141
4	0.030	0.027	0.030	0.000

Fig. 8.9 Importance and exchange network block matrices (China-EU). 1 = Chinese public actors, 2 = Chinese non-state actors, 3 = EU public, 4 = EU non-state (EnergyGov 2014; Piefer et al. 2015a: 43)

preference for private sector cooperation (Piefer et al. 2015a; Knodt et al. 2019: 29) (Fig. 8.9).

The EU-China energy dialogue has evolved considerably in the years since its inception, and despite different political prioritizations, indications of increased interest from both sides can be witnessed, such as the amplification of cooperation areas, the high-level participants (such as the Energy Commissioner or NEA Administrator), EU-Member State coordination within the High Level Meeting on Energy in 2012 and the establishment of joint centers and implementation of concrete projects. Nevertheless, Chinese actors prefer cooperation with EU Member States compared to with the EU. Interview partners referred to Member State cooperation as just 'another cup of tea' (Interviews in Beijing and Brussels, February–May 2012; Knodt et al. 2015a: 42f, 2019: 29).

Another dimension of our analysis explores the extent to which the two actors perceive each other's willingness and ability to cooperate with each other. In the elite interviews and the survey, our respondents were asked to assess different roles of the EU and China, for instance, "being an agenda-setter" or "playing with a hidden agenda". The two indicators of evaluations and roles are argued to identify the emotive charges of mutual perceptions and visions of agenda-setting qualities, mutual learning, change of standpoint as well as having an open agenda (Knodt et al. 2015b: 335).

The self-perceptions of the EU are positive (Fig. 8.10). The EU actors themselves tend to view the EU as an active agenda-setter and emphasize the EU's high compromise-building qualities. They also see the EU as taking an interest in other

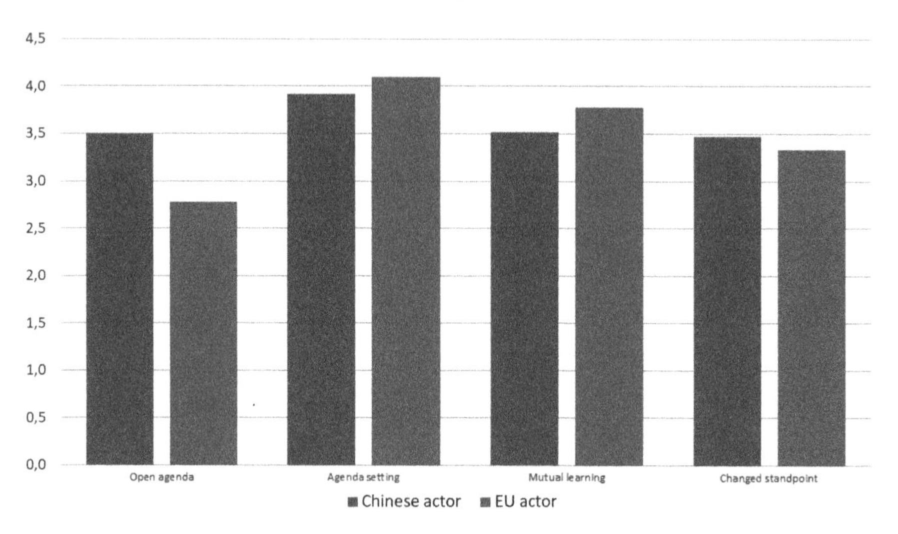

Fig. 8.10 Perception of the EU's properties as a dialogue partner in the EU-China energy dialogue, means of all answers; 1 = "Strongly disagree"; 5 = "Strongly agree" (EnergyGov 2014)

negotiation partners, i.e., being open to learning. This overwhelmingly positive self-image is however slightly less supported by non-state actors, who also stress that the EU has been increasing pressures on emerging powers during the course of negotiations (EnergyGov 2014; Knodt et al. 2017: 174, 2019: 30).

Perceptions of China in the EU-China energy dialogue by the Chinese and European actors converge only on some aspects (Fig. 8.10). Chinese elites see China's involvement in the EU-China energy dialogue as being rather transparent. In contrast, the EU considers China playing with a "hidden agenda". Furthermore, China sees itself willing to compromise, whereas the EU considers China as an actor less inclined towards compromises (Fig. 8.11). Such diverging perceptions may undermine the feeling of trust between the two actors (Zha and Lai 2015: 139; Knodt et al. 2017: 187, 2019: 30).

Concerning the learning effect, interviews showed that Chinese interviewees perceived themselves as learning—specifically, in the areas of technology, energy management and regulation—more from the EU than vice versa. The most prominent perception was the one of misunderstandings and misperceptions of the EU by Chinese interviewees. They referred to the EU's framing of Chinese enterprises as competitors, e.g., within the solar panel trade dispute. Hence, trade frictions were perceived as likely to increase. Restrictions on market access are, according to the interviewees, hindering the flow of technological innovations and expertise. Blocking European companies from entering the Chinese energy market means that European advanced technology and skills are also blocked from reaching China. Many European enterprises already reluctant to sell their technology to China are aware of this issue. Interviewees remarked that such a market access problem could only be settled at the EU level, between the EU and the Chinese government. Another frequently

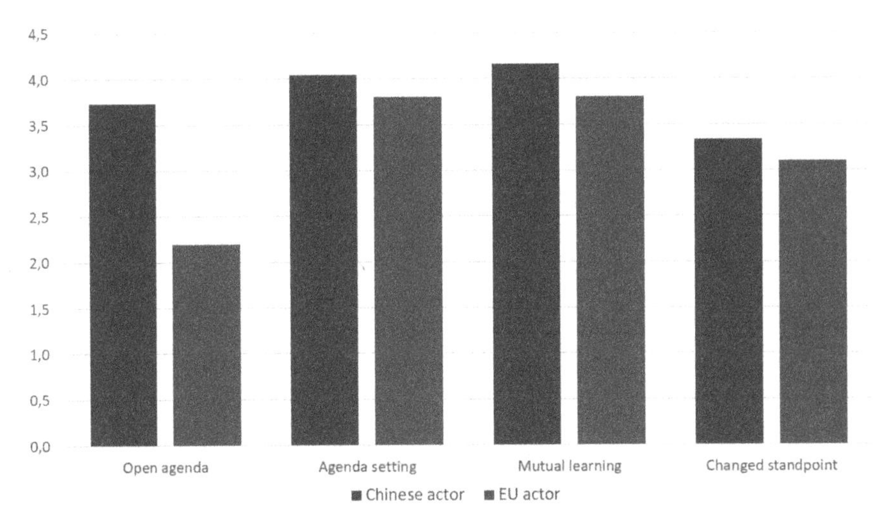

Fig. 8.11 Perception of China's properties as a dialogue partner in the EU-China energy dialogue, means of all answers; 1 = "Strongly disagree"; 5 = "Strongly agree" (EnergyGov 2014)

named problem was the lack of openness of the Chinese and EU energy markets to foreigners, particularly the former market. Respondents also agreed that due to different stages of development in the economic and energy spheres, the interests of the EU and China differ. To improve cooperation, the interviewees proposed more and better communication among the parties. Various communication mechanisms already established were perceived as positive attempts to help the situation (Zha and Lai 2015: 140; Knodt et al. 2017: 188, 2019: 30).

Germany is China's most important trade partner from all EU-27 member states. Of all EU-27 trade with China, the German share is approximately 48%. Following the UK's withdrawal from the EU, this also applies to trade with India (31%).[10] The same holds true for investment. Since July 2010, China is no longer considered a developing country that could receive Official Development Assistance (ODA) from Germany. Instead, Germany and China have converted their relationship into a strategic partnership in 2010.[11] Within the partnership, BMWi is responsible for the overall coordination in Germany, while the NDRC is the responsible counterpart on the Chinese side. As with the EU strategic partnership with China, the German-Chinese strategic partnership is having an energy dialogue. The dialogue takes place within the German-Chinese Forum for economic and technological cooperation. Two action groups are responsible for energy issues. Within the "AG Energy", the German Federal Ministry for Economic Affairs and Energy (BMWi) and the Chinese NEA are

[10] https://ec.europa.eu/eurostat/statistics-explained/index.php?title=China-EU_-_international_trade_in_goods_statistics#Trade_with_China_by_Member_State and https://ec.europa.eu/eurostat/statistics-explained/index.php/India-EU_%E2%80%93_international_trade_in_goods_statistics#Trade_with_India_by_Member_State (accessed 30-Mar-2020).

[11] https://www.bmz.de/de/laender_regionen/asien/china/index.jsp (accessed 30-Mar-2020).

talking on renewable energies. In a second group (AG Energy Efficiency), the BMWi and the NDRC are working on energy efficiency and strategies for energy saving. In October 2014, the Forum agreed on a common action plan for innovations where energy strategies in both groups are mentioned and an intensification of the effort was announced. At the moment, issues such as energy efficiency, biogas, e-mobility and climate protection are on the agenda. Thus, the Bundesministerium für Verkehr, Bau und Stadtentwicklung—BMVBS (now Bundesministerium für Verkehr und digitale Infrastruktur—BMVI) had signed a common declaration on cooperation on sustainable mobility, energy efficiency and emission reduction, and innovative transport technology with the Chinese Ministry for Science and Technology. With regard to the two last issues, the German Ministry of environment has signed a declaration with the Chinese Ministry of Science and Technology to intensify cooperation on environmental issues and a declaration of intent with the Chinese Ministry for environmental protection to expand the Chinese-German environmental partnership. The agricultural ministries of the two countries have signed a Memorandum of Understanding on biogas cooperation. E-mobility seems to be another major interest that has become a major field of cooperation where the German Ministries such as for Transport and Digital Infrastructure and Education and Research are involved. Linked to the EU-China Energy Dialogue, the German-Chinese strategic partnership in energy issues, technical cooperation in energy efficiency and low-CO_2 urban construction are also of great interest for the two countries (Knodt et al. 2017: 151f). In 2017, the Sino-German energy partnership dialogue was dealing with electricity market reforms. As China is undertaking a comprehensive electricity market reform that aims to establish an efficient, competition-based market model and a market-based electricity price-setting mechanism, Chinese partners were interested in German experiences. In a meeting of Rainer Baake, State Secretary in the Federal Ministry for Economic Affairs and Energy, with his colleague, Hu Zucai, Vice Chairman of the National Development and Reform Commission (NDRC) of the People's Republic of China, further cooperation in issues on electricity market designs were decided (Federal Ministry for Economic Affairs and Energy 2017). The Chinese government is interested in the German Energiewende and keen on learning about best practices and sorting out possibilities for investment, such as the investment of the China State Grid Corporation in the Desertec Industrial Initiative in 2013. The China State Grid Corporation is very active in approaching all German research institutes and universities dealing with the Energiewende. At the same time, the two countries are competitors on solar panels and RE technologies (Westphal 2015: 95f; Knodt et al. 2017: 152).

Energy cooperation between the UK and China is relatively new, with the first UK-China high-level energy dialogue meeting held only in 2010. The focus in the bilateral relations is on low carbon energy, including among other components exchanges of technologies, low carbon development, and emissions trading and services (Zhang 2015). This focus is intended to secure the supply of energy, to prevent climate change and to boost business cooperation and investment opportunities in low carbon technologies (Kim 2015: 103–104). In this field, the transfer of carbon capture and storage (CCS) technology is considered the first example of successful cooperation between China and the UK. Furthermore, the UK's own energy reforms, namely, moving from coal to gas, are of interest to China due to China's own high reliance on coal in its energy mix (discussed above). The UK also supports China in enabling a more eco-friendly coal consumption (ibid.: 104–105). Another area of cooperation is nuclear power development. Here, China actively seeks closer ties with the UK. In October 2015, a strategic investment agreement was signed between China General Nuclear Power Corporation (CGN) and Électricité de France (EdF) (the French electricity giant (owned primarily by the French government) that owns and operates almost all nuclear power plants in the UK). This agreement was part of Xi Jinping's agenda during his visit to the UK. This agreement will provide certain financial support to new British nuclear power plants, e.g., at the nuclear power station Hinkley Point. Such participation of a Chinese state-owned company taking stakes in nuclear power plants in Europe is a very new and predictably controversial phenomenon. On the one hand, the participation enables future UK developments in domestic nuclear power production and boosts economic opportunities for China in this field. On the other hand, it has triggered many critical voices concerned with security issues. Here, an open investment environment appears to clash with national security interests (Nakano 2015). In 2016, Amec Foster Wheeler, a British engineering company, and China National Nuclear Corporation (CNNC), China's other major nuclear power company, agreed on mutual cooperation (both are known to address high temperature reactors) (World Nuclear News 2016). A third central area of China-UK collaboration is climate change and renewable energies. In 2013, the Offshore UK-China Wind Cooperation project was launched. Its aim is to create an offshore wind energy industry able to compete with other energy producing industries. Thus far, the costs tend to be higher than the expected benefits (Kim 2015: 105). These concrete measurements of closer cooperation between the two actors are argued to lead to a rather positive image of the UK as an energy partner in China. This image contrasts with the past; previously, the UK was viewed as an actor lacking concrete actions in its energy interactions with China (Kim 2015: 106, Knodt et al. 2017: 152f).

Rather predictably, energy cooperation between France and China focusses primarily on matters surrounding nuclear energy. Most recently, in May 2016, the French Foreign Minister Jean-Marc Ayrault and the Chinese Premier Li Keqiang confirmed their commitment to closer cooperation between China and France, including cooperation in the field of nuclear energy (Xinhua 2016). The agreement to cooperate on the peaceful use of nuclear energy is not new, however—it dates back to an agreement between the two countries in 1982, which was the first agreement ever for China in this field (Yi 2014). Since then, closer ties have been established in this area, with the agreement in 2010 initiating an encompassing partnership covering all stages of the nuclear fuel cycle, followed by a joint statement on civil nuclear energy cooperation in 2015. This cooperation is a result of the French industry's strong and ongoing involvement in the Chinese nuclear energy power sector. Over the last decade, French nuclear companies such as Areva, Alstom and EDF have been key actors in China's construction of nuclear power plants. Importantly, one of the pillars in China-France energy cooperation is nuclear safety (France Diplomatie 2016). China and France also cooperate in the field of climate change; they launched a joint presidential statement on climate change (it was agreed upon in the margins of the climate summit in Paris in 2015). Finally, the French Development Agency has been involved with China since 2004 through the framework of a partnership with China's Ministry of Finance and National Development and Reform Commission (NDRC). The aim of this engagement is to support China's transition to a low-carbon and environment-friendly economy. In total, 24 projects have been implemented since 2004 (France Diplomatie 2016). During Hollande's first visit to Beijing in 2013, China and France agreed on bilateral energy cooperation in the areas of urbanization, industrial energy saving and renewables (Yi 2014; Knodt et al. 2017: 153f).

8.6 EU-India Energy Dialogue

Until signing a Strategic Partnership Agreement with India in 2004, the EU's focus in Asia was primarily on China to the neglect of acknowledging the potential of the rising Indian subcontinent. As the EU classified all South Asian countries as "non-association" partners, a previous already negotiated India-EU Association Agreement couldn't be realized (Abhyankar 2009, cited in Joshi and Gansehsan 2015: 153). Nevertheless, India and the EU had already cooperated in an institutionalized way in the area of trade and signed two Commercial Cooperation Agreements in 1980. In the 1990s, the relations between the two partners were broadened by several political agreements, such as the political cooperation statement (1993), a Joint Political Statement (1994) and a Cooperation Agreement (1994) (Joshi and Gansehsan 2015: 153; Knodt et al. 2017: 185). Explanations of the sluggish moving forward in cooperation have to be seen in the colonial and post-independence of India and a long-time disinterest of the EU. The Indian government for long time did not perceive signals from the EU side on a positive note due to its colonial past (Jaffrelot 2006). In addition, only in the 1990s has the EU become more aware of India being a strategic partner. The Europeans increased their efforts to keep the focus of their diplomatic effort on not only China but also India. Thus, the EU enhanced cooperation with India in several sectors, among others in energy and climate issues (Jaffrelot 2006; Bava 2008; Noronha and Sharma 2009; Knodt et al. 2019: 16). As part of the Strategic Partnership Agreement between the EU and India 2004, an energy dialogue was initiated as one of the sectoral dialogues. Its key priorities are the following: development of clean coal technologies, increasing energy efficiency and savings, promoting environment-friendly energies and assisting India in energy market reforms.[12] Since 2006, the energy dialogue panel and its working groups have held regular meetings. Working groups meet on four different thematic areas

[12]https://ec.europa.eu/energy/topics/international-cooperation/key-partner-countries-and-regions/india_en (accessed 30-Mar-2020).

and report back to the panel. The four working groups are (1) Coal/Clean Coal Technology, (2) Renewable Energy/Energy Efficiency, (3) Petroleum/Natural Gas and (4) Fusion/ITER. Special attention is given by the coal working group to the prominence of coal as the main source of energy production in India. At first glance, this focus might not be consistent with the EU's priorities on climate change and renewable energies, but interviews in New Delhi (September 2013) revealed the rationale behind this engagement; coal will remain the largest energy source in India in the near future; it is thus better to engage on clean coal technologies than only in sectors that might have minimal influence on India's overall energy situation. A greater impact can therefore be achieved by cleaning up coal, which is also where the greatest Indian interest lies (Piefer et al. 2015a: 47f; Knodt et al. 2019: 16).

In contrast, despite a strong sustainability rhetoric from both partners, India and the EU within the dialogue, the Working Group on Renewable Energy and Energy Efficiency has rarely met. Among the rare outputs was the Joint Work Program on Energy, Clean Development and Climate Change from 2008. Overall, the focus of the group was on a large range of areas of cooperation such as sharing experiences on running public awareness campaigns, standard setting for energy-using products, developing technologies on energy efficiency and establishing efficiency standards for buildings, photovoltaics, solar thermal, wind energy, biomass combustion and gasification, and biofuels. However, the outputs of this Working Group remain limited. An additional explanation for this underperformance might be that the topic of renewable energy within the cooperation is primarily situated in the development cooperation portfolio, which is phasing out within this budget period. Moreover, as in all the EU-BIC cooperation, Member State cooperation, e.g., between India and Germany or Spain, played an important role.

Meetings of the EU-India Energy Panel and its working groups have a strong business orientation, and the working groups are instrumental in paving the way for cooperation between European and Indian energy firms. At the 6th EU-India Energy Panel, the decision was made to extend the dialogue to include regular exchanges on energy security matters and to enhance the security of supply and the coordination of positions in international organizations and initiatives on energy. This strategic dialogue deals with EU-India coordination of relations with Russia, Turkmenistan and other Central Asian countries. At the EU-India Summit in February 2012, the energy agenda was refocused to give it a new kickstart with seven new priority areas, which have been embedded in the Joint Declaration for Enhanced Cooperation on Energy; it reiterated the importance of the development and deployment strategies for clean energy production, inter alia clean coal technologies and advanced coal mining. Both partners agreed to improve the energy efficiency of products and energy efficiency in the building sector. Cooperation on renewable energy sources and the development of smart grids was decided, as was the focus on energy safety (nuclear and off-shore drilling safety). The EU and India endeavor to advance in developing fusion energy as a future sustainable energy source (Piefer et al. 2015a: 48). However, personnel constraints—in the sense of limited and frequently changing staff—on both sides and the phasing-out of development cooperation with India are obstacles in this area. In the future, these efforts will be incorporated under the Energy Panel, but the question of how to replace budgets that are no longer available remains. Interviewed stakeholders hoped that the new Partnership Instrument would provide bridge funds (Piefer et al. 2015a: 48), because development is now addressed by the new Clean Energy and Climate Partnership agreement (Knodt et al. 2017: 188).

Recently, the cooperation became a new impetus through the success of COP 21 in December 2015. At the 2016 EU-India Summit in Brussels, the EU and India agreed on a 'Joint Declaration between the EU and India on a Clean Energy and Climate Partnership' on the promotion of clean energy generation and increased energy efficiency for climate action. The aim is to continue dialogue and cooperation on clean energy, energy efficiency and climate action (European Council 2016b). The partners underlined the positive contribution that clean energy generation and increased energy efficiency can make to global energy security (European Council 2016a: 1; Knodt et al. 2017: 187). Miguel Arias Cañete, European Commissioner for climate action and energy, noted that India, as a major world player, has to be seen as a crucial partner for the EU on energy and climate matters (European Commission 2016d). In the declaration, India and the EU perceive each other as "global partners and the world's largest democracies" and "reaffirmed their commitment to strengthen the EU-India Strategic Partnership based on shared values and principles" (European Council 2016a: 1). This new effort is clearly linked to the 2015 Paris Agreement. The EU's hopes for a renewed climate dialogue with India include reinforced energy cooperation, primarily on renewable energy sources, and the promotion of clean energy generation and increased energy efficiency (Knodt et al. 2017: 181, 2019: 17). The 14th India-EU summit in New Delhi 2017 stressed this effort. The partners at the summit 'welcomed the establishment of the South Asian Regional Representative Office of the European Investment Bank (EIB) in India and noted that its investments, especially in urban mobility and renewable energy projects, will support India-EU collaboration on the Climate Agenda' (European Council 2017: 7). Partners adopted a Joint Statement on Clean Energy and Climate Change, reaffirming their commitments under the 2015 Paris Agreement. They acknowledged the progress on the Clean Energy and Climate Partnership, adopted at the 2016 EU-India Summit. In relation to energy and climate change, the partners highlighted their interest in reducing primary resource consumption and enhanced the use of secondary raw materials. Both sides 'welcomed the contribution of the International Resource Panel, the Indian Ministry of Environment, Forests and Climate Change (through the Indian Resource Panel) and of the National Institution for Transforming India (NITI Aayog) to developing strategies for this crucial economic transition. Both sides agreed that the newly established G20 Resource Efficiency Dialogue will be an ideal platform for knowledge exchange and to jointly promote resource efficiency at a global level' (European Council 2017: 8).

As in the case of the EU-China energy cooperation, a network analysis was carried out in the case of the EU-India energy dialogue. However, in the case of India, only the aggregated data of the importance and exchange networks can be presented due to the request of the Indian respondents.[13] Thus, qualitative data and background

[13]For a methodological explanation of the networks, see Subchapter on EU-China energy dialogue.

information will be used, but institutions will not be named or shown in network graphic or tables. If we look at the overall network data, the Indian case is not very dense and central—density of the importance network, 0.14; density of the exchange network, 0.10; centrality of the importance network, 15.4; and centrality in the importance, 13.7. These outcomes reflect the fragmentation of the Indian governance system. No European actor is listed among the five most important actors within the importance network again, and only DG Energy is listed as the second-most important actor. As brokers in the exchange network, the EU Delegation in India and DG Energy are rated among the five most important brokers. The EU Delegation was even perceived as the most important broker. If the network data are aggregated, the exchange among actors of different groups is less than in the other dialogues. The Indian public and non-state actors engage in regular reciprocal exchange and EU non-state actors mostly address the EU. The exchange between the Indian public sector and EU non-state actors as well as between Indian non-state actors and EU public actors is very low and thus, under the average. Additionally, the EU only addresses the Indian public actors, but not vice versa. The interaction within the importance network seems more balanced, but still the EU public regard the Indian public actors to be important but not vice versa. Significantly, all actor groups regard the Indian public actors as important and try to gain access to them (Piefer et al. 2015a: 48–50; Knodt et al. 2019: 19) (Fig. 8.12).

The overall assessment of the EU-India dialogue through interviews in New Delhi (September 2013), showed the dialogue to resemble a talk-shop. Compared to the Indo-German Energy Forum, which is a dialogue platform where joint projects are

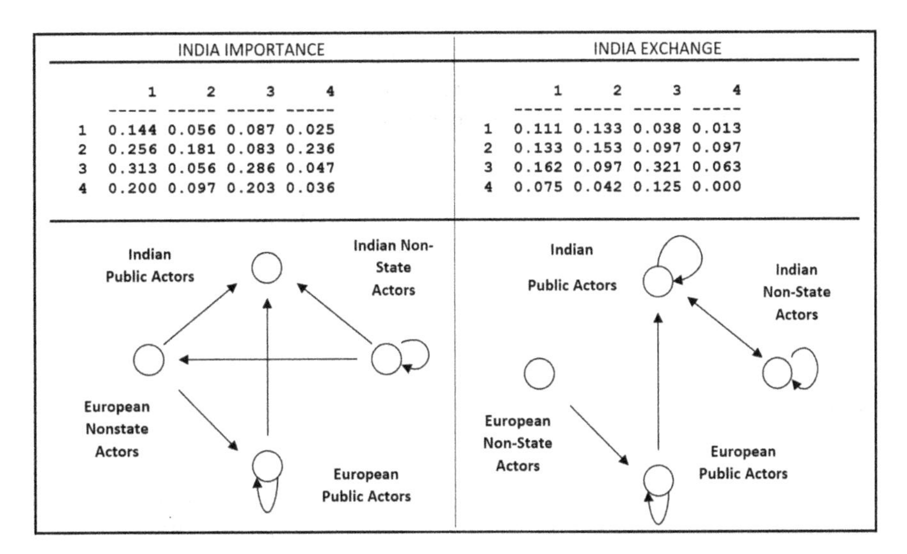

Fig. 8.12 Importance and exchange network block matrices (India-EU), 1 = Indian public actors, 2 = Indian non-state actors, 3 = EU public, 4 = EU non-state (EnergyGov 2014; Piefer et al. 2015a: 49)

being planned and implemented, the lack of concrete cooperation projects was criticized. When asked why there are no joint projects, as in the China dialogue, respondents from both sides gave the same answer—the EU expects concrete inputs on what India would want, and India expects the EU to come up with concrete proposals of what it could offer. Thus, it seems as though discussions are going in a circle, and due to personnel constraints on both sides to follow-up and engage in discussions with various stakeholders, no concrete proposals materialize. For the energy dialogue to remain relevant, this blockage needs to be overcome. The interviewees called for a jointly financed India-EU Energy Policy Centre (September 2013). One step towards this idea could be seen in the portfolio of the regional office of the European Investment Bank (EIB) opened in New Delhi in March 2017, which is financing, e.g., joint solar power parks. Overall, for such joint projects and cooperation, the EU and India have to cope with horizontal or dispersed ministerial structures in India and the complex structure of European Directorates General—and vertical coordination problems—such as EU coordination with their 28 Member States or Indian state coordination. As in all EU-BIC energy dialogues, the value added of the EU-BIC cooperation in comparison to its Member States is not obvious for the actors involved but is asked for by Indian interviewees (Piefer et al. 2015a: 50; Knodt et al. 2019: 20).

There are tensions between the EU and India concerning closer economic cooperation, and the long-lasting and still ongoing negotiations on a bilateral free trade agreement between the EU and India might also impact energy cooperation. While most of the controversies between the two actors deal with other major issues, energy policies are also affected because 'green energy'/'clean energy' is part of the FTA negotiations and would be covered in a specific section of the FTA. In effect, the FTA should cover the elimination of tariff and non-tariff barriers to clean energy technologies as part of a general liberalization strategy for environmental services, which would for example affect investment conditions in the Indian energy sector. Here, FDI regulations do not allow 100% foreign capital but rather are restricted to only 74% (Sawhney 2008: 19). Furthermore, there is much pressure, especially from the EU and specifically Germany, to install a regime on intellectual property rights that would go even beyond the WTO's TRIPS agreement (The Hindu 2013a, b). As long as the EU-India FTA negotiations are experiencing a series of standstills due to different liberalization schedules and the long-standing dispute on investment conditions and IPR, this difficulty will thwart agreements in the energy sector, might hinder more substantial cooperation, and would also affect questions of technology transfer, as far as IPR issues, such as licensing and patenting are affected (Piefer et al. 2015a: 50).

The analysis of the TU Darmstadt data also included evaluations on the perceived roles and characteristics of each other within the dialogue. The analysis of elite interviews and the survey aimed to identify properties assigned to the EU and India as dialogue partners by stakeholders involved in EU-India energy dialogue. As already observed in the Chinese case, differences between the perceptions of the EU and India could be found. In the case of India as a dialogue partner, Indian respondents considered India acting with an open agenda in this dialogue, whereas the EU actors

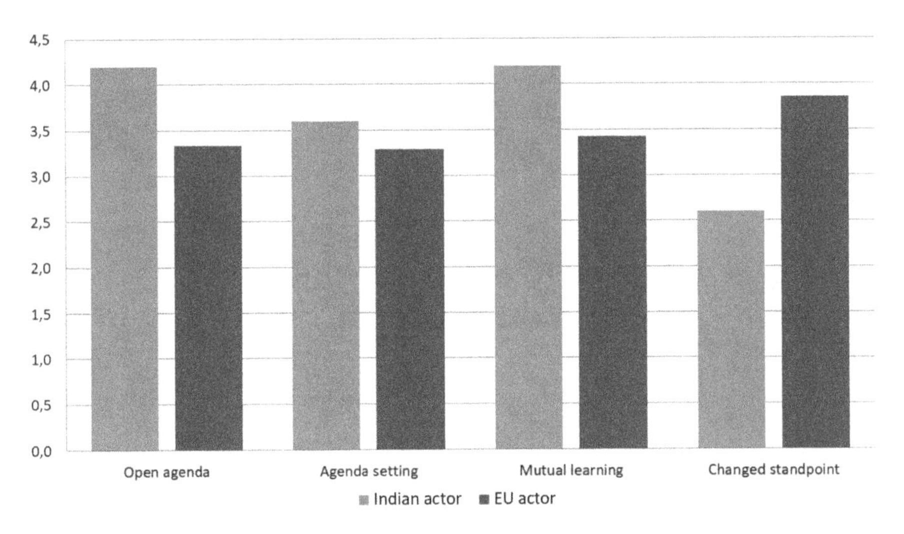

Fig. 8.13 Perception of India's properties as a dialogue partner in the EU-India energy dialogue, means of all answers; 1 = "Strongly disagree"; 5 = "Strongly agree" (EnergyGov 2014)

perceived India as acting with a hidden agenda. The same pattern can be seen in the case of mutual learning. Whereas Indian respondents show India being open for mutual learning, European actors perceived the Indian side to be less willing to engage in learning processes. EU respondents but not Indians perceive Indian actors as not insisting on their point of view but as changing their stand (Knodt et al. 2017: 209, 2019: 20, 22).

Concerning the EU as cooperation partner, European actors saw themselves as having more influence on the agenda setting in contrast to how Indian stakeholders perceived the situation (Fig. 8.13). Indian actors credited the EU with being more open for mutual learning than did the EU actors themselves. Concerning the open agenda, the perception of the EU's properties as a dialogue partner show the same mistrust as in the Chinese case. Thus, the Indian respondents are more reluctant to perceive the EU as having an open agenda than are EU actors (Knodt et al. 2017: 209f, 2019: 22) (Fig. 8.14).

The analysis of the mutual perceptions leads to the conclusion that each side has a rather negative perception of the other side, which potentially could disturb fruitful cooperation between the two partners (Knodt et al. 2017: 210, 2019: 22).

Within the EU-27, Germany plays a crucial role for India. In 2019 it has been at first place in imports from India and first place in exports of the European Union towards India.[14] At the heart of the German-India energy governance, the Indo-German Energy Forum was set up. It was regarded by our Indian interview partners as a success model (Interviews EnergyGov in New Delhi, September 2013). In 2006, the Indian Prime Minister Manmohan Singh and the German Chancellor Angela

[14]https://ec.europa.eu/eurostat/statistics-explained/index.php/India-EU_%E2%80%93_international_trade_in_goods_statistics#Trade_with_India_by_Member_State (accessed 30-Mar-2020).

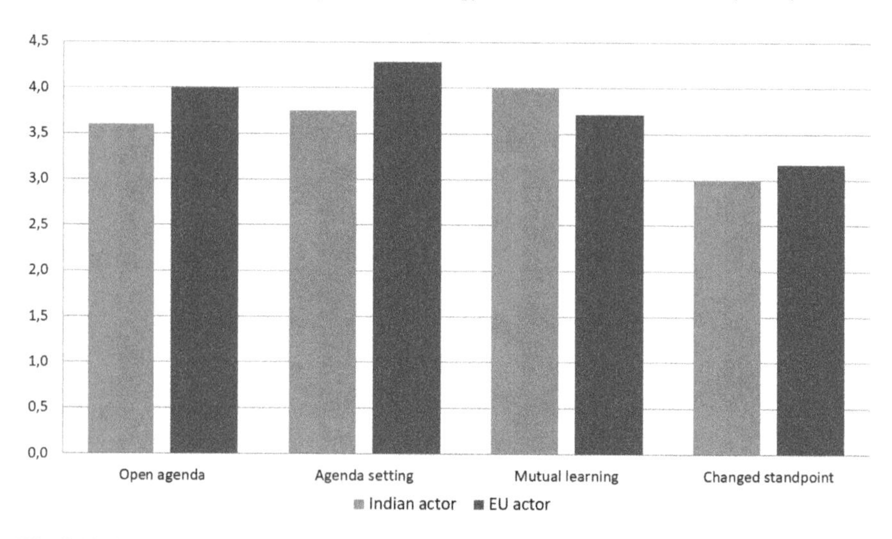

Fig. 8.14 Perception of the EU's properties as a dialogue partner in the EU-India energy dialogue, means of all answers; 1 = "Strongly disagree"; 5 = "Strongly agree" (EnergyGov 2014; Knodt et al. 2017: 222)

Merkel agreed on creating the Indo-German Energy Forum with the objective to enhance cooperation in the fields of energy security, energy efficiency, renewable energy and investment in energy projects. The current planning and implementation takes place on the implementation level between Indian Ministries and GIZ. Participants in the dialogue are the Indian and the German Governments, public institutions and the private sector. When the Forum started in 2006, the already since 2003 existing bilateral Indo-German Energy Program (IGEN), which supported the implementation of the Indian Energy Conservation Act from 2001, served as an example for the improvements that can be achieved through the cooperation of both countries. It is carried out by GIZ on behalf of the German Federal Ministry for Economic Cooperation and Development (BMZ), coordinating activities with KfW Development Bank and working together with the Indian partners. In 2008, another bilateral program was created aiming at supporting concrete cooperation projects that emerge from the dialogue. The connection is two-sided: On the one hand, the bilateral cooperation projects provide inputs on problem constellations and possible solutions on the ground to the dialogue. On the other hand, the dialogue results are transferred to the bilateral programs to promote their implementation. The results so far achieved seem to confirm the adequateness of this approach. One project can be seen in the Indo German Solar Energy Partnership to support and intensify the energy transformation in India towards the enhanced use of solar energy. The partnership was signed in 2015 by the partners. In the same year, in October 2015, the partners agreed to continue cooperation towards the common aim of developing climate-friendly, efficient and sustainable solutions for India's expanding energy needs and other areas of sustainable development under the heading of the Indo-German Climate and Renewables Alliance. That alliance is constructed as an overarching alliance between India and

Germany with the objective to give recognition to ongoing collaboration of various stakeholders on energy and climate change and to enhance cooperation and synergies in these fields (Government of India, Ministry of External Affairs 2017; Knodt et al. 2017: 191).

The UK and India also cooperate in the energy field; however, energy was initially discussed in a general context of bilateral relations with a focus on technology transfer and investments. Energy was not treated as a separate area of cooperation, a change that occurred due to the civilian nuclear accord signed in 2010 to enhance nuclear energy cooperation (Kim 2015: 106). After the UK-India Summit in November 2015, bilateral energy relations were strengthened and intensified through a 'Joint Statement on Energy and Climate Change' (Department of Energy and Climate Change 2015). This document evinces the willingness for closer cooperation in the field of climate change and shared commitments ahead of the then-upcoming UN Climate Summit in Paris (ibid.: 2). Tellingly, in Paris, the UK-India climate and energy partnership was showcased at the Indian Pavilion. The then British Prime Minister Cameron and Prime Minister of India Modi agreed on regular meetings at the ministerial level to advance current cooperation projects and explore the potential for cooperation. Both actors signed a Memorandum of Understanding to enhance cooperation on electricity market reform, energy efficiency, offshore wind energy, solar, smart grids, energy storage and off-grid renewable energy services (ibid.). The energy market in India is of particular concern to the British private sector because the Indian government has set a price ceiling for energy (particularly on natural gas), a move that hinders UK companies in turning a profit but protects consumers (Kim 2015: 107). In the Joint Statement, these aspects are to be combined to "promote energy security and economic growth and deliver a cost-effective low carbon transition" (Department of Energy and Climate Change 2015: 4). It remains to be seen whether the announced reforms of the Indian gas market will boost UK investments. At the ministerial level, it is expected that UK-India energy relations will receive less attention in the aftermath of Brexit and Cameron's resignation (Knodt et al. 2017: 191f).

Nuclear energy is at the core of the energy relations between India and France, with France being the first country to sign a civil nuclear agreement with India (2008). In 2009, a Memorandum of Understanding was signed between the two partners to pave the way for six nuclear reactor units in Jaitapur (in progress) (Kumar 2013). France's expertise in this field makes it an attractive partner for India. Conversely, France expects significant economic benefits and to gain a global ally in the nuclear power energy field from this partnership. France strongly supports India's efforts to become part of the global export control regime. To both actors, nuclear energy is of great importance for ensuring energy security and combating climate change. Following the UN Climate Summit in Paris, India and France sought to implement the Paris Agreement through cooperation in sustainable development and energy, covering governments, regions, cities and companies. This approach was emphasized in their Joint Statement from January 2016 during Hollande's visit to New Delhi during the 67th Republic Day celebrations. Both partners are also members of the International Solar Alliance (ISA) (2015), seeking to enhance the use of solar energy on a global scale and cooperating on joint research, development, financing and technical innovation. French companies are heavily engaged in not only the field of nuclear energy but also the area of renewable energy. Additionally, the French Development Agency (AFD) is a strong supporter of India's development of cleaner and more efficient energy (e.g., through technical assistance in the establishment of Smart Cities in India). Although France and Germany continue to provide financial development assistance to India, such development cooperation was phased out in the UK in 2015. As discussed above, the EU's development cooperation with India is also being phased out, and renewable energies are no longer being financed through DG DevCo (Piefer et al. 2015a: 48; Knodt et al. 2017: 192f).

8.7 EU-Brazil Energy Dialogue

Brazil and the EU began exchanging formal diplomatic missions in the 1960s. Initially, cooperation was based on strictly commercial ties, and the bi-regional agreement with Mercosur was the cornerstone of EU-Brazil relations until the initiation of the Strategic Partnership in 2007. The partnership builds on regularly held bilateral summits including sectoral dialogues. The energy dialogue focusses on energy supply, renewable energies, energy efficiency, technology and infrastructure (European Commission 2007). The EU-Brazil energy dialogue in November 2009 decided to exchange experiences and technical consultations on regulatory issues for competitive energy markets, including investment opportunities, as well as on energy efficiency and demand management (Council of the European Union 2009). Special focus is on second-generation biofuels and the promotion of EU-Brazil industrial cooperation on low carbon technologies such as clean coal and nuclear energy and cooperation on nuclear safety. Furthermore, as new areas, the dialogue was expanded on offshore safety, ocean energy and trilateral cooperation with African countries on bioenergy (Piefer et al. 2015a: 34; Knodt et al. 2019: 7).

The EU-Brazil Dialogue Support Facility claims that the EU and Brazil share much common ground as far as energy is concerned due to their large internal markets with high energy demand and the acknowledgement of the potential of renewable energies by also spearheading standards for countries in other regions. However, it also alludes to the major pitfalls of the dialogue by stating that the dialogue is not a priority for DG Energy, which is more focused on issues related to adjacent countries in the European neighborhood and on issues related to energy supply security, especially in the cases of cross-border gas transport (EU-Brazil Dialogue Support Facility Homepage, 2013). Biofuels can be regarded as the main topic and at the same time main controversy of the dialogues; this notion is consistent with the great priority that Brazil attributes to biofuels in its external energy relations. Brazil is aiming for its second-generation biofuels to be certified according to EU standards to have better market access opportunities not only in Europe but also in Africa and Asia. Reluctance to the full certification process had been observed from the EU side, so that Brazil aimed for an equivalence agreement (Interviews Brasilia,

December 2013). In the latest Joint Action Plan (2011) as well as in the declaration of the EU-Brazil Summit in 2013, both envisage joint trilateral cooperation projects in the bioenergy sector in Africa. A first feasibility study has been conducted for cooperation with Mozambique, but the project was suspended due to EU concerns in terms of biofuel standards. Furthermore, the EU side criticized that the feasibility study was conducted under the lens of private sector interests. It was financed by Brazil and jointly conducted by the Getulio Vargas Foundation and Vale, one of the largest Brazilian private companies aiming at expanding its investments and projects on the African continent. Thus, the value added for Brazil of cooperating with the EU engaging in the energy dialogue and joint trilateral cooperation projects was to implicitly have its standards certified. In the last summit statement of 2014, reference was only made to support for trilateral cooperation of EU Member States with Brazil (Piefer et al. 2015a: 34f).

To analyze the interaction, importance and exchange of the actors involved in the EU-Brazil energy dialogue again, a network analysis will be run.

Compared to the other networks analyzed above, the Brazil-EU energy dialogue network is the most dense and central one (density, 0.29 in the importance and 0.14 in the exchange networks; centrality of 25.8 in the importance network). The most important actors (see Fig. 8.15) are the Ministry of Mines and Energy (MME) and the Ministry of External Relations (MRE). These results are not surprising as the Department of Energy of the MRE is formally responsible for energy issues within

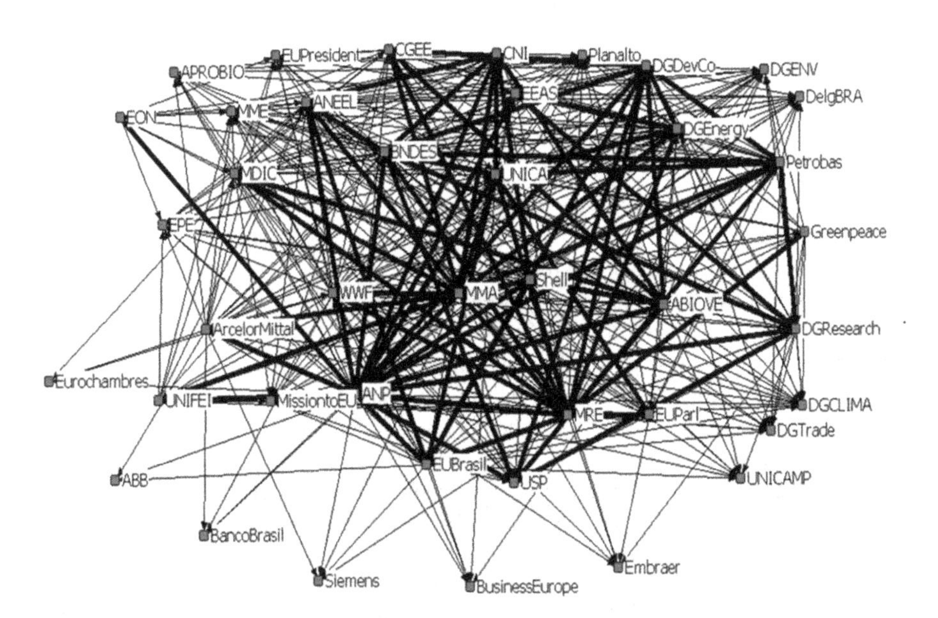

Fig. 8.15 Importance network for Brazil-EU energy cooperation, importance data: unilateral (thin ties) and reciprocal relations (bold ties) (EnergyGov 2014; Piefer et al. 2015a: 35; for actors please see Table B.3 in Appendix B)

Table 8.3 Top five actors in Brazil-EU energy dialogue (EnergyGov 2014; Piefer et al. 2015a: 36)

Importance network Five most important actors (indegree centrality)	Exchange network Five most important actors (indegree centrality)	Broker function (exchange) Five most important actors (betweenness centrality)
MME 53.7	PETROBRAS 27.5	MDIC 14.6
MRE 51.2	MDIC 25.0	UNICA 8.5
UNICA 48.8	UNICA 25.0	MRE 5.6
DGEnergy 46.3	MRE 25.0	ANEEL 5.1
Planalto 46.3	MME 22.5	PETROBRAS 3.6

Brazil and—in cooperation with the MME—the dialogue. The latter has an interesting position in the institutional set-up, being responsible for mines and energy. In our qualitative interviews in Brasilia (December 2013), some respondents alluded to the distinction of MME rather focusing on mining, that is, resource extraction and the exploitation of energy sources, and thus a more fossil fuel-oriented agenda. Besides, MME and MRE, the Ministry for Development, Industry and Commerce (Ministério do Desenvolvimento, Indústria e Comércio Exterior—MDIC), the Sugarcane Producers' Union (UNICA), Planalto/the Brazilian Presidency and PETROBRAS are among the top five of the importance and exchange networks (Table 8.3) (Knodt et al. 2019: 11). Although formally not the main responsible partner, the Ministry of Environment (MMA) is also one of the key actors in the Brazil-EU energy dialogue on issues of renewable energies and energy efficiency. MMA ranks 6th in the betweenness centrality and is thus also one of the main actors. The first energy-related issues in Brazil-EU cooperation were initiated by MMA within the scope of the dialogue on the environmental Dimension of Sustainable Development (Piefer et al. 2015a: 35f).

Table 8.3 clearly shows that the organization with highest indegree centrality values in the EU-Brazilian exchange network on energy issues is PETROBRAS, closely followed by MDIC, UNICA, and MRE. In other words, these four organizations are chosen by at least one-quarter of the respondents as important exchange partners. MME also shows up as forming part of the five most important organizations in terms of exchange, although this organization's position turns out to be much less pronounced. As already within the other dialogues, nearly no EU actor appeared upon the top five actors in the network, except DG Energy. In the Brazilian case, DG Energy could be found as the only EU actor among the five most important actors in the importance network (not in the exchange network). The External Action Service (EEAS) and DG Trade are also near the center of the network. In this result, the Brazilian case is unusual but could be explained by the development of the EU Brazil relationship. Before the Strategic Partnership was signed in 2007, the focus of the relationship was on trade diplomacy within the scope of the EU-Mercosur dialogue. Interestingly, DG Development and Cooperation (DevCo) and the Delegation of the European Union to Brazil (DelBRA) are regarded as important and are involved in some regular exchanges, but are not as central as the other three EU actors. The importance of DG DevCo results from the fact, that until the phasing out of bilateral development cooperation of the EU and Brazil, all energy-related activities were financed by DG DevCo and not by DG Energy. The coordination between DG DevCo and DG Energy at times was suboptimal, as interviewees in Brasilia (December 2013) explained (Knodt et al. 2019: 11).

Important roles in the EU-Brazil network are taken by PETROBRAS, as the country's largest state-owned energy company, and the Sugarcane Producers' Union (UNICA), with many other dialogue participants regularly exchanging information, which, due to the strong focus on biofuels in the Brazil-EU energy dialogue, is not surprising. The important position of those actors and of the Brazilian Confederation of National Industries (CNI) can be explained by the role the private sector plays in most issues of the dialogue, e.g., on biofuels, renewables, and offshore safety. In addition, UNICA and PETROBRAS play a role as brokers within the exchange network (Piefer et al. 2015a: 37). Regardless, the most important broker has to be seen in the Ministry for Development, Industry and Commerce (Ministério do Desenvolvimento, Indústria e Comércio Exterior—MDIC), which is interesting, as it has not been one of the most important actors but is rather situated in a position of being responsible for cross-cutting issues dealing with energy and foreign investment. The Ministry of Planning was not included in our network analysis; otherwise, it might have been identified as a broker because it is responsible for the Dialogue Support Facility, which is financed from the development cooperation budget of DG DevCo and aims at enhancing all sectoral dialogues that are currently stagnating (Piefer et al. 2015a: 37; Knodt et al. 2017: 115, 2019: 12).

A glimpse at the aggregated data of network block matrices (Fig. 8.16) on importance and exchange reveals interesting observations about the EU's role and behavior in energy cooperation with Brazil. Four categories of organizations (domestic public, domestic non-state, and EU public, EU non-state) maintain relatively dense relationships among each other in terms of the exchange of strategically important information, but this level of relationship does not exist for non-state EU-level organizations. Brazilian domestic actors are the most active in the network, possibly due to the specific time of our analysis. As mentioned previously, Brazilian actors attempted to engage with the EU on the certification of Brazil's second generation of biofuels according to EU standards in the years 2012 and 2013. Thus, public EU actors are addressed by Brazilian public actors, which is the exemption in the three cases. It seems that Brazilian public and non-state actors regard the EU public actors to be important and exchange information with them. Generally, and beyond this special time frame, interviewees confirmed that the EU is not the most important partner for Brazil (Interviews Brasilia, December 2013). As in the other dialogues, the bilateral relations with EU Member States have always been more important, a fact mirrored in impressions of the EU being not very visible among Brazilian actors and even some EU Member States representatives in Brasilia. EU public actors cooperate most with other public EU actors and do not interact with the outside to a greater extend. As Brazilian energy companies are partially state-owned and the others also maintain close linkages to Ministries, the network shows a dense reciprocal relationship between Brazilian public and non-state actors. Thus, it will be interesting to

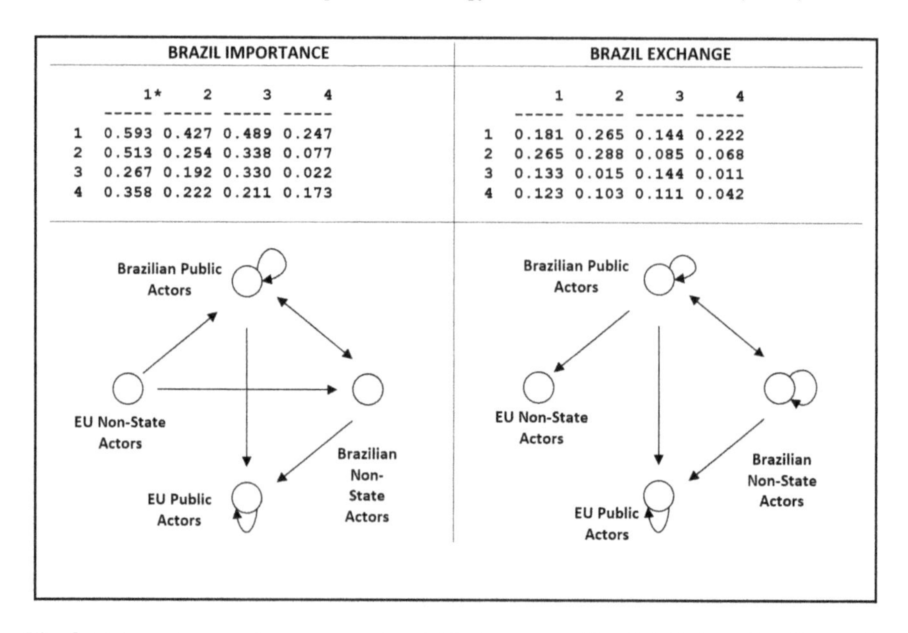

BRAZIL IMPORTANCE

	1*	2	3	4
1	0.593	0.427	0.489	0.247
2	0.513	0.254	0.338	0.077
3	0.267	0.192	0.330	0.022
4	0.358	0.222	0.211	0.173

BRAZIL EXCHANGE

	1	2	3	4
1	0.181	0.265	0.144	0.222
2	0.265	0.288	0.085	0.068
3	0.133	0.015	0.144	0.011
4	0.123	0.103	0.111	0.042

Fig. 8.16 Importance and exchange network block matrices (Brazil-EU), 1 = Brazilian public actors, 2 = Brazilian non-state actors, 3 = EU public, 4 = EU non-state (EnergyGov 2014; Piefer et al. 2015a: 38)

look for further explanations for the EU's role in the network and for opportunities to improve cooperation and mutual understanding (Piefer et al. 2015a: 38f; Knodt et al. 2017: 116, 2019: 12).

The overall assessment of the EU-Brazilian energy dialogue, gained through the network analyses as well as interviews and survey data, shows the relationship is performing under its potential. One important factor for this shortfall might be the mutual perception of Brazil and the EU within the EU-Brazilian energy dialogue. Overall, Brazil is a very self-confident cooperation partner for the EU. The country has an internal market almost the size of the EU. In addition, in the 2010s, it followed a clear strategy of 'BRICSalization' (Gratius 2012) and South-South cooperation that comes with a reorientation away from the US and EU. The country's growing role in international as well as South-South arenas and its status as a regional power brought a very positive self-image of Brazil's role as a dialogue partner in energy cooperation concerning Brazilian actors and an even more positive image when we only look at Brazilian public actors (Fig. 8.17). Brazil has always had a clear vision of how the energy dialogue can serve broader national interests. The EU on the other hand seems to lack such a clear strategy but continuously showed that the country still represents one of its closest strategic partners for the EU in terms of normative orientation and foreign policy goals. Referring to the perception of the actors' properties as dialogue partners, EU actors perceived Brazil as more of an agenda-setter than Brazil perceived itself, which might be a result of Brazil's image as a self-confident, interest-driven actor. Both groups perceived Brazil as acting with an open agenda. Interestingly,

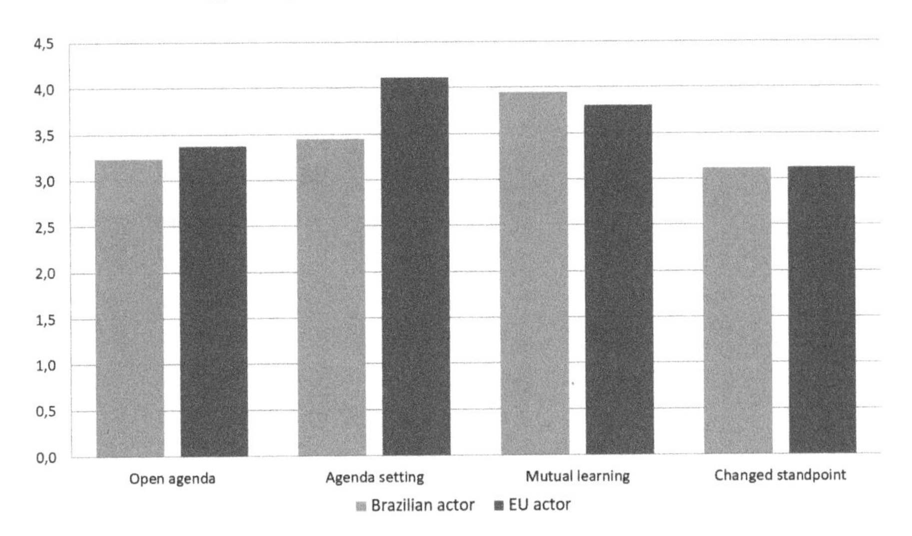

Fig. 8.17 Perception of Brazil's properties as a dialogue partner; means of all answers; 1 = "Strongly disagree"; 5 = "Strongly agree" (EnergyGov 2014)

this perception doesn't hold true for the perception of the EU as a dialogue partner. Brazilian actors perceived the EU as not playing with an open agenda, which comes with a certain degree of mistrust from the Brazilian side. Brazil is seen as being open for mutual learning in the dialogue and able to change its standpoint from both sides (Ribas and Schaeffer 2015: 186f; Piefer et al. 2015a: 68; Knodt et al. 2017: 137f, 2019: 13f).

Brazil on the other hand has more negative perceptions of the EU as a dialogue partner (Fig. 8.18), especially concerning the hidden agenda mentioned previously. Here, the EU sees itself as having a rather open agenda. Brazil actors also see the EU as setting the agenda, which is seen as less pronounced by the EU itself. In addition, they interpret the EU as an actor that is open for mutual learning, able to change its standpoint and thus open for compromises. In both cases, the EU actors perceive themselves as not as open for mutual learning (Knodt et al. 2015a: 68). The mistrust that appears in the analyses together with different perceptions might be an obstacle for further cooperation (Knodt et al. 2017: 137, 2019: 14).

Energy relations between Germany and Brazil have changed radically since their first energy agreement in 1975 (on the peaceful use of nuclear energy). Today, renewable energies and energy efficiency take central place in the Strategic Partnership (established in 2008) between the two countries.[15] Several official documents and agreements serve as a basis for their cooperation in the areas of energy and climate. Among these are the German-Brazilian Energy Agreement/German-Brazilian Energy Partnership (2008), the German-Brazilian Memorandum on Climate Change (2009), the Brazilian-German Committee on Climate Change, the

[15] http://www.bmz.de/de/laender_regionen/lateinamerika/brasilien/index.jsp (accessed 30-Mar-2020).

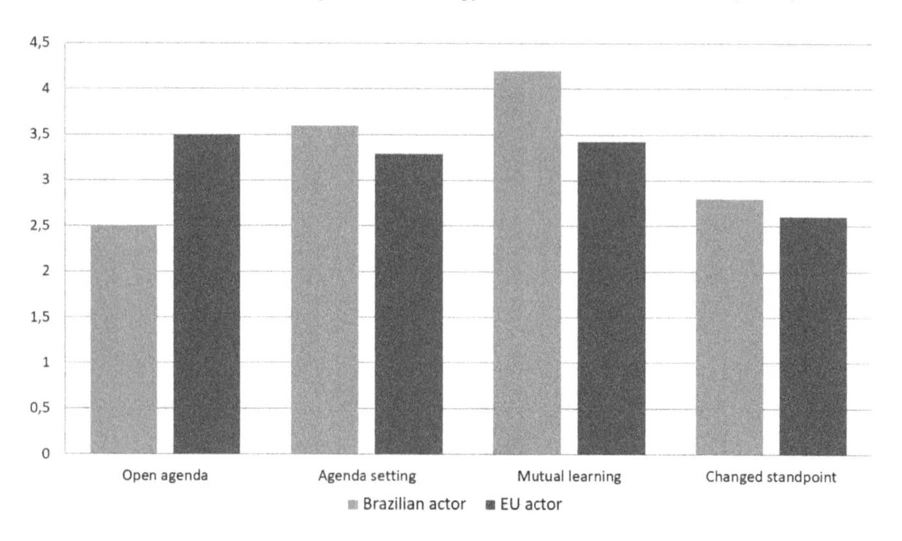

Fig. 8.18 Perception of the EU's properties as a dialogue partner; means of all answers; 1 = "Strongly disagree"; 5 = "Strongly agree" (EnergyGov 2014)

Cooperation for Sustainable Development, and the Brazilian-German Joint Statement on Climate Change signed before the UNFCCC Climate Summit in Paris 2015 (Westphal 2015: 94; BMUD 2015). The German-Brazilian Energy Partnership in 2015 was not yet filled with life, especially not in the case of renewable energies. The working group on bio-fuels showed the differences between both partners. The Germans' wish for the installation of working groups on other renewables failed due to Brazilian resistance, although it was the focus of the German-Brazilian development cooperation. Thus, the German KfW and the Brazilian BNDES are financing the construction of wind parks in Brazil. One reason for the log-in of the bilateral negotiations has to be seen in the different priorities in promoting renewable energies in 2015. While Germany focused on electricity transformation and the development of solar and wind energy, Brazil emphasized hydro, transport and ethanol (Röhrkasten 2015). Nevertheless, both showed a joint commitment to a binding agreement at the then upcoming climate summit and further emphasize bilateral cooperation in fighting deforestation, ensuring low-carbon development and promoting renewable energies (BMUD 2015). Concrete projects and regular meetings indicate substance and mutual commitment to these agreements. Specifically, the Brazilian energy ministry has become more active in this respect. The German Society for International Cooperation (GIZ), the Banking Group Kreditanstalt für Wiederaufbau (KfW), and the International Climate Initiative (IKI) all have projects in Brazil (Westphal 2015: 94). Despite this booming interaction, Brazil and Germany still view sustainable development and renewable energies differently—a difference that is challenging their energy cooperation. Whereas Germany promotes renewable energies to replace fossil fuels, Brazil focusses more strongly on the economic and agricultural development attached to renewable energy sources (Westphal 2015: 94; Knodt et al. 2017: 117).

Brazil's rising importance in global energy flows and its reputation as a frontrunner in biofuel production and an increasingly significant energy exporter make it an attractive energy partner to the UK. The first high-level energy dialogue occurred in 2011, with the then Prime Minister David Cameron confirming the enhancement of this dialogue in 2012 (Kim 2015: 107). Relations between the two are in general viewed as friendly, including their energy relations. The dialogue focusses on sustainable development (e.g., low carbon development, renewables, and sustainable bioenergy). Both sides also aim at stronger commercial relationships. The UK's energy relations with Brazil are thus largely driven by the private sector. This sector has a strong interest in entering the Brazilian energy market, despite difficulties presented by Brazil's local content policy. Conversely, Brazilian energy companies are also interested in accessing the newest energy technology in Europe; e.g., Petrobras has a London office (Kim 2015: 107–108). The UK-Brazil Economic and Financial Dialogue (EDF) in 2015 confirmed its willingness for a stronger economic relationship, including transport and energy issues (HM Treasury and Osborne 2015; Knodt et al. 2017: 117f).

The Strategic Partnership between France and Brazil was established in 2006, followed by an Action Plan in 2008. The main area of cooperation lies in defense and the common border (between Brazil and French Guiana) shared by the two countries. Strong energy relations have thus far not been developed. Predictably, nuclear energy is the main cooperation area. It was a part of the Action Plan, together with sustainable development (Ministry of Foreign Affairs 2016a). In the area of nuclear cooperation with France, transfer of technology, particularly concerning the development of nuclear submarines, is a key interest of Brazil (Vaz 2014: 11). In 2013, the Franco-Brazilian Economic Forum (2013) occurred, focussing on science and technology and on the interests of the private sector (Ministry of Foreign Affairs 2016b; Knodt et al. 2017: 118).

The potential of Brazil as an energy partner for the EU and its Member States is obvious; Brazil is likely to become the fifth largest economy within the next 20 years, and its internal market is almost the size of the EU's. In this context, the EU has acknowledged that Brazil remains one of its closest strategic partners in terms of normative orientation and foreign policy goals. The Strategic Partnership initiated in 2007 can be considered a clear response of the EU to Brazil's economic rise and its increasing political presence on the international scene. However, at the same time, Brazil has been reorienting its foreign policy. Traditionally, Brazil has focused on the US and the EU in its external relations. Today, Brazil's foreign policy stresses its direction towards an increasing "BRICSalization" and South-South cooperation (Gratius 2012; Santander 2015: 186–187). In the energy field, Brazil moved towards being a very self-confident cooperation partner for the EU, with clear visions of how the energy dialogue can serve its own domestic interests in a broader geopolitical context (Piefer et al. 2015a: 39). Worryingly, Brazil has not been mentioned in the new

EU Global Strategy (EEAS 2016), perhaps indicating that the EU is not considering Brazil among its top strategic partners in the world (Knodt et al. 2017: 118).

8.8 Recommendations for Future International Energy Relations

There are two major challenges for future cooperation in energy matters. First, there are challenges in carrying out multilateral cooperation at the international level. Second, there exist challenges to cooperating on a bilateral basis, especially between the EU and emerging powers such as Brazil, China and India.

To address the first point, international Energy Governance is executed in a highly fragmented environment of a highly diversified network of international governmental organizations that deal with the regulation of energy supplies. For the transformation of the national energy systems, this heterogeneous structure leads to "forum shopping" by the states in energy matters and to suboptimal outcomes. Multilateral cooperation at the international level should be streamlined. A new initiative for an encompassing multilateral energy organization at the international level in the wake of the Paris Agreement would be one possibility to coordinate energy policy through an international platform. Either a completely new organization could be launched to overcome legacies of the past, or one of the existing platforms such as the UNEP could be used.

The current development of on the one hand strengthening the climate change-energy transformation nexus and on the other hand the withdrawal of the US from multilateral agreements on climate change and a reorientation away from renewables towards fossil fuels, demonstrates the need for a new multilateral international forum on energy transition. If it does not seem possible to create one energy platform at the global level, then at least it seems to be necessary for some of the existing organizations to expand their membership towards the BIC countries. Bilateral cooperation could help to pave the way in some of the organizations.

Reflecting the lack of international energy governance, bilateral energy dialogues nonetheless gained momentum in the international realm for the discussion of energy transformation. However, and the following will refer to our second point, bilateral dialogues, at least those between the EU and the BIC countries, are not tapping their full potential. They are limited in scope and characterized by mutual distrust and diverging perception of norms and goals. It appears that partners only show minor interest in the dialogues. We recommend that bilateral dialogues should be enhanced both in scope and quality.

Dialogue partners should devote more commitment and develop a long-term vision for the dialogues and the expected results of engaging in this cooperation mode. Long-term visions should be linked to international agreements such as the Paris agreement and sound possible joint contributions and strategies to global goals.

Bilateral dialogues should be enhanced in their status as a valuable tool to develop common strategies for energy transformation. At the moment, it looks as though the partners do not devote sufficient commitment to the energy dialogues. As the EU does not seem as important for BICs as the Member States, commitment is less intensive. Thus, dialogues vary in intensity, frequency of meetings and outcomes. Dialogue partners do not devote enough commitment to develop a long-term vision for the dialogues and the expected results of engaging in this cooperation mode, which implies the need to build up mutual trust between partners, and multi-directional learning, equal partners and joint ownership are required as a basis for successful political communication. Dialogues should also be carried out more reflectively. Learning from past experiences and collaboration should be more central, which in essence calls for systematic knowledge management by all actors involved. Especially on the EU's side, the permanent rotation of personal is hindering learning processes and is a large obstacle for the proceeding of the dialogues.

Importantly, bilateral dialogues are still characterized by mutual mistrust, minor interest of the emerging powers in the dialogue and divergent perceptions of the underlying norms and goals. To overcome this deficit, we recommend conducting a systematic evaluation/assessment of the sectoral dialogues under the Strategic Partnerships in Brussels and the EU Delegations in Beijing, Brasilia, and New Delhi. The results should shed light on lessons learned and best practices that can support in steering the energy dialogues towards deeper, more efficient and trustful cooperation (Piefer et al. 2015b: 349).

Actors' motivations to enter into the dialogue mostly do not converge. Nor do the high expectations they maintain. While the EU in all the dialogues is motivated mostly by combating climate change and less by technology transfer, the BIC countries focus mostly on technology transfer and on the promotion of private sector cooperation or else market entry. Thus, it might be important for EU actors to ensure how these interests might actually be met. Possibly, the interest in technology transfer exhibited by some of the emerging powers is not always echoed by heightened EU interests in market entry in these countries. Consequently, in the long term, they might prefer to engage in technological cooperation with other countries or Member States, instead of the EU. The EU might therefore prefer to focus more on specific arenas dealing with issues of technological transfer or private actors' cooperation. A stronger understanding of ownership as the capability of both partners to conceptualize and implement policies based on a common understanding of demands and needs can be the key to intensified energy cooperation between the EU and emerging powers. Currently, the dialogues are driven by competing ownership; each partner wants to be the lead in terms of designing policies, setting the agenda, or suggesting concrete outcomes (Piefer et al. 2015b: 347). Joint ownership would instead raise innovative forms of energy cooperation. Joint ownership and mutual understanding can also be motivated by joint research among the partners of the bilateral dialogues. We recommend the EU and BIC countries to develop joint research strategies with new joint funding opportunities, which would empower both partners to set up budget lines for energy cooperation and would allow for transcontinental research cooperation for the era beyond Horizon 2020. With the phasing out of traditional development cooperation funds, the need for more-innovative funding models becomes evident. As such, joint funds provided by the EU and Brazil, India and China bilaterally could be thought of. Both sides are eligible to hand in project proposals for joint cooperation projects in one of the countries or in a third country. A steering committee would then decide on the implementation of projects. Thus, projects that are within the strategic interest of both sides could be carried out.

Another aspect is the mode of cooperation. Specifically, the EU has shown poor performance within these dialogues (Knodt et al. 2015a). It seems that the EU is very good at 'talking at' instead of 'talking with' external partners. The EU engages in top-down, one-way communication projecting interests, norms and values rather than developing a horizontal, dialogue-led, two-way communication process between equal partners (Chaban et al. 2016). To overcome this pattern, mutual trust and multi-directional learning, equal partners and joint ownership are required as a basis for successful political communication and new initiatives for an encompassing multilateral energy organization at the international level in the wake of the Paris Agreement.

With the phasing out of traditional development cooperation funds, the need for more innovative funding models becomes evident. As such, joint funds provided by the EU and Brazil, India, China and South Africa on a bilateral—or even plurilateral—basis of the BICs group could be thought of. Both sides are eligible to hand in project proposals for joint cooperation projects in one of the countries or in a third country. A steering committee will then decide on the implementation of projects. Thus, projects that are within the strategic interest of both sides can be carried out, and the possibility for trilateral cooperation with third countries is a given. This model on the one hand ensures joint ownership of the dialogue and its step towards collaboration; on the other hand, it encourages more horizontal cooperation among equal partners and thus enhances interest and trust in the dialogue. For each side, it is important to see that some concrete outcomes of the dialogue are measurable. Joint projects are one way to achieve this goal. The issue of funding usually being a problem could be overcome by modes of joint funding from the beginning (Piefer et al. 2015b: 348).

To overcome the deficit described, we recommend conducting a systematic evaluation and assessing the sectoral dialogues within the EU and the BIC. The results should shed light on lessons learned and best practices that can support steering the energy dialogues towards deeper, more efficient and trustful cooperation. At this point, EU Member States should also be involved to obtain better vertical coordination between the EU and national level. In addition, Member States should make use of the possibility of the regional cooperation opportunities mentioned in the Governance regulation to enhance cooperation on energy issues that are enhancing consensus on decarbonization strategies. Thus, there should be regional cooperation between Germany and Poland on the cutback of the coal mining industry.

Bilateral dialogues should be embedded within international multilateral fora. For instance, the Brazil-EU energy dialogue envisions coordination within the International Partnership on Energy Efficiency Cooperation (IPEEC). Additionally, other international organizations could be thought of, such as IRENA or UNEP. Concerning international forums, IRENA is rapidly growing, and European Member States, not least Spain and Germany, have been active in shaping the organization and spreading its message. Increasingly, emerging powers are also having a say in and making strategic use of IRENA. As IRENA's mission includes the development of an open-access knowledge base on energy transition worldwide, BICs and the EU could devote more attention to this process to bundle bilateral dialogues under its roof.

A strong stumbling block of a successful bilateral energy dialogue is a strong commitment from the EU side. Strong commitments can only be made if the EU position builds on a broad consensus among the EU Member States, as energy policy within the EU is not based on exclusive EU competences. In the last 10 years, European energy policy witnesses a confrontation between the Central and Eastern European (CEEC) and the Northern and Western EU Member States within EU energy policy. On the one hand, Member States such as Germany and Denmark led a group of environmentally and climate friendly governments within the EU and pushed for the decarbonization strategy of the Commission. On the other hand, especially the Visegrád states (Poland, Slovakia, Czech Republic and Hungary) and Bulgaria and Rumania under the lead of the Polish government set a priority for the goal of energy supply security. This lack of consensus impacts the EU's role in the bilateral energy dialogues, as the EU is seen as not speaking with one voice and not performing as "one" trustful partner. Thus, we recommend that the EU work towards a broader consensus in its energy and climate policy. The EU should make use of all possible instruments to find a common position, especially on the issue of its decarbonization and CO_2 reduction strategy.

The European Commission has to obtain leverage on the national energy mixes without—according to Art. 194(2) TFEU—having European competences in this issue. We recommend that the soft governance instruments, as proposed by the Commission in its Regulation on "Governance of the Energy Union" (Governance regulation) as part of its newest legislation act, the "Clean and Secure Energy for All Europeans" or so-called "winter package" of November 2016, should be strengthened.

We also recommend the linkage of energy policy with other EU policies, especially the regional policies and those such as the European Structure and Investment fund (ESI). For this purpose, the efforts within the Energy Union and the specific cooperation program with the priority "low CO_2 economy" in the European Fund for Regional Development (EFRE) should be linked more closely. Furthermore, a program for cooperation between regions for the joint conversion of coal mining fields could be set up to support especially the conversion of coal mining regions and other decarbonization strategies, which could be easily added to the new initiative of the "Green Deal" of Ursula von der Leyen (European Commission 2019). Structural funds could be "earmarked" to force Member States to use the money for decarbonization strategies and serve as a conditionality tool for those Member States underperforming in their energy transformation goals.

If there is no consensus to be reached within the EU, there should be a coalition of the frontrunner Member States within the EU and beyond, such as the "Powering Past Coal Alliance", with more than 25 nation states and a further 22 sub-state and non-state members. Initiated by EU members France and Great Britain, as well as non-EU states such as Canada, it establishes the goal of reducing CO_2 emissions though a phase-out of coal-fired power stations. The declared target is a phase-out of coal by 2030 in the OECD and EU nations and for the rest of the world by 2050.

Chapter 9
Questions of Distributive Justice

9.1 Introductory Remarks

Technically produced energy is a "basic resource": a supply of cheap and safe energy is indispensable for mankind to maintain physical subsistence. Energy is needed to process and provide drinking water and sufficient food. Large parts of the world would not be habitable without a constant flow of energy to heat or cool the living environment. Technically generated energy is an essential prerequisite for transport and communication and thus the basis for the creation and preservation of prosperity by division of labor, specialization, and participation in markets and society.[1] The reliable availability of energy everywhere has a technical and economic leverage effect for increasing productivity and thus for achieving a standard of living that goes beyond the elementary securing of subsistence. And only where they have energy at their disposal people can secure the resources with which they are able rise above the state of nature; energy, therefore, is the basis for all culture and all prosperity—with all the opportunities it offers, the generation and consumption of energy inevitably is associated with risks, too. These range from the hazard potential of a simple fireplace to the hazard potential associated with malfunctions in a nuclear power plant. In addition, the intensive use of energy "consumes" the environment in which people live. Degradation and extraction of energy sources and the landfill of residues consume soil, the use of hydropower and wind energy transform the landscapes, emissions pollute water and air, and the accumulation of combustion residues in the atmosphere gradually changes the climatic conditions for life on Earth. And the intensified use of energy also has social consequences: The benefits and burdens of a technical energy supply are unequally distributed, and in an increasingly networked world, conflicts are increasingly arising. That gives rise to questions of

[1] While the central contribution of division of labor, specialization and trade to the development of prosperity since Adam Smith's "Wealth of the Nations" (1779) is undisputed, the influx of energy indispensable for this has only recently received the necessary attention. Cf. for instance Hall and Klitgaard (2012) and Smil (2017).

C. F. Gethmann et al., *Global Energy Supply and Emissions*,
Ethics of Science and Technology Assessment 47,
https://doi.org/10.1007/978-3-030-55355-5_9

fair distribution on a global scale and to the consideration of ethically legitimized strategies that allow people to meet their respective requirements. In the following, from the point of view of philosophical ethics, the prerequisites are discussed which must be created in order to develop what may be called "energy justice".

The starting point of ethical considerations of justice is always a principle of impartiality. Claims are to be judged "without regard for the person"—in judiciis non est acceptio personarum habenda, as it is called in the old formula—regardless of the abilities, talents, resources of the individual, regardless of its gender, race and religion, and regardless of where or when he or she lives. However, attention must be paid to what I. Kant calls the practical necessities of our actions in his "Groundwork of the Metaphysic of Morals" (Kant 1785; cf. also Stemmer 2000). An equal distribution of all resources over all—potentially infinitely many—generations e.g. would reduce the usage rights of each individual generation to infinitesimally small and uselessly small units. The increased use of resources by those who, because of special capabilities, can use them more productively than others and who are willing to exchange products may be the preferred option for all concerned. And if public goods such as legal institutions, which many consider a necessary condition for the emergence and persistence of larger, more efficient exchange markets, can only be made available if territorial units are formed that grant privileges to some (their "citizens") and not to others, then such inequalities can also be justified. Accordingly, wherever in philosophical ethics a fundamental equality of all is mentioned, the corresponding aim is not an equal distribution of all goods or services or the equality of all living conditions, but the equal treatment and consideration of all raised claims.[2] Where claims compete with each other and cannot be satisfied at the same time, unequal treatments that are inevitable due to practical necessities require a rational justification of practically necessary unequal treatment towards everyone involved. The claim of a "rational" justification means that there is not one justification presented to this and another justification to the other person, and that they do not merely unfold their persuasive power under certain situational conditions and in a certain linguistic guise. Rather, the claim is made that ethical justifications are understandable and convincing for everyone at all times, i.e. taking into account all aspects [technically speaking, it has to be justifiable invariantly with regard to addressees and situations; for more details, see Gethmann and Sander (1999)]. Thus, justice in this ethical sense does not mean "equality", but "rationally justified inequality".

In view of the magnitude of the task, given the vast complexity of practical necessities relevant in the field, no definitive ethical answer can be expected for questions

[2]It is controversial to what extent Marx's understanding of a communist society constitutes an exception. In his programmatic writings, however, he is unambiguous: his prominent distribution principle "From each according to his ability, to each according to his need" (Engels 1847), for example, seeks merely to define abilities and needs as the yardstick for the social distribution of goods—instead of class affiliation and capital ownership. Cf. also F. Engels' Preface to Marx' "The Poverty of Philosophy" (1847): "Justice and equality of rights are the cornerstones on which the bourgeois of the eighteenth and nineteenth centuries would like to erect his social edifice over the ruins of feudal injustice, inequality and privilege". For a more detailed discussion, cf. Buchanan (1982).

of global energy justice. Rather, a problem analysis and discussion of aspects to be considered serve to prepare and structure a debate that must be conducted publicly and that may produce a collective understanding of what is necessary for a fair distribution of energy and the consequences of its usage on a global scale. Debates of this kind are, however, typically burdened with a high potential for misunderstanding, because relevant concepts are understood in different ways and because the different contexts in which they are used are associated with different worldviews. In particular it must be assumed that the parties involved, on the basis of their respective preconditions, unreflectively have hidden pre-conceptions on what the means and the outcomes will be, which will be an obstacle for mutual understanding. And because people tend to expect their self-evident, quasi second nature conviction systems to be shared by others and tend to perceive deviant practices as provocation, there is a high potential for conflict. Therefore, in order to prepare the debate, an exploratory process is necessary that contributes to clarifying the concepts, identifies essential presuppositions of such a discourse of justice, and sensitizes to the risk of culturally determined presumptions about other people's beliefs and convictions.

9.2 Challenges of a Global Energy Justice

While in political or economic contexts questions of distribution on a global scale are always related to an external framework and preconditions of action that transcend organizational units, ethical questions of distributive justice are usually posed as questions of communication and interaction between individuals, groups or institutions within a social corporation (family, company, state, etc.). The long-standing discussion on questions of distributive justice in the environmental sector, for example, is predominantly taking place in a domestic context (cf. Gethmann 1995). Grosso modo, until a few years ago there has been only one question dimension that transcended this perspective, namely the problem of distributive justice between generations with a view to long-term responsibility. However, this question—which has only been dealt with systematically for a few decades—has a clear diachronic perspective.[3] If questions of distributive justice that surmount the narrower framework of states were discussed at all in a synchronous perspective, then mostly under a dualistic presupposition: rich affluent societies verus "third world", developed industrial nations versus regions with catching up development, donor countries versus recipient countries. However, there are a number of questions that imply the problem of distributive justice between states and state-like corporations in a synchronous and multilateral perspective. These concerns, for example, rules of martial law including the distribution of the costs of war, diplomatic relations, international treaty law, the proliferation of certain weapons, etc. But it is presumably only with the climate discussion that the question of justice is explicitly addressed in such a synchronous

[3]Cf. the discussion using the example of nuclear waste disposal in Streffer et al. (2011).

perspective (Gethmann and Kamp 2020). Here the well-known questions of distributive justice, as they exist between individual and institutional actors, are delegated to states or state-like actors. How for example does the inequality of distribution due to the factual inequality of interests and needs relate to the normative equality of the actors? If justice is "justified inequality": what are the adequate justification criteria, what are the relevant benchmarks? How are historical merits balanced in relation to those who are added later and those who profit from them ("historical justice")? How are natural or fateful advantages (e.g. wealth of mineral resources) or disadvantages (e.g. desolate political conditions due to failures of colonial powers), that are not the respective nations' own merits or failures, to be compensated for? But with the application of the justice issues on states or state-like actors, it must be examined whether and in what sense organized large corporations could be subjects and objects of distributive justice or whether criteria of justice, as they are defined in ethical theory, are only applicable on individuals.

So, for instance, a simple idea, though not easy to operationalize from a legal point of view, is that universal human rights can at least be used as distribution criteria. Human rights focus primarily on the protection of the individual and at best indirectly allow conclusions to be drawn on issues of distributive justice. Furthermore, the reference to human rights is associated with the notion that economic globalization could prove to be a vehicle of normative universalism, so to speak. On the whole, however, the discussion of the last decades shows that economic globalization by no means goes hand in hand with the globalization of universal norms, such as human rights. Rather, the appeal to universalistic moral concepts, including concepts of justice, are often interpreted as a kind of "propaganda trick" of Western exploitation strategies ("Eurocentrism"). On the other hand, particularistic moral concepts are used as an expression of special traditions and cultural understandings (cf. Gethmann 2006). Not least, the balancing of the universal principles of distributive justice ("an equal share for everyone", "each according to his ability", "each according to his performance", "each according to his needs", etc.) takes place in affinity to social traditions as well as to the status and role of the negotiators.

The energy issue exacerbates the above-mentioned problems of global distributive justice to an extent beyond what is known in the field of small-scale individual and institutional actors. At the same time, the energy issue, apart from individual aspects, is not yet the subject of an appropriate discussion on a global scale, keywords such as "energy" or "electricity" etc. are not found in the relevant articles of philosophical encyclopedias, not even in a specialized 1200-page "Encyclopedia of Global Justice" (Chatterjee 2011). However, many conflicts that are already acute or are to be expected in the near future require ethical reflection. In particular, at the level of principle, considerations must be made as to what can be regarded as a fair distribution of benefits and burdens, of opportunities and risks of an energy supply that is beneficial and adequate for all. At the same time, the question must be asked as to how institutional framework conditions could be designed to make the formation of fair conditions promising.

A particular difficulty with regard to state and quasi-state actors lies in the fact that, from a global perspective, there are a number of states that do not want to or cannot carry out government tasks in the traditional sense. The reasons for this are internal or regional conflicts that partly result from colonization, social conflicts caused by population growth, economic weakness or poverty, etc. In extreme cases of state disorganization, the term "lost" or "failed states" is used.[4] In any case, discussing justice issues in the context of collective problems on a global scale involves including the debate on global political self-organization. Since there is currently no tendency towards a world state construction and a world state may not be desirable for fundamental reasons (Höffe 2007), considerations of justice on a global scale can only be combined with considerations of a supranational, legally constituted political world system.

The term "energy justice" expresses the fact that due to being intensively interlinked decisions of the national energy economies are not possible without retroactive effect on the interests of other energy regions. These decisions concern aspects of the choice of an energy policy, the commitment to a particular mix of technical strategies, the requirements of resource extraction, the erection of an infrastructure for distribution and provision, and the associated emissions. In principle, all these decisions are ethically relevant, and whenever justice is a relevant criteria the consequences of options taken are as relevant as the consequences of rejected ones, and there is no difference of principle between executions and omissions. A few central points are briefly outlined below:

Technical Strategies—Including Their Research and Development

The choice of national energy production strategies has implications for others. Scientific basic and application research can open up to other new options, testing and further development of technical strategies creates gains in safety and efficiency, use in the field and corresponding upscaling in production ensures easier and more favorable availability.

Resource Extraction

Extraction strategies affect environmental aspects and the living conditions of those who mine substances, some of them at risk to health, in oil-producing countries and in rare earth mining regions more often than not under exploitative social conditions and in de facto enslavement.[5] At the same time, possible subsistence protection and development potential through trade are given. The members of future generations are always affected by investment controls and environmental impacts, both here and there.

---

[4]The definition of failed states focuses partly on economic and partly on legitimatory aspects. Uniform definitions and indicators have not become established, so that the numerical data also fluctuate. An assessment of the current state of affairs in 178 compared nations with regard to relevant factors can be found in the "Fragil State Index" (http://fundforpeace.org/fsi/data/ accessed 13-Dec-2019).

[5]Cf. the impressive descriptions in Wenar (2015) and Yergin (2011).

Infrastructure and Distribution

Energy itself is a commodity. Due to changeability and shape-specific transportability, the place of production and the place of use can be far apart; transport can cross national borders and even continents (as in the planning for the failed major project Desertec, for example[6]). Risks associated with the extraction, processing and transport of energy sources can easily be transferred to others (creation of externalities). Unused potentials (such as comparative location advantages) can go hand in hand with withdrawal of benefits (for producers there and consumers here) and acceptance of avoidable risks (with regard to security of supply, operational risks, emission risks).

Security of Supply—With Users in Mind

Modern energy production allows for a relative independency of natural conditions and enables users to focus their lives on other advantageous aspects (for example, they no longer have to settle in the river valley to operate the mill wheel with the power of water and can settle anywhere operating the mill with electricity from a power plant). This creates, on the other hand, social dependencies that can be exploited (for example, if the miller working on the hill is dependent on regular supplies from an electricity supplier due to his specific investments in an electric drive). Here, organizational precautions must be taken. The same applies to major players such as states as importers and exporters of energy. The omission of action is to be considered here as well: The situation in which citizens do not use options because they do not dare to become dependent due to a lack of organizational framework conditions, such as a lack of contract protection, is suboptimal.

Emissions

The transformation of energies is accompanied by emissions, some of which are directly responsible for hazards and health risks, while other risks are very much mediated and oblique, as the example of climate change shows. The risks associated with emissions can be influenced in many ways: through the choice of energy production options (gas, sun, lignite), through the choice of technical means (old and new coal-fired power plants), avoidance (e.g. filters or sequestration technique, but also increased efficiency or reduced use). The cross-border and global relevance of national decisions as well as the repercussions of international agreements on regional living conditions have to be considered, especially with regard to the omission side. In general—and especially with regard to developing countries—restrictions can delay development and thus deprive people of benefits and pose avoidable risks. A particular challenge lies in the fact that the risks associated with so-called climate change are directly related to the development already achieved in the industrialized countries. This raises the obvious question of how historical benefits or merits of some societies (e.g. prosperity due to scientific and technical proficiency or a high and constant degree of organization) are balanced in relation to those who come later

[6]For an overview of the project and an analysis of its failure cf. Schmitt (2018).

and profit from these merits as well as those who's development has been hampered by others ("historical justice"). If the outcomes of historical developments as well as future needs and potentials should be fairly balanced, the question must be answered as to which periods of time should be taken into consideration (should, for example, the state identity apply or at least a rough territorial identity?).

In order to answer these and other similar questions, proposals and recommendations for rules must be developed according to ethical standards, which have the prospect of being accepted by everyone, including all aspects. This would be impossible in a context where it is completely unpredictable what reaction one might expect to a particular action taken. At the same time, everyday experience shows time and again how fallible human assumptions and expectations about the reactions of people in their own immediate environment are and how complex and confused the chains of social action are. In the thematic framework of a global energy supply set here, the complexity of the interrelations on the one hand, and the diversity and disparity of the forms of life to be considered on the other, increase these difficulties to the point of over-complexity.

In such contexts, any attempt to demonstrate an ethically appropriate procedure based on supreme principles to those concerned is doomed to failure. Where the general ability to consent is not to be bought by losing oneself in the abstract general and thus not contributing to an operable solution of the problems, the effort to achieve acceptability must be sensitive to the specific preconditions of those affected and their respective way of life. In order to develop appropriately differentiated approaches to solutions that do justice to the manifold disparate requirements and prerequisites, one will rather think about organizational forms and procedures through which an appropriate clarification can be brought about by the participants themselves. On the one hand, this requires the transparent and comprehensible processing of the required factual information, the technical options on the one hand, and the cause-effect relationships that provide information about the associated opportunities and risks on the other. There is a need for a public debate in which goals and expectations are compared, motivation and reasonableness are weighed against each other and converted into formal entitlements and obligations. The first cornerstones have been laid at the international level for the debate on a global scale. With the increase in mobility and the expansion of communicative opportunities, which in the first place opened the view to the global range of the task, the prerequisites for its processing have also improved considerably. At the same time, however, the scope and magnitude of the challenge can only be assessed. In order to contribute to the preparation of the upcoming discourses, to prevent misunderstandings, to clarify some preconditions and to warn against inappropriate and exaggerated expectations, some central questions from the point of view of philosophical ethics are raised and dealt with below.

According to Aristotle and most modern ethicists, just social conditions are those in which only justified inequalities exist.[7] Such inequalities can exist, for example,

[7] Aristotle, Ethica Nicomachea, Book V; cf. e.g. Rawls (1971), Gauthier (1982), Hinsch (2002) and Gosepath (2015).

with regard to the qualitatively and quantitatively different availability of raw materials, different levels of technical development, different political or legal structures, different cultural framework conditions, levels of development and many other criteria. Justified inequalities are any—given or manufactured—allocations of such advantages or disadvantages for which there is a justification. Thereby, it is irrelevant whether the justification was explicitly submitted or is merely assumed by the relevant parties. However, not all inequalities are per se justifiable and sometimes the justification of an inequality will have to be assessed before its justification can be assessed:

The fact that some have more rain, others less is an unequal distribution, possibly even one that the affected people complain about. However, they will not expect any justification as long as it is beyond human control to control local precipitation.[8] At the moment when technical possibilities become conceivable with which deposited quantities of water can be transported from one place to another, the unequal distribution of water—at least in principle—changes from a condition that does not require justification to one that requires justification. If, for example, someone now makes the demand to be allowed to alleviate an existing shortage with the water available in quantities elsewhere, the need for justification becomes practically manifest. If the distribution of water proves to be unfair (because there is a justification for a differing distribution) on the basis of an argumentative review, it may already be possible to justify the imperative of exploring the technical possibilities through which a redistribution can then take place. However, the subsequent question of burden sharing—who invests which financial and cognitive resources in research, who carries them out, etc.—also becomes a question of fair distribution. This question, too, did not need to be justified until the changing framework conditions made it so. However, the question as to whether the framework conditions are (already) such that they make a distribution need justification or not can be quite controversial: For example, there may be different, even erroneous, opinions about the realism and feasibility of conceivable technical intervention possibilities, and even convincing proof that options for changing a situation are not yet in prospect is not yet a viable justification for a lack of need for justification—this can always only be the result of a negotiation process and an agreement between the parties involved. The need for justification must be determined accordingly by the parties involved and cannot be ascertained by third parties. This also applies to science, which can, however, provide a weighty argument for such negotiations with references to the existence or absence of technical prerequisites, the higher or lower "readiness level" and the shorter or

[8]The fact that such inequalities can be regarded as divine providence and that in the context of a theistic world view it could certainly be deplored as injustice by those concerned that the Creator has put them "into misery", while others enjoy their happiness, will indeed play a role for their reaction to the perceived injustice. The constellation between the complaining person and the divine "distributor" could also be described as a conflict. Conflict resolution efforts on the part of the person concerned could then be presented in the form of magical or religious practices. Here and in the following, however, injustice is always addressed as a type of conflict that exists between conflict parties acting according to human standards. Religious convictions are only taken into account to the extent that they play a role in the discussion of conflict resolution strategies.

longer expectation horizons of technical developments. And through the open provision of information about technical developments and the options for action that this opens up, it can also trigger debates about the need for justification—science and the consulting services of scientists are accordingly already an important point of reference in justice debates against this background. After all, it is essentially a consequence of scientific and technological developments that today, on a global scale, distributions are considered to be in need of justification, some of which just a few decades ago simply had to be accepted as "given". Indications such as the fact that the distribution in question had already been found, that it was anchored in tradition or that it "had always been like this" no longer represent a viable contribution to the debate on the need for justification: Such indications merely contest the need for justification and do not justify the need not to justify it.

In justice contexts, justifications can already become complex in trivial distribution constellations, and in the context of collective action problems of global range one can regularly expect a high degree of such complexities. The overriding question of a fair distribution is broken down into countless detailed questions on which, in case of doubt, a separate agreement must be reached if the forthcoming consultation is to lead to an amicable decision that is supported by the parties involved and accepted by the parties concerned. The most complete and consensual answer possible to such detailed questions becomes all the more urgent if the aim of the deliberations is to achieve a lasting conflict-management effect: By conducting the deliberations as completely and explicitly as possible and by making the presentation of the results as immune as possible to misunderstandings, the binding force of given approvals is increased and the decision is more robust against revisions. In some cases it will be possible to deal with the detailed questions in separate debates, so that the differentiation of the detailed questions can at the same time be understood as an aid to structuring the discourse organization. Often, however, the various questions are syndromically embroiled with each other, so that they are not treated separately, but can only be addressed in a similarly complex overall solution. What often already applies to simpler distribution issues applies here in particular: You will not always be able to count on the one, the final, the optimal, the decisive answer. Over-complex discourse constellations like this are "a labyrinth of paths" as Wittgenstein (1953) puts it talking about understanding language and philosophical problems in general: "You approach from one side and know your way about; you approach the same place from another side and no longer know your way about." It is not only the different answers parties give to the discourse that are in dispute, but also how the debate is to be organized, what role this or that participant should play, how and in which order the questions are to be asked, which contexts are relevant, how central expressions are to be understood and so forth. The following considerations can therefore be seen as contributions to the preparation of the discourses to be conducted.

9.3 The Subjects of Distribution

Distributions need to be justified if there are conflicts between the parties involved—i.e. purposes which the parties could achieve with full or wider availability of the distributed resources cannot be achieved if others use the resources to achieve their goals. Accordingly, the standard case in the debate on distributive justice is competition for scarce resources which, if one uses them, are not available to the other at the same time so that he either couldn't reach his or her purposes or at least has to change plans. Hereby the term "purpose" is generally to be understood in such a way that the more favorable, faster, more efficient, more sustainable… achievement of a certain purpose can also be presented as a purpose. Equally, the term "resources" is to be understood in its broadest way: Whatever helps an actor to reach his or her purposes is called a resource. It would be especially inadequate to reduce the concept of "recourses" to a materialistic understanding and to use it in the same way as e.g. "goods": In this understanding, the prehistoric Neanderthal man would have had more energy resources than the current inhabitant of the Neanderthal—especially in the amount of coal, gas, oil, uranium, etc. that he and his human successors burnt down so far. If, however, resources are not understood as goods, but more adequately in functional terms as that which is necessary to achieve a certain purpose, then today's human being, with its enormously expanded possibilities for action, has at its disposal many thousands of the resources that the Neanderthal man had.[9] Resources, so understood, are generated by inventing and discovering techniques and strategies for action and are constantly increased by innovation—and so, for example, what was a smelly, oily black mass, which made soil worthless wherever it appears at the surface, became the most important resource of our mechanized civilization and the fuel with the largest share in the world's energy mix (34% in 2018). It's important to keep this difference in mind, for two reasons: In its correct, functional understanding it is not reasonable to speak of a finite totality of resources to be distributed.[10] And if distributive justice is about whatever counts as a means to an end (and not about finite amounts of material), than innovative achievements and the capacity to innovate are part of the equation. As e.g. the use of knowledge by one person does not exclude others from its use, there are problems of justice of a different and unique nature with regard to the distribution and accessibility of such practical knowledge.

---

[9]The calculation by Nordhaus (1996) impressively proves this: according to his findings the Stone Age man had to invest approx. 50 h of work around 9,300 b.c. in order to obtain a thousand lumen-hours of light (corresponds to the approx. one-hour light output of a conventional 75 W lightbulb), whereas the members of Western industrial nations today only have to work approx. 0.00012 working hours—calculated in the mid 1990ths, before the LED lamps became widespread.

[10]This thesis is emphatically supported by Simon (1996). For a historical outline of resource multiplication through innovation from antiquity to the present, see Mokyr (2017). In this context it is worth noting, that the same applies to information: Since information are the result of information gathering efforts, and since one never can know ex ante whether and when further efforts might be fruitful, there is no such thing as "complete information". By principle reasons therefore decisions and resolutions about just distributions are to be made with incomplete information (Kamp 2015a).

Even more far-reaching in this direction is a proposal which does not so much define the possibility of realizing purposes (in the sense of a disposition of resources and technical know-how) as the object of distributive justice, but rather the actual capabilities to freely define purposes and to strive for the means of pursuing them. A just distribution then also includes measures of freedom or, in a negative delimitation, restrictions of freedom: To what extent is the individual granted the political or economic freedom to obtain the goods necessary for his or her purposes? To what extent does he have access to education and knowledge? To what extent can he freely dispose of his labor and property in order to make reliable plans for the future? To what extent is he protected against health, social and other risks in order to be able to strive for new challenges? And last but not least: To what extent is he able to appropriate and to use energy as a resource that allows for more productive and efficient action and enables him to use the time and energy thus released to achieve further purposes? Approaches of this kind then also lead further into fundamental welfare economic debates, which is why they are primarily discussed in debates on the appropriate organization of the welfare state or on development aid. Contrary to the question with which they were brought into the discussion—"Equality of What" was the title of a lecture by Sen (1979), to which large parts of the discussion can be traced—"capabilities" and similar concepts are most likely to be a measure of the need for action of national and supranational organizations and thus only indirectly the subject of distributive justice: Since, for example, there is no power in anyone to "distribute" political and economic freedom fairly among the nations (and even within states the granting of freedom can only be reconstructed with difficulty as a distribution problem), the comparison of the capabilities of different persons or average capabilities of regions or societies offers organizations such as the UN, WHO, UNICEF and other indications of where their resources should be primarily directed, how they should be directed, and for what goals they should be used. Accordingly, the so-called capabilities approach was also proposed by its representatives as an alternative construction to economic standards such as gross domestic production (see in particular Nussbaum 2011). Where—as with the issues at stake here—the distribution of rights, freedoms and institutional resources is not the primary concern, this approach seems less appropriate. Nevertheless, the mere effort to achieve a fair distribution of resources without a simultaneous effort to develop possibilities and abilities to use them would be a questionable undertaking (strictly speaking it would be a mere allocation of material). In this respect, questions of a just global energy supply must always be dealt with in connection with questions of political and economic freedom of action, the stability of institutions and the fair distribution of public goods, the freedom to acquire education and to develop one's personality. Any debate on a fair allocation of energy resources therefore will be oriented to these overarching criteria.

However, in addition to the category of energy resources, the category of opportunities and risks associated with the consequences of action is also important for coping with collective problems of action on both a regional and global scale. Here, for example, inequalities of distribution regularly occur when large-scale facilities such as pipelines or power plants are built. For advanced technical civilizations, conflicts among those affected by risks and the beneficiaries of such plants are a structural feature: the operation of large-scale facilities or possible incidents poses risks for the immediate surroundings, while the opportunities are distributed more widely across the area. At the same time, those affected by risks often have a higher organizational interest and potential to assert their interests.[11] However, with regard to less developed regions of the world, risks are often of greater significance that arise precisely because technical potentials to concentrate and thus reduce risks do not materialize due to a lack of material, cognitive or organizational resources. One example is the health risk that stems from pollutant emissions from domestic heating already mentioned above in Sect. 3.2.8 in Chapter "Aspects of Environmental Compatibility of Energy Systems" above which for three billion people is still the everyday matter of course for the requirements of cooking and heating, and to which the WHO attributes 3.8 million premature deaths a year.[12] Last but not least, collective action problems such as climate change, in which the consequences of decisions taken here and now can occur in distant places around the world and only in the near future, place special demands on the distribution of risks and opportunities. These requirements are structurally different from those in local conflicts and must be treated fundamentally differently: Those affected are often able to stand in for themselves in overcoming local conflicts through direct participation, or they are able to fall back on organizational precautions and procedures whose legitimization they recognize and accept factually. Such conflict management instruments are not available where conflict parties are separated in space and time. Instead of a balance of opinions and interests, in which the pacifying consensus is essentially determined by the actual acceptance of the participants, much more emphasis must be laid on the acceptability of proposed solutions: Reasons are required which will still be valid even if other persons who have not had a say in the debate so far join it, and if further data and facts are presented which have not yet been taken into account in the perspectivations taken up so far. One will tend to expect such proposed solutions to have the potential to win the approval of everyone and under all circumstances.[13] In such a data- and fact-based ideal discourse, for example, the risks associated with the one or other technical option would then be methodically assessed according to the best available evidence, and, beyond all subjective estimates, rhetorical refinement and the factual distribution of power, the risks and opportunities would be compared

[11] The assertiveness potential of smaller groups with a higher density of interaction (such as those that would be directly affected spatially by a large-scale installation) in the face of the "silent majority" of beneficiaries, who would have to make far greater efforts to overcome their mutual anonymity and represent their interests in an organized manner, was exemplarily presented by M. Olson as a "logic" of collective action (1965).

[12] https://www.who.int/news-room/fact-sheets/detail/household-air-pollution-and-health, accessed 13-Dec-2019.

[13] In the scientific and epistemological debates, there is then talk of a situation- and addressee-invariant defensibility of assertions that form the ideal of scientific discourse. The same applies analogously to the defensibility of calls that want to be regarded as ethically justifiable (cf. e.g. Gethmann and Sander 1999).

with those caused by greenhouse gases elsewhere. Such risk/opportunity considerations can be more convincingly justified if risks and opportunities are made more operationally tangible as the possibility of gains and losses in prosperity, gains and losses in life years for society as a whole or for specific groups, possible health impairments under certain prior burdens or the like. The choice of such standards is associated with considerable abstractions, and, depending on the standards and the way in which and to which they are applied, one can expect a more or less large degree of approval on the part of the distribution addressees. Nevertheless, the structuring contribution of such abstractions to methodologically clean application is enormous, so that they should be used as orientation if acceptable decisions are to be taken that are not supported by sentiments, opinions and majorities, but by rational argumentation.

As soon as one leaves the level of concrete, presentable material goods and more adequately broadens the perspective so that the significance of distributions for the persons concerned is included, the determination and evaluation of the object, the "substratum" of fair distribution soon becomes an enormously complicated task: Whether something is a resource for someone, a pure but worthless material, or even an object of annoyance, how he perceives and values risks and opportunities of technical measure, what purposes one sets and what suits him as an appropriate and affordable means, is highly dependent on the circumstances he or she lives under. And where, due to a degree of technical and organizational development already achieved, alternative means could serve to achieve the purpose, in case of doubt one will base demands for justice less on claims for a certain type of resource than on risks and opportunities for everyone concerned. One therefore always has to take the factual constellations of distribution into account.

9.4 Distributional Constellations

Because justifications are basically linguistic procedures, the constellations where distributions are in question are given in verbal descriptions. And there always is more than only one way to describe these constellations. This description dependency of any justification leads to a further question: the invested description of the distributional constellation will by no means always be the same between the participants, whether they are author, addressee or in some observer role interpreter of a justification. The question of who is affected and who is entitled to claim can be just as controversial as the question of whether the need for justification is to be limited to the current distribution constellation or whether this distribution constellation is to be regarded as one within a sequence of distribution constellations. Only then the very common form of conflict resolution of a time-shifted compensation would be possible ("You this time, I next time"). The question of whether future or past distributions, where different actors will be or were involved, must be posed separately once again. This question is of particular importance, for example, in view of the conflict in which developed industrial nations, which have already contributed to the emergence of greenhouse gases in the atmosphere for more than two centuries, are confronted with the demands of those nations that are striving for such a development, but should not use available resources in order not to exceed the overall CO_2 budget and not to miss the target (2 °C-target or other). Factual historical distributions trivially are not based on joint decision-making—in many cases the historical actors will not even have understood the appropriation of resources as a distribution constellation—partly because scarcity was simply unthinkable,[14] partly because historical anti-humanist ideas meant that "the other" was not perceived as a relevant actor, whose action planning must be taken into account and whose objectives must be coordinated with one's own. However, this does not exclude the possibility that, within the framework of current distribution discourses, the outcomes may be interpreted as the result of distribution constellations and determined as fair or unfair.

Individual and collective perspectivations and interests (some of which are culturally predetermined) shape the reconstruction, interpretation and evaluation of social constellations.[15] In this respect, distribution constellations that do not transcend the boundaries of a culture of interpretation and move within a framework in which homogeneous moral concepts or effective legal institutions provide decision-making routines for balancing different perspectives and partial interests are fundamentally different from those that extend beyond the boundaries of homogeneous cultures and overarching institutions. So e.g. the estimation of the stability and continuity of the given social, political and economic framework plays a crucial role for the perception of the relationship in which the present see themselves to the future. The words "perception" and "estimation" are intended to underline the significance of subjective refraction: For example, expectations of economic stability and thus the willingness to make decisions today in anticipation of future consequences are not only dependent on the actual circumstances, so that the willingness to cooperate and invest in anticipation of future earnings is generally higher where, for example, stable legal institutions and effective control of powers also provide a reliable prospect of access to earnings. The willingness to base justifications for distributions on the anticipation of future returns to cooperation will very much depend on such perceptions and estimations. Where immediate and urgent needs for action to secure life and ward off acute shortages characterize everyday life, longer-term planning will not be possible, so future cooperation gains will hardly determine decisions and planning horizons hardly extend beyond the present day. In many Western societies, on the other hand, sustainability discourses and the assumption of responsibility for future generations have increasingly determined long-term social planning in recent decades.[16]

---

[14]D. Hume's in 1751, for example, stated, that no one can commit an injustice by using the abundantly available air wastefully, although it belongs to the "most necessary of all things" (Hume 1751:Part 3, para. 4). To think of climate as a resource and to think of emissions in terms of a budget is a development that took place only over the very last decades.

[15]This was prominently emphasized in the approaches of philosophical hermeneutics. Cf. Gadamer (1969) as representative for the debate.

[16]The great importance attached to the issues of sustainability and intergenerational justice when setting the political agenda, especially in Europe, is not necessarily to be seen as an indication of the existing need for protection and rescue of civilization and its environment. Rather, it can also be

Cultural, ideological and religious attitudes in particular must also not be neglected in the global discourse on justice: The expectation that powers of destiny or a divine hand will provide a balance in a future life, the expectation that unpretentiousness, modesty and renunciation will be rewarded in the reborn life or by a place in paradise, determine the general conditions of the discourse situation considerably and will make participants with this background have different ideas of the object of discourse than those who strive for a fair balance between the prevailing "profane" conditions and for this, for example, make prosperity economic considerations oriented to measurable variables. Neither in the disputes about historical burdens of guilt nor in questions of the appropriate determination of planning horizons, which will also do justice to future generations, can one reckon with reasons for decision which are understandable from every perspective and can be accepted by everyone. Insofar as the underlying conflicts impair individual planning security and the potential for cooperative action, the participants will have to examine their willingness to agree to second- or third-best solutions, which are not completely free of assertiveness and negotiating power. Supranational organizations that are designed according to the thought model proposed by J. Rawls in such a way that every participant, regardless of his preconditions, imagines them as an instrument for reconciling interests and as a "cooperative enterprise for mutual benefit" (Rawls 1971:105), certainly offer the best preconditions here—though they are, of course, in the end a leading ideal that can be approached, but whose permanent and static implementation at the international level is as little to be expected as at the national level. As experience has shown, compromises and agreements made are subject to the risk of changing requirements and expectations and are subject to temporal stress everywhere. Just as, however, a more satisfactory effect can generally be expected from a state border along mountain ranges and river courses than from an arbitrary demarcation of the border in the area, so efforts to compensate for historical burdens should be oriented towards more recent historical upheavals, such as the reorganization of states after the end of the Second World War.

9.5 Distribution and Participation

As little as the distribution constellation is given from the outset, as little it is determined from the outset who belongs to it, whose goals and purposes play a role and whose action plans a just distribution has to take into consideration. The question has a synchronic and a diachronic component: Being confronted with challenges of a global range does not mean that every human being currently living on earth is affected. At the same time, problems of action of this scope usually also affect members of future generations who have undertaken or failed to undertake coping

---

seen as an indicator of the high degree of stability and planning certainty that these countries have achieved. Contrasts between nations that plan for the long term and those that depend on immediate solutions to existential threads are impressively described in Deaton (2013).

efforts—be it because of long-term consequences of the actions taken or be it because their coping "consumed" considerable resources, so that members of future generations have fewer potentials available to cope with their challenges, be it that innovative options are not available because research resources were directed in this and not in that direction. The question must therefore be asked as to who—ideally conceived—is to be involved in the discourse and decision-making process and who—as is more likely to be realistic in the case of the problem constellations in question here—is to be regarded as the party affected, so that in the case of a distribution, consideration would have to be given to its action planning. If the present and future world communities deserve the same consideration, then conflicts emerge, for example with regard to the allocation of resources, between those who first and foremost strive for the prosperity of a planned, self-determined life and the possibility of reliable planning for the future and those for whom existing prosperity is to be secured and maintained. From an ethical perspective, it is basically irrelevant which epoch or nation a person benefiting or disadvantaged by a decision belongs to, even if the pragmatic necessities for prioritizing those close to the decision and for orienting action towards delimited economic communities, common legal spaces or solidarity communities are to be recognized. If, however, in an energy policy decision situation, there are prosperity risks for members of future generations, who are expected with a certain probability on the basis of certain scenarios, in competition with manifest life risks and the withdrawal of an otherwise highly probable prosperity development of the now living, then, from an ethical perspective, the avoidance of immediate dangers generally deserves to be preferred to coping with long-term risks. However, the constitutional conditions of planned cooperative action must always be taken into account: Organizations of a size that can also develop a sufficient impact for the decisions to be taken here require considerable investments in material and temporal resources, which only those who recognize a longer-term perspective will be prepared to spend. In particular, the willingness to cooperate, which is indispensable for the long-term preservation of such institutions, presupposes confidence in the future benefits of current cooperation. In the ideal case, institutions are therefore justified in the long term, i.e. without a definite time limit, and always keep an "open horizon", so that the participants do not cease their willingness to invest if their collapse appears to be foreseeable. Basically, institutions are forms of organized cooperation. But rational actors will withdraw their willingness to cooperate if they have to fear that others will do so in anticipation of a foreseeable collapse of the institution, when, if they won't, they only had to bear the burden of cooperation without getting anything in return.[17]

[17]To put it in R. Axelrod's words, institutions live "in the shadow of the future": Future benefits from cooperation motivate the individual to cooperate now (e.g. in the preservation of institutions), even if it would be more profitable for the individual not to cooperate (e.g. to exploit the institution as a freeloader) in the respective upcoming decision-making situations with an exclusive view to the directly expected returns. If, however, the shadow of the future dissolves (and an end to the institution becomes foreseeable), cooperation collapses. In game theoretical description the actors play so-called endgames. Cf. detailed Kamp (2015). The argument has a structural kinship with I. Kant's, when he sees the categorical imperative based, among other things, on the practical necessity

In another respect, the question of who is to participate in the discourse, whether it is an ideal or a real one, is relevant. When it comes to distributing a bundle of resources, it is regularly to be expected, that not only those have an interest in the outcome, that for lack of alternatives inevitably need some of the resources, but also others, that aren't interested in the resource itself. If an actor C is interested in a cooperation with B, but B only becomes capable of action and cooperation if he also receives a certain share from a forthcoming distribution of a resource between A and B, then C will certainly also have an interest in the outcome of the distribution, even if he himself has no interest in the resource to be distributed. If one now adds the idea that the cooperation between A and B can result in a product that can be enormously advantageous for a D, then it becomes completely clear that the question of whose interests are to be taken into consideration in a distribution no longer has a trivial answer once the degree of complexity of the distribution constellation corresponds the complexity of life-worldly interrelationships: Since also third parties often have preferences for the division of a resource without having interest in the resources itself, it would be shortened to include only the potential recipients of a resource in the considerations of justice. Questions of distributive justice therefore deserve a more holistic view in tasks of the size and complexity of a fair global energy supply and cannot be answered with a focus on individual resources.

However, the talk of preferences must not always be reduced to "selfish" preferences. People can have a preference for the welfare of their own children, for the welfare of children in general, for future generations, for workers in sweatshops far away or for species-appropriate animal husbandry. Accordingly, the terminology used in economics, action theory, decision theory and planning theory, from the concept of interest to that of purposes and goals to the concept of utility, should not be reduced to an understanding that includes only the satisfaction of one's own needs and one's own prosperity. The utility of an action for an actor A, for example, is precisely what allows A to achieve his purposes, whatever they may be and whoever maybe the beneficiaries.

9.6 Distribution Postulates

Another question leads to the principles that can be invoked as justification for distributions. A seemingly obvious answer—often suggested by textbook examples and in educational contexts—is not possible here: equal distribution. The principle of equal distribution (it must be distinguished from the principle of equality of rights and from the principle of equal opportunities), however obvious it may seem at first, loses its power of persuasion as soon it is applied to larger collectives which, oriented towards

to preserve the institutions where they are condition of the possibility of purposeful and effective action (e.g. personal benefits from a broken promise depend on the existence of a social institution of promising—which would not exist if everyone would try to exploit the institution and breaks promises at will).

the different interests and the different capacities of individuals, base their supply of resources on a division of labor and thus an unequal distribution of competences and responsibilities. In order to achieve an equal distribution, a considerable authoritarian intervention in the freedom of the individual would be required in order to create and maintain the necessary preconditions. The same applies mutatis mutandis to any effort, to level different inclinations, motivations and requirements as well as the different capabilities and prerequisites to use a given framework for one's own purposes and to enforce one's own interests against the interests of others. Nevertheless, an understanding of justice as equality—as equality of citizens before the law and of equal rights and opportunities as market actors—offers an ethically and pragmatically justifiable yardstick for the quality of existing organizational forms and for decisions on the further development of organizations. But equality must not be the sole yardstick here either: Those who adhere in principle to the ideal of equality will have to balance out social equality and the advantages that a free development of individual talents and potentials can have for the whole of society.

If one wants to realize distributions that are sensitive to the different concerns and prerequisites of the participants and have the aim of balancing interests, then the common formation of framework conditions in a free discourse of all participants will be the desired ideal (Habermas 1982). Such a discourse will either directly address the solution of concrete distribution tasks ("material justice discourse") or rules, procedures, and possibly an institutional framework through which just distributions are to be ensured or better ensured than before ("procedural justice discourse"). In the factual realizations of such ideal discourses, it is to be expected in any case that the parties will invoke principles for the defense and examination of distribution or procedural proposals that are at the center of justice-theoretical considerations since the antiquity and that determine theoretical and practical debates to this day. They have their regular place in local and global debates on questions of equitable supply of resources of all kind, including energy, as well as in questions of the equitable distribution of family rights and duties, a fair tax system and the equitable distribution of public goods. They are all mentioned in a much-quoted passage by John Stuart Mills—exemplarily related to the fair distribution of a collectively generated labor income:

> Some… consider it unjust that the produce of the labour of the community should be shared on any other principle than that of exact equality; others think it just that those should receive most whose wants are greatest; while others hold that those who work harder, or who produce more, or whose services are more valuable to the community, may justly claim a larger quota in the division of produce. And the sense of natural justice may be plausibly appealed to in behalf of every one of these options. (Mill 1864: 68)

Mill already indicates the finding of a detailed philosophical examination of these principles, which is widely accepted today: Which distribution principles are proposed in a distribution constellation and which can be enforced in the discourse is not decided in advance—for an equitable distribution of resources no direct general rules can be established ex ante, because the justification depends, among other things, on the subjective perception of the distribution constellation, the subjective evaluation of options and on the individual and collective weighing of needs in discourses. In the context of a fair energy supply and the associated question of a fair distribution of risks and consequential burdens, they are all familiar in one form or another and have already been addressed in part above: Is it rather understood occasionally as a singular distribution or as part of a whole sequence of expected distributions? Is maybe the distribution understood as compensation of previous distributions? Can future distributions be included in the current assessment of needs? Is there a state of emergency that forces actors into certain minimum requirements? Questions like these and many more are open for negotiation in an ideal discourse, and because there are no defining standards or unquestionable criteria to answer them with unanimous consent, no valid statements can be made ex ante about the outcome of the forthcoming discourses. The resulting distribution will be not least a matter of factual morality, co-determined by traditions and well-rehearsed practices, and less a question of rational examination and ethical rules. The assessments of needs and how they are weighed against others e.g. are not subject to any moral "jurisdiction". In the same way, there is no such thing as an appropriate yardstick for assessing risks or the willingness to live in the vicinity of technical installations. And practical outcomes often even succeed without requiring consistency and general validity of the solutions achieved. In the opposite, what people are willing to accept and are willing to take into account is very much dependent on what is socially established and what they are used to (Renn 2008). But while in most areas a distribution of energy resources has prevailed and is widely accepted that is based on the willingness of the demanders to pay, in a globally perspective the demand for a just energy supply is very much based on the intuition of an affordable and secure energy supply for everyone in the manner of a human right.[18] Those, however, who make concrete proposals and try to justify them specified with the one or other distribution principle, regularly face counter-arguments that give preference to other distribution principles, that—to cite Mill again—"the sense of natural justice may be plausibly appealed to", trained by tradition and well-rehearsed customs.[19] In any case, those who propose a distribution or are unwilling to accept a completed distribution and do not simply want to impose their position with power and force will have to justify their claims. Such justification cannot, however, be based on arguments that are not already universal or universalizable. Statements of the kind 'A (I) shall get something (more)' merely formulate a claim and are no more an argument as statements of the kind 'I need more', 'I have the older rights' or also 'We have been suppressed for centuries, now it's our turn'. An argument that wants to have prospect of examination and consideration on the participants in discourse though must be of the form 'Everyone who has property E shall get …'. 'E' thereby may stand for a particular ability or productivity, the degree of need, inherited class or family privileges, the historical suffering of a group, or any other potentially

[18]For example, the UN Secretary-General heads a "Sustainable Energy for All initiative", the aim of which is to ensure universal access to modern energy services (http://www.un.org/sustainabledevelopment/energy/, accessed 13-Dec-2019).

[19]The public debates are full of examples of more or less quickly withdrawn proposals, from flat tax to income-based tiered pricing for access to public goods, income-related fines for traffic offences and much more.

claiming quality. Anyone who claims a certain right for him- or herself or another person or group characterized on the basis of certain properties must then, for reasons of pragmatic consistency, grant this right to anyone else who possesses this characteristic. In moral communities, entitlement rights are thus "always already" claimed and justified with universal arguments—ultimately because otherwise the discourse community refuses to accept them.[20] Whether and to what extent then universalized principles such as "everyone according to his need", "everyone according to his possibilities", "everyone according to his abilities", "everyone according to his commitment", "everyone according to his merit", "everyone according to his achievements" are regarded as decisive by the parties to the discourse, and which principle flows into the decision-making, is not predetermined by any external circumstances. Here, the sciences and humanities can offer support services and provide valuable advisory services, for example with references to historical distribution effects, statistically collected or theoretically derived estimates of expected consequences of any kind of distribution, or simply with linguistically sharpened differentiation services. In no way, however, can the question of which distribution principles are to be applied be answered by pure reasoning or scientific analyses "in abstracto". According to the well-known "suum cuique" any consideration will have to take individual interests into account, as they can be ideally expressed in a free and open discourse. Therefore the requirement for universalizability is rather to be understood as a formal principle, not directed towards distribution outcomes, but towards the distribution procedure, as an requirements for its design: design distribution procedures in a way so that everyone can assert his or her claims and that everyone's individual prerequisites are taken into account! Therefore, in the modern era, the philosophical debate on justice has increasingly turned away from the so-called material questions, i.e. questions directed at the concrete distribution, and towards the more formal questions of the design of framework conditions and procedures that lead to just distributions.

From a historical point of view, questions about criteria, structures and procedures have increasingly been included in the debates on ethics, particularly with the emergence of anonymous large societies. Promising proposals and recommendations have been developed on these questions and their ability to consent can be effectively determined by the actual participation of those involved ("political philosophy"). Thus, the thinkers of the so-called Scottish Enlightenment discuss questions of justice on the basis of the phenomenon of the spontaneous self-organization of interacting communities and the allocation principles of free markets, Montesquieu proposes to protect the individual's freedom of action and expression by division of power and mutual control of governmental branches, and Kant's categorical imperative formulates the principle of rule of law against the absolute claim to power of the state ("Act only according to that maxim whereby you can at the same time will that it should become a universal law"). Where the behavioral expectations that people have of each other become insecure and volatile, a debate about the (further development of) procedures for the (re-)establishment of relationships that everyone

[20]This is a variant of Hare's (esp. 1952) universalizability argument, reformulated for distributive justice and relativized for moral communities.

(sufficiently) finds fair can, on the one hand, have the prospect of general agreement and, on the other, have a regulative effect. In this sense, Rawls (1971, 1999) and the discussion that followed him then endeavored to show criteria for the assessment of social orders that above all emphasize their suitability to establish conditions that are justly presented. In the local context, these can be required to ensure adequate access to an affordable and reliable energy supply. It cannot be expected that all scarcity problems will be eliminated—but it can be expected that all rivalry arising from scarcity conditions will be overcome in the most efficient and conflict-free way possible. Ideally, a pareto-optimal distribution is realized: There would then be no alternative to the implemented distribution that would place one participant in a position he prefers without having to place another participant in a position he does not prefer.

9.7 Universal Principles and Their Global Application

Where a justified choice from the discussed set of principles is up for discussion, the existing forms of organized interaction, as they are given by the formation of traditional communities, legal communities, territorial states, markets, etc., are an essential frame of reference. This framework sometimes consists of more informal-implicit as well as more formal-explicit rules, is sometimes well reflected by those who obey the rules, but often enough obeyed only intuitively and without recognizing, that the behavior is a rule-following and a reaction to contextual framings at all. Often enough these rules are not consistent with each other but vary from context to context.[21] Which rules people apply in everyday behavior depends very much on the interpersonal relation they have, especially whether they are closely connected or anonymous to each other, and whether they interact on a regular basis or just once, so that the actors cannot assess the reactions of the others and positive sanctions for supportive actions as well as negative sanctions for exploiting advantageous situations are not to be expected. Therefore, an extension of ethical considerations over given social frames of reference to a global scope is not merely an "upscaling" or extension which otherwise leaves all the general conditions unchanged. Although humanism and the Enlightenment have been the starting point for thinking on ethical issues in a way that includes the "whole human race", it is still unclear how this abstract formula is to be translated into promising proposals and recommendations for applicable concrete agreements and regulations.

The challenge mankind is facing here can be illustrated by an analogy that makes use of the possibility assumed in some science fiction films to coordinate actions across galaxies by means of a common definition of an intergalactical time frame

[21]This insight is well established in philosophical debates at least since L. Wittgenstein's Philosophical Investigations (1951). For an application on ethical consideration cf. McDowell (1981). For insights from behavioral research cf. Ariely (2008), Kahneman (2011) and the papers collected in Kahneman and Tversky (2000).

like a "stardate" that constitutes unambiguous reference points in order to compare temporal relations over very long distances in cosmic dimensions: At the latest since the chronometers at the stations of major cities were "synchronized" in the 19th century, we have had a concept of simultaneity at different locations and equivalent time units that is unproblematic in everyday contexts: if it is 12 o'clock in Greenwich, then also in Exeter (cf. Hylton 2015, Chap. 9). But if we ask the question: "If it is 12 o'clock in Greenwich, what time is it on the moon?" one falls into confusion: Just as the concept of simultaneity, applied to the cities of Greenwich and Exeter, is based on conventions that have been established and on considerable technical prerequisites for their implementation and control, so, of course, corresponding conventions could also be fixed for the time relations between Greenwich and (a place on) the moon. We certainly would also have a vague idea of the technical means needed to create the conditions for the necessary intersubjective control to comply with the convention—perhaps a further development of caesium atomic clocks, which time stamp the signals from GPS satellites. But up to now there are neither the necessary conventions established nor the necessary technical infrastructure. And while it might seem simple on first glance, on a closer inspection many questions arise: How should the conceptual reference systems be designed? In view of the distances (which can only be overcome by any means over time) and the physical requirements, how can technical systems be constructed with the appropriate reliability and their trouble-free functioning controlled over time? Although, when we talk about something like "stardate", the idea of a "simultaneity", spreading homogeneously and space-filling (with ourselves as a point of reference of course), seems to be clear on the first glance, nevertheless the frame of reference necessary for an answer is missing. And although such a system can theoretically be created, coordination of action over cosmic distances always requires a technically controlled convention that regulates simultaneity for two related locations (such as the location of two coordinated clocks), as is already the case for the coordination of railway timetables.[22]

Humanity finds itself in an analogous situation when it is confronted with the question of how to "create" justice and ensure a fair energy supply on a global scale—not only in thought experiments in principle, but also in real and effective ways. Neither is there a global "frame of reference" established culturally and historically or by explicit agreement across borders. The reference to one's own intuitions, however stable they may be when one thinks, for example, of distribution constellations within one's personal environment, is just as little a reliable point of reference as the spontaneous reaction people show when they recognize a drowning child in a pond nearby that they can save by taking only minor risks for themselves. Intuitions that are built up and trained in constellations and with examples like this would not suffice as an argumentative basis for this or that distribution ration on a global scale or to demonstrate an obligation towards all people to save them from need.[23] If one takes a closer look and examines—as illustrated by the example of "stardate"—the

[22]For a resilient control, at least two clocks are required, see Janich (1985).

[23]Singer (2009), for example, undertakes this thoroughly popular but misleading attempt.

supposed self-evidences with regard to the conditions of their possibility, then the validity of the prerequisites of such transfers from the immediate environment into the global scale becomes apparent: Moral intuitions, as already indicated above, are learned, habitualized dispositions of judgement that are formed in an environment of action that is characterized by direct or indirect acquaintance of the actors and, if necessary, by a direct or indirect acquaintance of the actors. The moral community in which the intuitions are developed, is constituted and consolidated by regular social control and the possibility of direct, personally addressed sanctions. Often even supported by mutual sympathy the actors pursue similar goals and share attitudes and expectations that arise from the same tradition the same social environment and are supported by homogenous educational ideals. But these intuitions quickly fail and are misleading where they are applied to anonymous large societies in which the interactions are not supported by emotional or firm social ties and where one cannot address stable behavioral expectations to unknown interaction partners. Where partners of interaction change frequently and interaction is volatile and anonymous, where the attribution of actions and their consequences is insecure and where positive reactions to supportive behavior cannot help to stabilize a common moral and where damaging or exploiting behavior cannot be sanctioned efficiently, expectations that stem from experiences in smaller, interconnected groups will be disappointed. There is a fundamental gap between interaction that takes place in small, mutually connected groups, and that in larger societies and different tools and measurements are required for making it fruitful and keep it functioning. So the attempt to tie the division of labor within the family or the distribution of food during a hike with friends to the willingness to pay a certain amount of money and to bind distribution to a price signal would be just as misguided as the expectation that among the citizens of a municipality or a state, through mutual goodwill, a just and satisfactory distribution of goods and burdens would "naturally" result. At least when it comes to building up and maintaining common goods together with a fair distribution of the necessary efforts and the outcoming benefits fairly, the limits of a solution based solely on good will or right intuition become apparent. The "Tragedy of the Commons" has become proverbial for the exploitation and freeriding problems that arise quickly in constellations like this where common resources tend to be overused on the on hand side and burdens of care and conservation are neglect on the other.[24] Larger communities, from village communities to modern nation-states, therefore, have developed numerous strategies to escape this common dilemma, with the formation of an identifiable group identity playing an important role. This includes, on the one hand, the creation of binding forces within the group by providing public goods and granting participation rights to those and only those who belong to the group, and, on the other hand—as the flipside of the coin, so to speak—the sharp exclusion of "the others" who do not belong to the group (Ostrom 1990).

[24]Hardin (1968). The literature structure underlying the problem, especially in numerous cases of overuse of resources. Cf. representatively for many of the systematic investigations Trapp (1998).

Within the groups, moral rules form the common basis for a successful, because largely uniform interaction, which is lowered into stable intuitions and fixed dispositions for action. At the same time, a certain uniformity of interaction is a prerequisite for moral rules to exist as an "ensemble of practices" (Marquard 1986) and to constitute the practices and communities that the individual presupposes as a frame of reference for his planning and action. And even if all Enlighteners since antiquity have rightly lamented failures or limitations of traditional morals and called upon people to work on their rational further development, countless developments can be listed in history, all of which were brought about in the name of justice, but which have remained in memory above all because of their unjust consequences, the destruction of prosperity and peace. Thus, overtaxing morals can contribute to their erosion and thus also to the dissociation of communities constituted by common customs. Without the potential for action that arises from a successfully cooperating common purpose, however, services that are expected to have a global impact cannot be provided.

In order to overcome the fundamentally group-related frameworks the evaluation of distribution principles as appropriate and just dependent of, to avoid inappropriate projections of one's own ideas and to "globalize" an idea of justice that can then be implemented systematically without endangering the necessary basis, suitable, practically effective forms of organized regulation must be built. For being effective and having a steering effect, it is required that they are known to all participants, that they can be observed and controlled in an appropriate manner, that they can be adapted and improved in an orderly manner if circumstances change, and that the participants are willing to orient themselves to these framework conditions at all. This willingness can be generated by perceptibly sanctioning deviant behavior or by creating a normative framework that is "incentive-compatible", so that it is in the interest of every individual to act in accordance with it whatever purposes he or she may be pursue. Very quickly, however, the point is reached at which visible, formally founded and permanently established institutions must take over the tasks that in non-anonymous small communities can often still be carried out with support of a common moral and without further organizational effort. Last but not least, the emergence and functioning of markets that want to ensure efficient cooperation even among mutually anonymous participants, especially for certain forms of interaction, requires such institutional framework conditions. In order for market interactions to take place, for example, players offering their services must be able to rely on the fact that in the case of a freely negotiated and agreed exchange, the consideration will also be provided. Under anonymous conditions between actors who enter into negotiations with very different prerequisites and under who differ widely in their share of power, abilities and resources, this trust will not easily be established. But those who are concerned about one-sided exploitation and have to fear that they cannot reliably acquire by exchange what they need for their way of life will not be prepared to specialize in activities for which they find favorable conditions and in which they can be productive. It will also have to devote a significant proportion of its time and goods resources to defending itself against the threat of assault by others. Institutions that, for example, by providing immediate, reliable and efficient sanction instruments for those who refuse consideration after receiving services, are therefore an essential prerequisite for the formation and operation of markets—and thus for the development of prosperity that is associated with the possibility of being able to concentrate on activities with which one can be particularly productive. A market dependent on a technical infrastructure in which energy, energy sources or energy services are efficiently exchanged would not be conceivable under such circumstances.

In the same way that clocks synchronized over distances by technical means first of all create a time regime in the area and a resilient idea of simultaneity, which, for example, allow timetables and flight plans to be aligned with sufficient accuracy to the place of departure and destination, such institutions then enable a regime of justice, in which reactions to collective problems of action and just solutions for colliding claims are founded. Only such institutions will be able to provide effective procedures to balance submitted claims in an acceptable manner and to provide effective strategies for conflict resolution. And because obligations are created first and foremost by agreements, are made permanent through conventions, and are made explicit through decision-making, the conclusion of contracts and the formation of institutions can build durable standards that include consideration for the other on a global scale.[25]

Even though the technical and organizational requirements for a uniform time regime in cosmic dimensions have not yet been met and it remains unclear how the synchronization between clocks that are light years apart can be controlled (especially when acceleration affects the running behavior of clocks), the locally developed practices of temporal coordination here on earth can guide the considerations as a "guiding idea" of cosmic simultaneity as to what a solution should achieve, though it may be far away from having a concrete idea of what it may look like in realization and how the many detailed problems may be solved. Analogously, the previous ideas of justice offer everyone a kind of guiding idea of globally just relations. In particular, they are the basis for the perception of existing relationships as unjust and thus the starting point for activities to change these relationships in line with the guiding idea. If, however, one directs one's attention to the questions of detail, to the question of the more precise determination of the objects and constellations of just distributions, to the question of who should actually be involved and which principles should be used for assessment, then the picture becomes more diffuse. Who believes that globally valid and recognized standards for just relations could easily be developed according to the pattern of strategies and rules, some of which have evolved over centuries and form the often invisible cement of their own relatively homogeneous

[25]Th. Hobbes (1651) describes a state in which there is competition for the elementary resources, but in which obligations and claims have not already been created by agreements, as an original state. Until the standards for this are created with the constitution of the first rules, there is no such thing as justice or injustice in this state—people in the state of nature virtually only follow the laws of nature as long as they are the only standard. Such an original state is conceivable both at the level of the individual, under suitable marginal assumptions, e.g. a biological disposition to instinctive care for one's neighbors or the necessity to form defense communities against organized enemies, etc., and also at the level of the hordes, settlement communities or the territorial states.

and relatively stable communities (Elster 1989), is rather misguided by this idea. But what can be achieved—continuing Aristotle's line on to the Scottish Enlightenment, to Montesquieu, Kant and Rawls—is the critical assessment of existing strategies and rules, led by criteria, and the construction of organizations, institutions and markets according to their suitability to work towards this guiding idea.

9.8 Questions of Justice-Oriented Rationality of Action

Especially with regard to ethical considerations that will helpful in transcending cultural-historical borders on a global scale, ethics is thus confronted with basic challenges, and only a few approaches, if any, have been developed to counter them.[26] In principle, the question would have to be answered as to what proposals and recommendations for "universally valid" and at the same time practically implementable rules would look like if, on the one hand, the implementability is not independent of what the addressees are prepared to accept on the basis of their respective moral presettings, and, on the other hand, it is not the factual, but often arbitrary agreement ("acceptance") on the basis of given moral presettings that serves as the yardstick, but the general acceptability, that is rationally founded in that it includes all aspects.

The orientation towards general acceptability aims at generalizability in a multiple sense:

Non-arbitrarity

As is expressed by the most basic legal principle, same cases should always be treated equally and unequal unequally. In jurisprudence, this rule is referred to as the prohibition of arbitrariness: Only if same cases are treated equally, only if action (or action planning) A and action (or action planning) B, which corresponds to A in all relevant points, consequences included, are judged equally, actors can establish stable mutual behavioral expectations. If within a group resources are allocated today according to principle p_1, tomorrow, in an otherwise identical situation, according to principle p_2 and the next day in an again similar situation according to principle p_3, those who support this allocations and do not give a justification that refers to relevant differences in the cases will not be seen as a reliable interaction partner. If today's distribution of joint earnings from work favors A while yesterday's distribution favored B without there being any justification for the unequal treatment, if the behavior that has been applauded today is punished tomorrow, then actors cannot plan social interaction for the future and cannot anticipate the other's action in the future. In case of doubt, therefore, it seems rational to refuse to cooperate further on or to consume today what one might otherwise have traded tomorrow in expectation of a greater benefit. Non-arbitrarity, based on generalizable justifications, therefore is essential for the establishment and preservation of a social framework that allows

[26]Cf. the overviews of the diversity of approaches in Moellendorf and Widdows (2015) and the thematic compilations in Stückelberger (2016).

for plannable cooperation. It is what I. Kant in his “Groundwork of the Metaphysic of Morals” (1785) identifies as a “condition of possibility” for stable social relations and not the least for moral (“Sittlichkeit”) itself. Generalizability thereby aims at the assessment of entire classes (sets) or types of identical cases and considers individual subjects of assessment in the light of these similarities. In this light, individual distributions for which there would have been a more favorable alternative for certain actors concerned may appear acceptable, even if they find it difficult to accept the alternatives spontaneously and only a reference to the overriding rationality of equal treatment is needed in order to achieve reflected acceptance.

Robustness

General statements of the type: “whenever…, then…”, logically analyzed as “for all cases x: if x, then …”, do not only claim validity for previously observed applications, but also for future applications. In particular, if the connection established with such statements is to become the basis for future action planning, the cases will be differentiated and specified until the statements are as robust as possible against the manifold variance of the circumstances. If, after countless successful cases, the means M does not always prove to be successful when actor A wants to achieve a certain purpose, then the if-clause may have to be supplemented by further conditions (“if x, then …, unless…”, resp. “for all cases x, for all cases y: if x and y, then …”). The same also applies to agreements or rules designed to ensure future coordination of the action plans of various actors with as little disruption as possible. The fundamental fallibility of scientific statements corresponds to the concept of incomplete contracts: All eventualities cannot be anticipated.

Resilience

The target systems on which people plan are never complete and consistent, they change in the course of life, but also in different life situations. Actors sometimes subordinate goal a to goal b, sometimes goal b to goal a, be it that the framework conditions change (e.g. slippage of prosperity into poverty), that roles or perspectives change (e.g. between that of beneficiary of public goods and that of the taxpayer), be it that social moods fluctuate.[27] Agreements should be agreeable for each of the participants among all the changes that arise.

Inclusion

Agreements should be acceptable to each of the participants and should not be at the expense of third parties. Those whose action plans are affected should be able to give their approval, whether directly affected or only by distant consequences in time or space.

In this context, it is worth mentioning, that generalizability and acceptability are not characteristics that can be attributed once and for all to an agreement, convention, rule or standard. As in the case of scientific statements, whose claim to general

[27]Cf. the contributions in Lichtenstein and Slovic (2006).

validity must always be put to the test and occasionally rejected, here too the reservation of fallibility always applies. In particular, the equality of cases cannot simply be stated—for example by simply ostending to the cases. Correspondingly, the claim for an equivalent treatment could not simply be made. Textbook examples as they are often cited in ethical debates are deceptive here. A typical case is again the one of the passer-by, witnessing how a child threatens to drown in a pond who he could save while having to accept only small risks for himself, which is referred to in order to show that in any case, where there is a helpless person in need one could save with small risks for oneself, one it obliged to do so, regardless of where the person in need is, whether in close proximity or at a greater distance.[28] Even if one acknowledges—which is by no means trivial—the existence of an obligation to save the child in a situation as described, this would not imply an obligation of equal scope for those cases in which not a fellow human being is visibly threatened by drowning within reach, but a person in the distance, maybe on another continent, is threatened by starvation, who could be saved for the equivalent of only a dollar. In both cases a life-threatening situation exists and the risk or the effort for the provider of assistance to remedy this situation may well be similar. In this respect, the cases are initially to be regarded as cases of the same type, so that, prima facie, by referring to the requirement of non-arbitrary nature from the given (or assumed) obligation towards the drowning child, an obligation could be derived towards all children in life-threatening situations, as long as they only could be saved with similarly assessed risks. But that the information about the situation in the distance is a mediated one, that other actors stand between the actor and the child threatened with fatal danger, that he has to make an effort to inform himself about sufficiently reliable help options that can only be implemented indirectly and with time delay, that he cannot react to immediate visual and acoustic stimuli, but if necessary, that his reaction to the situation in the distance requires deriving a state of emergency from words and numbers etc. makes the assumption of an equality of constellations, on which the conclusion of analogy is based, already problematic. Above all, however, the argument assumes that (i) generalizability would be a demand directed quasi from the outside to the actions of the individual and that (ii) the equality of cases, which any requirement of generalizability presupposes as given, is the subject of a statement, not of a (joint) determination. In both cases this would be a normativistic fallacy.[29] The orientation towards generalizability and the joint development of an understanding of the same cases to be treated in the same way serves the organization of cooperation and the creation of a reliable environment for action, which provides security for individual

---

[28] See Singer (2009). The analogy argument is often quoted and is—especially outside professional philosophical circles—often cited in order to promote financial transfers from developed regions to regions in need of development. Cf. Peter Singer's own online platform https://www.thelifeyoucansave.org (accessed 13-Dec-2019).

[29] A normativistic fallacy is the missed (but often superficially convincing) attempt to derive concrete requests or recommendations from general principles without due consideration being given to situational circumstances that stand in the way of the application of the principle to the given case. The definition of the normativistic fallacy goes back to Höffe (1981:16), cf. for instance Gorke (2003, Chap. 14).

planning. So, it is perhaps a praiseworthy and supportive project to promote equal treatment of cases and to fight politically for framework conditions which then also favor de facto equal treatment. The effort to demonstrate a general duty for everybody from the individual case, on the other hand, is as misguided as the effort to infer a general law of nature from a single observation. What is to be combined into a class of cases that require equally treatment and what is to be treated as unequal is rather the object and—if successful—the result of incessant negotiations and constant readjustment. And it is not even clear from the outset who in this negotiations bears the burden of proof: traditions and the rules included in them often form a kind of "default attitude" here, so that the burden of justification is attributed to those who want to follow deviating rules. Not only do traditional practices occasionally prove to be inconsistent, unsuitable for new situations or even disruptive for new challenges—for fundamental reasons the mere reference to existing practices can never be considered a rational answer when an existing practice is called into question. One argument that should not be underestimated, however, is the reference to the evolutionarily developed functionality of existing practices, even if further development is required in order to adapt the practices to new and changed framework conditions. In spite of all the restrictions on action that would result from the binding to the existing, established—fundamentally new, "tailor-made" solutions presuppose that the framework conditions for their design and implementation would have been fully grasped, or at least the essential sections would have been grasped. History gives clear indications that it is highly risky to make such prerequisites. With regard to the entitlement to participate in such distribution discourses and the obligation to comply with the discursive agreements reached there, it is, however, possible to formulate an equality rule at a "meta-level". A reliable conflict resolution will only be available if all those who are potentially able to make a claim are given the same opportunity to participate in discourse, and if, in addition, all those who participate in discourses are equally committed to a consensus of distribution. The individual, where this participation in the discourse of the individual takes place in an organizational framework through mandate holders and representatives, is not released from responsibility. It remains his or her duty to incentivize the incentivators to set fair, appropriate and effective incentives (Trapp 1998).

An energy supply system (in general: a distribution of resources) that is justified in orientation to the principle of ethical universalism by a discourse equally accessible to all (its result may lie in equal or unequal distribution) is called "just".[30]

This shows that the prescriptive content of "justice" has two clearly distinguishable moments: firstly, a moment of equality, which, however, refers to the pre-discursive entitlements and obligations, such as equal access to discourse, equal treatment of all participants by the "rules of procedure", etc. Secondly, there is a moment of justified inequality, which refers to the intra-discursively found distributions, i.e. to the result

[30] As altogether in this contribution here "justice" is used only in the sense of distributive justice (iustitia distributiva), not in the sense of exchange justice (iustitita commutativa) or justice before the law (iustitia legalis).

of the distribution. Accordingly, two postulates can be formulated as an explication of the prescriptive content of "justice":

Equality Postulate

Act in such a way that everyone who makes relevant claims receives the same rights and obligations when participating in discourses!

Distribution Postulate

Distribute so that each allocation is justified by a discourse of equal participants!

These postulates refer to the scheme of distribution discourses and can therefore be described as "formal". The more "material" criteria of distributive justice are obtained by looking at the arguments put forward by the participants to justify their claims. By definition, actors try to create opportunities and to avoid risks. Regardless of the question of whether there can be purely individual opportunities and risks at all, a fair distribution of opportunities and risks has to do with the question of what should be expected of others and what others should expect of us. For the decision of this question, individual or collective preferences of the respective individual or collective actors cannot therefore be (alone) decisive. This also applies when entire generations are assumed as actors. Ultimately, therefore, only procedures can be determined according to which it is decided whose claim is to be taken into account and whose "will" is to be realized in whole or in part.

For the distribution substratum "opportunities" and "risks", the demand for pragmatic consistency results first of all directly in the

Rule of Risk-Taking

Be prepared to take risks if you have already accepted or expected others to accept similar risks and therefore consider them acceptable!

Further rules, which then supplement such a general rule for concrete cases—derived from basic considerations—and develop it in a practical manner, are then possible further candidates for justification, to which the discourse community can commit itself:

Rule of Opportunity Sharing

Act in such a way that you let the risk takers share in the opportunities as much as possible!

Rule of Risk Allocation

Decide risk options in such a way that those who so far have benefited least from opportunities have the greatest relative advantage!

Risk Provisioning Rule

Act in such a way that you can compensate those who bear the risks of your opportunities as much as possible in the event of damage!

Such rules, however plausible they may be, immediately lead to considerable operational difficulties when considering interactions of a rapidly reached level of complexity, but especially when considering the interaction network of an entire society. Even if one makes optimal assumptions for the individual actors with regard to their insight and intentions for action, it is therefore quite unclear at the collective level how an equitable distribution of opportunities and risks can be thought out. Philosophers, economists, lawyers and others have been working for several years on formal models that can make clear how distributional discourses function in relation to promising and risky actions.

9.9 Supranational and Internal Conflicts

Management of internal conflicts, the peaceful reconciliation of claims between citizens in the countless questions of individual lifestyles, and the maintenance of a willingness to cooperate on all sides in a large collective are only some of the numerous tasks and functions of state organizations. Ideally, the state is legitimized as a "cooperative enterprise" by serving the mutual advantage of its citizens wherever possible (Rawls 1971, p. 105) and enabling them to reap the fruits of peaceful cooperation without requiring and spending substantial resources on it. Where the rule of law is established and institutionalized, state agencies themselves are subject to the law, and conflicts between state representatives are manageable as efficient as conflicts between citizens. The state protects the citizen against attacks by others on property and life, thereby creating the planning security required for investment in the future and sustainable action. Ideally, supranational organizations and their relationship to the individual states will also be designed precisely in accordance with these requirements. However, since it is hard to imagine anything other than tiered responsibilities and multi-level systems (see Sect. 8 above), the coupling is also associated with challenges that have ethical relevance and can entail considerable burdens for the establishment of equitable conditions.

In order to maintain the willingness of individual citizens or like-minded citizens to cooperate, who unite to achieve common goals, internal conflict regulation can have a binding effect that is precisely contrary to supranational cooperation interests. This becomes particularly clear when, for example, a volume of CO_2 emissions defined by its accumulated effects on the climate is defined as a budget that is to be distributed among the global community as a whole. Measures that are adopted at the national level to ensure internal cooperation and serve to reconcile the interests of actors who pursue incompatible goals due to different expectations of prosperity, different risk appetites or different conditions for action are often at the expense of third parties: Whether it is regional economic interests that allow one nation to hold on to the recycling of the locally cheap and abundantly available coal, or whether it is the rejection of risks associated with the capture and storage of CO_2 or the continued operation of largely CO_2-neutral nuclear power plants—compromises that are found on this basis might be an effective conflict resolution at the national level but at the same time might be unfair in view to third parties insofar they do not exploit available potentials for emission reduction and requiring an unnecessary share of the overall available residual budget. This is especially the case if there was little or no scope for others to switch to other energy options—e.g. because of the level of social or technical development or because of other regional circumstances. The reduction of the budget by one side would then either mean that they would have to compensate for the use by others by reducing energy consumption or would have to accept the risks of climate change to which others would be exposed. The examples chosen above make it clear that this is above all a question of the fair distribution of risks: the low technical and economic risks associated with the maintenance of coal-fired power plants, for example, and the considerable use of CO_2 quotas, are offset by the risks to which others are exposed as a result of the remaining options. Anyone who fails to exploit the savings potential associated with the operation of nuclear power plants or other low-emission technologies will expose others to a certain amount of risks that goes along with climate change. Given the scientific assessment of the respective risks (see Chapters "Strategic Energy Requirements—Technological Developments" and "Aspects of Environmental Compatibility of Energy Systems"), this would not be in accordance with the rule of risk allocation.

In western industrial nations—precisely because the question of who is allowed to impose what risks on whom is the subject of social compensation—decisions on the energy infrastructure of such scope are taken by representatives of society and often with the involvement of the affected part of society. In particular, the binding effect of such decisions, which are the result of intensive direct citizen participation, may be regarded as particularly high. A representative of the state would have hardly any room for maneuver in international negotiations to even relativize the decisions thus found. However, as the arguments about the risks of nuclear power or about installations such as power lines or wind turbines, which are intended to make energy supply with renewable energy sources possible, show, the requests for participation are generally not raised with a view to the available emissions budget, but rather with a view to quasi local risks, which are then passed on to inhabitants of other states as "externalities".

Such externalities arise not only because of the global interconnectedness due to an atmosphere that is common to all. Such externalities are as well formed by the creation of common markets and global trade: waste is shipped to countries where few benefit from cash payments, while large sections of society, which are not involved in decision-making and do not have sufficient political power, have to bear the health risks. On international markets, raw materials are traded that help the consumer to spend a comfortable and healthy live while people of the region where they are mined accept considerable risks for themselves and the environment. More often than not, the cash flows allow for perpetuating a balance of power that stabilizes the exploitative conditions. While in countries which are already largely developed as "cooperative enterprises for the benefit of all" or which owe their existence to this idea, the discovery of profitable mineral resources finds good prerequisites for promoting general prosperity (example: Norway), such a discovery can easily lead to exploitative conditions in societies which are already extremely unequal and not protected by robust constitutions under the rule of law. In the literature, this phenomenon is referred to as the resource trap or resource curse (Sachs and Warner 2001; Wenar 2015). It is associated with dictatorship and oppression, with corruption and instability, with a high debt risk, lasting political instability and persistent violence in the above-mentioned failed states (Acemoglu and Robinson 2012). Supranational institutions alone cannot provide the necessary balance here. At the same time, individuals are overwhelmed by the diversity of information and the complexity of the interrelationships. It is therefore the responsibility of the individual states in particular not only to take up expressions of interest and to orient themselves towards the willingness of citizens to accept them, but also to promote acceptable decisions supported by rational argumentation in an international vote. Not least because of the potential offered by the new media to organized groups to assert their special interests, the state, as a cooperative enterprise, is called upon to the advantage of all to review the generalizability of the incentives given to it by citizens and domestic organizations (see Sect. 9.8 above).

Appendix A
Strategic Energy Requirements

C. F. Gethmann et al., *Global Energy Supply and Emissions*,
Ethics of Science and Technology Assessment 47,
https://doi.org/10.1007/978-3-030-55355-5

Table A.1 Environmental criteria and indicators established in the NEEDS project (Hirschberger and Burgherr 2015)

Criteria/indicator	Description	Unit
Environment	Environment-related criteria	
Resources	Resource use (nonrenewable)	
Energy	Energy resource use in whole lifecycle	
Fossil fuels	This criterion measures the total primary energy in the fossil resources used for the production of 1 kWh of electricity. It includes the total coal, natural gas. and crude oil used for each complete electricity generation technology chain	MJ/kWh
Uranium	This criterion quantifies the primary energy from uranium resources used to produce 1 kWh of electricity. It includes the total use of uranium for each complete electricity generation technology chain	MJ/kWh
Minerals	Mineral resource use in whole lifecycle	
Metal ore	This criterion quantifies the use of selected scarce metals used to produce 1 kWh of electricity. The use of all single metals is expressed in antimony-equivalents, based on the scarcity of their ores relative to antimony	kg(Sb-eq.)/kWh
Climate	Potential impacts on the climate	
CO_2 emissions	This criterion includes the total for all greenhouse gases expressed in kg of CO_2 equivalent	kg(CO_2-eq.)/kWh
Ecosystems	Potential impacts to ecosystems	
Normal Operation	Ecosystem impacts from normal operation	
Biodiversity	This criterion quantifies the loss of species (flora and fauna) due to the land used to produce 1 kWh of electricity. The "potentially damaged fraction" (PDF) of species is multiplied by land area and years	PDF * m^2 * a/kWh
Ecotoxicity	This criterion quantifies the loss of species (flora and fauna) due to ecotoxic substances released to air, water, and soil to produce 1 kWh of electricity. The "potentially damaged fraction" (PDF) of species is multiplied by land area and years	PDF * m^2 * a/kWh
Air pollution	This criterion quantifies the loss of species (flora and fauna) due to acidification and eutrophication caused from production of 1 kWh of electricity. The "potentially damaged fraction" (PDF) of species is multiplied by land area and years	PDF * m^2 * a/kWh
Severe accidents	Ecosystem impacts in the event of severe accidents	
Hydrocarbons	This criterion quantifies large accidental spills of hydrocarbons (at least 10,000 tones) which can potentially damage ecosystems	t/kWh

(continued)

Table A.1 (continued)

Criteria/indicator	Description	Unit
Land contamination	This criterion quantifies land contaminated due to accidents releasing radioactive isotopes. The land area contaminated is estimated using probabilistic safety analysis (PSA). Note: Only for nuclear electricity generation technology chain	km^2/kWh
WASTE	Potential impacts due to waste	
Chemical waste	This criterion quantifies the total mass of special chemical wastes stored in underground repositories due to the production of 1 kWh of electricity. It does not reflect the confinement time required for each repository	kg/kWh
Radioactive waste	This criterion quantifies the volume of medium and high-level radioactive wastes stored in underground repositories due to the production of 1 kWh of electricity. It does not reflect the confinement time required for the repository	M3/kWh

Table A.2 Economics criteria and indicators established in the NEEDS project (Hirschberger and Burgherr 2015)

Criteria/indicator	Description	Unit
Economy	Economy-related criteria	
Customers	Economic effects on customers	
Generation cost	This criterion gives the average generation cost per kilowatt-hour (kWh). It includes the capital cost of the plant (fuel) and operation and maintenance costs. It is not the end price	€/MWh
Society	Economic effects on society	
Direct jobs	This criterion gives the amount of employment directly related to building and operating the generating technology, including the direct labor involved in extracting or harvesting and transporting fuels (when applicable). Indirect labor is not included. Measured in terms of person-years/GWh	Person-years/GWh
Fuel autonomy	Electricity output may be vulnerable to interruptions in service if imported fuels are unavailable due to economic or political problems related to energy resource availability. This measure of vulnerability is based on expert	Ordinal
Utility	Economic effects on utility company	
Financial	Financial impacts on utility	
Financing risk	Utility companies can face a considerable financial risk if the total cost of a new electricity generating plant is very large compared with the size of the company. It may be necessary to form partnerships, with other utilities or raise capital through financial markets	€
Fuel sensitivity	The fraction of fuel cost to overall generation cost can range from zero (solar PV) to low (nuclear power) to high (gas turbines). This fraction therefore indicates how sensitive the generation costs would be to a change in fuel prices	Factor
Construction time	Once a utility has started building a plant, it is vulnerable to public opposition, resulting in delays and other problems. This indicator therefore gives the expected plant construction time in years. Planning and approval time is not included	Years
Operation		
Marginal cost	Generating companies "dispatch" or order their plants into operation according to their variable cost, starting with the lowest cost base-load plants up to the highest cost plants at peak load periods. This variable (or dispatch) cost is the cost to run the plant	€ cents/kWh

(continued)

Table A.2 (continued)

Criteria/indicator	Description	Unit
Flexibility	Utilities need forecasts of generation they cannot control (renewable resources such as wind and solar), and the necessary start-up and shut-down times required for the plants they can control. This indicator combines these two measures of planning flexibility, based on expert judgment	Ordinal
Availability	All technologies can have plant outages or partial outages (less than full generation) due to either equipment failures (forced outages) or maintenance (unforced or planned outages). This indicator tells the fraction of the time that the generating plant is available to generate power	Factor

Table A.3 Social criteria and indicators established in the NEEDS project (Hirschberger and Burgherr 2015)

Criteria/indicator	Description	Unit
Social	Social-related criteria source: NEEDS Research Stream 2b survey of social experts. Quantitative risk based on PSI risk database	
Security	Social security	
Political continuity	Political continuity	
Secure supply	Market concentration of energy suppliers in each primary energy sector that could lead to economic or political disruption	Ordinal scale
Waste repository	The possibility that storage facilities will not be available in time to take deliveries of waste materials from whole life cycle	Ordinal scale
Adaptability	Technical characteristics of each technology that may make it flexible in implementing technical progress and innovations	Ordinal scale
Pol. Legitimacy		
Conflict	Refers to conflicts that are based on historical evidence. It is related to the characteristics of energy systems that trigger conflicts	Ordinal scale
Participation	Certain types of technologies require public, participative decision-making processes, especially for construction or operating permits	Ordinal scale
Risk	Risk	
Normal risk	Normal operation risk	
Mortality	Years of life lost (YOLL) by the entire population due to normal operation compared with/without the technology	YOLL/kWh
Morbidity	Disability adjusted life years (DALY) suffered by the entire population due to normal operation compared with/without the technology	DALY/kWh
Severe accidents	Risk from severe accidents source: NEEDS Research Stream 2b for severe accident data	
Accident mortality	Number of fatalities expected for each kWh of electricity that occurs in severe accidents with five or more deaths per accident	Fatalities/kWh
Maximum fatalities	On the basis of the reasonably credible maximum number of fatalities for a single accident for an electricity generation technology chain	Fatalities/accident
Perceived risk	Perceived risk	
Normal operation	Citizens' fear of negative health effects due to normal operation of the electricity generation technology	Ordinal scale

(continued)

Table A.3 (continued)

Criteria/indicator	Description	Unit
Perceived acc	Citizens' perception of risk characteristics, personal control over it, scale of potential damage, and their familiarity with the risk	Ordinal scale
Terrorism	Risk of terrorism	
Terror potential	Potential for a successful terrorist attack on a technology. On the basis of its vulnerability, potential damage and public perception of risk	Ordinal scale
Terror effects	Potential maximum consequences of a successful terrorist attack. Specifically for low probability, high consequence accidents	Expected fatalities
Proliferation	Potential for misuse of technologies or substances present in the nuclear electricity generation technology chain	Ordinal scale
Residential ENV.	Quality of the residential environment	
Landscape	Overall functional and aesthetic impact on the landscape of the entire technology and fuel chain. Note: Excludes traffic	Ordinal scale
Noise	This criterion is based on the amount of noise caused by the generation plant, as well as transport of materials to and from the plant	Ordinal scale

Table A.4 NEEDS technologies for year 2050 (Hirschberger and Burgherr 2015)

Characteristics	Units	1	2	3	4	5
		Nuclear plants		Advanced fossil		
		EPR	EFR	PC	PC-post CCS	PC-oxy-fuel CCS
		European pressurized reactor	Sodium fast reactor (Gen IV fast breeder reactor)	Pulverized coal (PC) steam plant	Pulverized coal (PC) plant with carbon capture and storage (CCS) post combustion	Pulverized coal (PC) plant with carbon capture and storage (CCS) oxyfuel combustion
Type of fuel		U235, 4.9%	Mixed oxide	Hard coal	Hard coal	Hard coal
Electricity efficiency	%	0.37	0.4	0.54	0.49	0.47
Electric generation capacity	MW	1590	1450	600	500	500
Load factor (expected hours/year)	Hours/year	7916	7889	7600	7600	7600
Annual generation (expected)	kWh/year	1.26E+10	1.14E+10	4.56E+09	3.80E+09	3.80E+09
Construction time	Years	4.8	5.5	3	3	3
Capital cost (net present value)	€/kWe	1498	1900	983	1560	1560
Total capital cost (net present value)	M€	2383	2756	590	780	780
Plant life	Years	60	40	35	35	35
Average cost of electricity	€cents/kWhe	3.01	2.68	2.96	3.94	4.00

(continued)

Table A.4 (continued)

Characteristics	Units	6	7	8	9	10
		Advanced fossil			Integr. Gasification Combustion Cycle	
		PL	PL-post CCS	PL-oxyfuel CCS	IGCC coal	IGCC coal CCS
		Pulverized lignite (PL) steam plant	Pulverized lignite (PL) plant with carbon capture and storage (CCS) post combustion	Pulverized lignite (PL) plant with carbon capture and storage (CCS) oxyfuel combustion	Integrated gasification combined cycle (IGCC)	Integrated gasification combined cycle (IGCC) with carbon capture and storage (CCS)
Type of fuel		Lignite	Lignite	Lignite	Hard coal	Hard coal
Electricity efficiency	%	0.54	0.49	0.47	0.545	0.485
Electric generation capacity	MW	950	800	800	450	400
Load factor (expected hours/year)	Hours/year	7760	7760	7760	7500	7500

(continued)

Table A.4 (continued)

Characteristics	Units	6	7	8	9	10
		Advanced fossil			Integr. Gasification Combustion Cycle	
		PL	PL-post CCS	PL-oxyfuel CCS	IGCC coal	IGCC coal CCS
		Pulverized lignite (PL) steam plant	Pulverized lignite (PL) plant with carbon capture and storage (CCS) post combustion	Pulverized lignite (PL) plant with carbon capture and storage (CCS) oxyfuel combustion	Integrated gasification combined cycle (IGCC)	Integrated gasification combined cycle (IGCC) with carbon capture and storage (CCS)
Annual generation (expected)	kWh/year	7.37E+09	6.21E+09	6.21E+09	3.38E+09	3.00E+09
Construction time	Years	3	3	3	3	3
Capital cost (net present value)	€/kWe	989	1560	1560	1209	1505
Total capital cost (net present value)	M€	939	1248	1248	544	602
Plant life	Years	35	35	35	35	35
Average cost of electricity	€cents/kWhe	3.01	4.08	4.16	6.17	7.26

(continued)

Table A.4 (continued)

Characteristics	Units	11	12	13	14	15
		Integrated gasification combined cycle				
		IGCC lig	IGCC lig CCS	GTCC	GTCC CCS	IC CHIP
		Integrated gasification combined cycle (IGCC)	Integrated gasification combined cycle (IGCC) with carbon capture and storage (CCS)	Combined Cycle	Combined cycle with carbon capture and storage (CCS), post combustion	IC engine cogeneration
Type of fuel		Lignite	Lignite	Natural gas	Natural gas	Natural gas
Electricity efficiency	%	0.525	0.465	0.65	0.61	0.44
Electric generation capacity	MW	450	400	1000	1000	0.2
Load factor (expected hours/year)	Hours/year	7500	7500	7200	7200	5000
Annual generation (expected)	kWh/year	3.38E+09	3.00E+09	7.20E+0.9	7.20E+0.9	1.000E+06
Construction time	Years	3	3	3	3	1

(continued)

Table A.4 (continued)

Characteristics	Units	11	12	13	14	15
		Integrated gasification combined cycle				
		IGCC lig	IGCC lig CCS	GTCC	GTCC CCS	IC CHIP
		Integrated gasification combined cycle (IGCC)	Integrated gasification combined cycle (IGCC) with carbon capture and storage (CCS)	Combined Cycle	Combined cycle with carbon capture and storage (CCS), post combustion	IC engine cogeneration
Capital cost (net present value)	€/kWe	1209	483	440	615	879
Total capital cost (net present value)	M€	544	483	440	615	0
Plant life	Years	35	35	25	25	20
Average cost of electricity	€cents/kWhe	6.57	6.78	5.99	8.69	11.10

(continued)

Table A.4 (continued)

Characteristics	Units	11	12	13	14	15
		Integrated gasification combined cycle				
		IGCC lig	IGCC lig CCS	GTCC	GTCC CCS	IC CHIP
		Integrated gasification combined cycle (IGCC)	Integrated gasification combined cycle (IGCC) with carbon capture and storage (CCS)	Combined cycle	Combined cycle with carbon capture and storage (CCS), post combustion	IC engine cogeneration
Type of fuel		Lignite	Lignite	Natural gas	Natural gas	Natural gas
Electricity efficiency	%	0.525	0.465	0.65	0.61	0.44
Electric generation capacity	MW	450	400	1000	1000	0.2
Load factor (expected hours/year)	Hours/year	7500	7500	7200	7200	5000
Annual generation (expected)	kWh/year	3.38E+09	3.00E+09	7.20E+0.9	7.20E+0.9	1.000E+06

(continued)

Table A.4 (continued)

Characteristics	Units	11	12	13	14	15
		Integrated gasification combined cycle				
		IGCC lig	IGCC lig CCS	GTCC	GTCC CCS	IC CHIP
		Integrated gasification combined cycle (IGCC)	Integrated gasification combined cycle (IGCC) with carbon capture and storage (CCS)	Combined cycle	Combined cycle with carbon capture and storage (CCS), post combustion	IC engine cogeneration
Construction time	Years	3	3	3	3	1
Capital cost (net present value)	€/kWe	1209	483	440	615	879
Total capital cost (net present value)	M€	544	483	440	615	0
Plant life	Years	35	35	25	25	20
Average cost of electricity	€cents/kWhe	6.57	6.78	5.99	8.69	11.10

(continued)

Table A.4 (continued)

Characteristics	Units	16	17	18	19	20
		Fuel cells				Biomass CHP
		MCFC NG	MCFC wood gas	MCFC NG	SOFC NG	CHP polar
		Molten carbonated fuel cells, natural gas	Molten carbonate fuel cells, wood gas	Molten carbonated fuel cells natural gas	Solid oxide fuel cells, (tubular), natural gas	Steam turbine cogeneration, short rotation forestry poplar
Type of fuel		Natural gas	Wood gas	Natural gas	Natural gas	SRF poplar
Electricity efficiency	%	0.5	0.5	0.55	0.58	0.3
Electric generation capacity	MW	0.25	0.25	2	0.3	9
Load factor (expected hours/year)	Hours/year	5000	5000	5000	5000	8000
Annual generation (expected)	kWh/year	1.25E+06	1.25E+0.6	1.000E+07	1.50E+06	7.20+07
Construction time	Years	0.83	0.83	0.83	0.83	2

(continued)

Table A.4 (continued)

Characteristics	Units	16	17	18	19	20
		Fuel cells				Biomass CHP
		MCFC NG	MCFC wood gas	MCFC NG	SOFC NG	CHP polar
		Molten carbonated fuel cells, natural gas	Molten carbonate fuel cells, wood gas	Molten carbonated fuel cells natural gas	Solid oxide fuel cells, (tubular), natural gas	Steam turbine cogeneration, short rotation forestry poplar
Capital cost (net present value)	€/kWe	1544	1544	1235	1030	2280
Total capital cost (net present value)	M€	0	0	2	0	21
Plant life	Years	5	5	5	5	15
Average cost of electricity	€cents/kWhe	8.74	8.44	7.29	6.73	7.29

(continued)

Table A.4 (continued)

Characteristics	Units	21	22	23	24	25	26
		Fuel cells	Solar				
		CHP straw	PV-Si plant	PV-Si building	PV-CdTe building	Solar thermal	Wind, Wind offshore
		Steam turbine cogeneration, agricultural waste, wheat, straw	PV, monocrystalline Si, plant size	PV, monocrystalline Si, building integrated	CdTe, building integrated	Concentrating solar thermal power plant	Wind
Type of fuel		Waste straw	Sun	Sun	Sun	Sun	Wind
Electricity efficiency	%	0.3	0	0	0	0.185	0
Electric generation capacity	MW	9	46.6375	0.4197375	0.939475	400	24
Load factor (expected hours/year)	Hours/year	8000	984	984	984	4518	4000
Annual generation (expected)	kWh/year	7.20+07	4.59E+07	4.13E+0.5	8.26E+05	1.81E+09	9.60E+07

(continued)

Table A.4 (continued)

Characteristics	Units	21	22	23	24	25	26
		Fuel cells	Solar				
		CHP straw	PV-Si plant	PV-Si building	PV-CdTe building	Solar thermal	Wind, Wind offshore
		Steam turbine cogeneration, agricultural waste, wheat, straw	PV, monocrystalline Si, plant size	PV, monocrystalline Si, building integrated	CdTe, building integrated	Concentrating solar thermal power plant	Wind
Construction time	Years	2	2	0.5	0.5	3	2
Capital cost (net present value)	€/kWe	2280	848	927	927	3044	1130
Total capital cost (net present value)	M€	21	40	0	1	1217	27
Plant life	Years	15	40	40	35	40	30
Average cost of electricity	€cents/kWhe	6.51	6.30	6.92	7.15	6.31	7.27

Table A.5 Tendencies in the development of main variables in the scenario studies (Deutsch et al. 2011)

Study	Timeframe	Modeltype	Model mechanism	Main target	GDP p.c.	Oil price	Gas price	Coal price	CO_2 price
	Up to year x	BU/TD	Opt or Sim	GHG or RES					
WEO 450pppm	2030	BU (add. TD)	Opt + Sim	-80%GHG (2050)	⬀	⇨	⇨	⬂	⇧
ETP Blue	2050	BU	Opt	-74% GHG (2050)	⬀	⬂	⬂	⬂	⇧
EU DG TREN Ref	2030	BU/TD mixed	Opt		⬀	⬀	⬀	⇨	⇨
EU DG ENV NSAT	2030	BU/TD mixed	Opt	20-20-20 targets	⇧	⇨	⇨	⇨	⬀
ECF 80%RES	2050	BU with add. TD	Sim	-80% GHG (2050)	⇧	⬀	⬀	⇨	⇧
Advanced E[R]	2050	BU	Sim	-95% CO_2 (2050)	⬀	⇧	⇧	⬀	⬀
Eurelectric PowCH	2050	BU/TD mixed	Opt	-75% GHG (2050)	⬀	⬀	⬀	⬀	⇧
WITCH Ref	2100	TD	Opt		⬀	⬀	⬀	⇨	○

⇧ Sharp increase
⬀ Moderate increase
⇨ Almost stable or small increase/decrease
⬂ Moderate decrease
⇩ Sharp decrease
○ no information available or no development estimated in the scenario studies

(continued)

Table A.5 (continued)

ETS coverage	Final Energy demand	Power demand	Energy efficiency	CO_2 Emissions	Nuclear generation	RES	RES (focus)	CCS	EV	Grid investment
			EEV per GDP		TWh		TWh	TWh		Add. Grids
OECD+ (2013). CDM	⇨	⬀	⬂	⬂	⬀	⬀	Wind (offshore after 2020), solar, bio	⇨ (by 2030)	⬀	⇨
OECD+OME	⬂	⬀	⬂	⇩	⬀	⬀	Wind, bio, solare (PV, CSP after 2020)	⬀	⇧	⇧
EU-27, CDM	⇨	⬀	⇨	⬂	⬀	⬀	Wind (offshore after 2020), solar, bio	⇨	⇨	⬀
EU-27	⇨	⬀	⬂	⬂	⇨	⬀	Wind, bio, solar	⇨	⇨	⬀
OECD (2020)	⬀	⇧	⬂	⇩	⇩	⇧	Wind (offshore after 2020), bio, heat pump	⬀ (after 2020)	⇧	⇧
Global (long term)	⇩	⬀	⬂	⇩	⇩	⇧	Wind, solar, geo	○	⇧	⇧
Global (2020)	⇩	⬀	⇩	⇩	⬀	⬀	Wind (offshore after 2020), bio, solar, PV	⇧ (after 2020)	⇧	⇧
○	⇧	⇧	⇨	⬀	⬀	○	Wind, solar	⇨ (by 2030)	○	○

add. additional
BU bottom-up
TD top-down
Opt optimization
Sim Simulation
EV electric vehicles
bio Biomass
geo geothermal
OME other major economies
CDM Clean Development Mechanism
ETS Emission Trading System

Arrows describe tendencies in the development of variables over the timeframe applied in the scenario studies. Developments are evaluated in comparison with developments in other scenarios studies (e.g. sharp increase in one variable means that the variable increases relatively sharp compared to the development in other scenario studies)

Table A.6 Planned and realised geothermal plants (electricity generation) in Central Europe

	Geoth. capacity in MW	Electrical power in MW	Temperature in °C	Delivery rate in m3/h	Drilling depth in m	Planned commissioning (year)
Germany (German)						
Gross Schönebeck Research Project	10	1.0	150	<50	4294	in trial operation, currently no power generation
Neustadt-Glewe	10	0.21	98	119	2250	Power plant operation since 2003–2009, power generation discontinued in 2009
Bad Urach (HDR pilot project)	6–10	approx. 1.0	170	48	4500	Project cancelled due to expiration of financing/drilling problems
Bruchsal	4.0	approx. 0.5	118	86	2500	In power plant operation since 2009
Landau in the Palatinate	22	3	159	70	3000	In trial operation since 2007. Temporarily suspended due to slight earthquake. Resumption with reduced pump pressure
home		4–5	>155		3600	In power plant operation since November 2012
Brühl	40	5–6	150		3800	(Drilling currently suspended due to lawsuit; lawsuit dismissed), GT1 successfully tested
Schaidt			>155		>3500	The mining approvals granted in 2010 have expired. The future is open

(continued)

Table A.6 (continued)

	Geoth. capacity in MW	Electrical power in MW	Temperature in °C	Delivery rate in m3/h	Drilling depth in m	Planned commissioning (year)
Offenbach at the Queich	30–45	4.8–6.0	160	360	3500	Stopped due to drill hole instability
reed city	21.5	Approx. 3.0		250	3100	Unfamiliar
Speyer	24–50	4.8–6.0	150	450	2900	Abandoned in 2005 because oil instead of water was found (three wells in trial operation)
Simbach-Braunau	7	0.2	80	266	1900	District heating since 2001, ORC power plant in operation since 2009
Unterhaching	40	3.4	122	>540	3577	In operation since 2008
Sour salmon	Approx. 80	Approx. 5	140	>600	>5500	In operation since 2013
Dry hair	Approx. 50	Approx. 5.0	135	>400	>4000	In operation since 2013
Mauerstetten			120–130		4100	Almost no deep groundwater found; 2014 hydraulic stimulation approved
Kirchstockach	50	5	130	450	>4000	In operation since 2013
Laufzorn (Oberhaching)	50	5	130	470	>4000	In operation since 2014
Church canopy			120	470	>4500	Focus on heat for greenhouses and district heating
Baptisteries			120	470	>3000	From 2015 (currently power plant construction)

(continued)

Table A.6 (continued)

	Geoth. capacity in MW	Electrical power in MW	Temperature in °C	Delivery rate in m3/h	Drilling depth in m	Planned commissioning (year)
Traunreut	12	5.5	120	470	>5000	Expected 2015, currently expansion of district heating network
Geretsried	40	5.3	145	360	>4500	Drilling work interrupted, 1st drilling with too little debris
Bernried at Lake Starnberg			150	2 × 450	>4500	In preparation, (start of drilling planned for 2015)
Weilheim in Upper Bavaria			150	>500	>4500	(Start of drilling planned for 2015)
wooden churches			150	240	5500	(Start of drilling planned for 2015)
Neuried (Baden)		3.8				(Start of drilling planned for 2015. Postponed due to a lawsuit of the city of Kehl and a considerable citizen resistance)
Austria						
Altheim (Upper Austria)	18.8	0.5	105	Approx. 300–360	2146	In power plant operation since 2000
Bad Blumau	7.6	0.18	107	Approx. 80–100	2843	In power plant operation since 2001
aspern	40		150	360	5000	Drilling work aborted
France (France)						
Soultz-sous-Forêts	12.0	2.1	180	126	5000	Test operation since 2008
Switzerland						
Basel	17.0	6.0	180		5000	Project discontinued due to earthquake

(continued)

Table A.6 (continued)

	Geoth. capacity in MW	Electrical power in MW	Temperature in °C	Delivery rate in m3/h	Drilling depth in m	Planned commissioning (year)
St. Gallen	approx. 30	3–5	150–170		Approx. 4000	Project cancelled, high gas inflow and increased seismicity during production test

Table from Greenpeace (2005) with additions from Wikipedia, "Geothermie"—Wikipedia. [Online]. Available: https://de.wikipedia.org/wiki/Geothermie. [Accessed: 20-Jul-2009]

Table A.7 Hydropower plants with rated outputs above 5000 MW

Name	Installed capacity in MW	Year of completion	Country		River
Three Gorges Dam	22,500	2008/2012	China		Yangtze
Itaipu Dam	14,000	1984/1991, 2003	Brazil	Paraguay	Paraná
Xiluodu	13,860	2014	China		Jinsha
Belo Monte	11,233	2016–2019	Brazil		Xingu
Guri	10,235	1978, 1986	Venezuela		Caroní
Tucuruí	8370	1984, 2007	Brazil		Tocantins
Grand Coulee	6809	1942/1950, 1973, 1991	USA		Columbia
Xiangjiaba	6448	2014	China		Jinsha
Longtan Dam	6426	2007/2009	China		Hongshui
Sayano-Shushenskaya	6400	1985/1989, 2010/2014	Russia		Yenisei
Krasnoyarsk	6000	1967/1972	Russia		Yenisei
Nuozhadu	5850	2014	China		Mekong
Robert-Bourassa	5616	1979/1981	Canada		La Grande
Churchill Falls	5428	1971/1974	Canada		Churchill

Wikipedia (https://en.wikipedia.org/wiki/List_of_largest_hydroelectric_power_stations?oldformat=true) (accessed 13-Dec-2019)

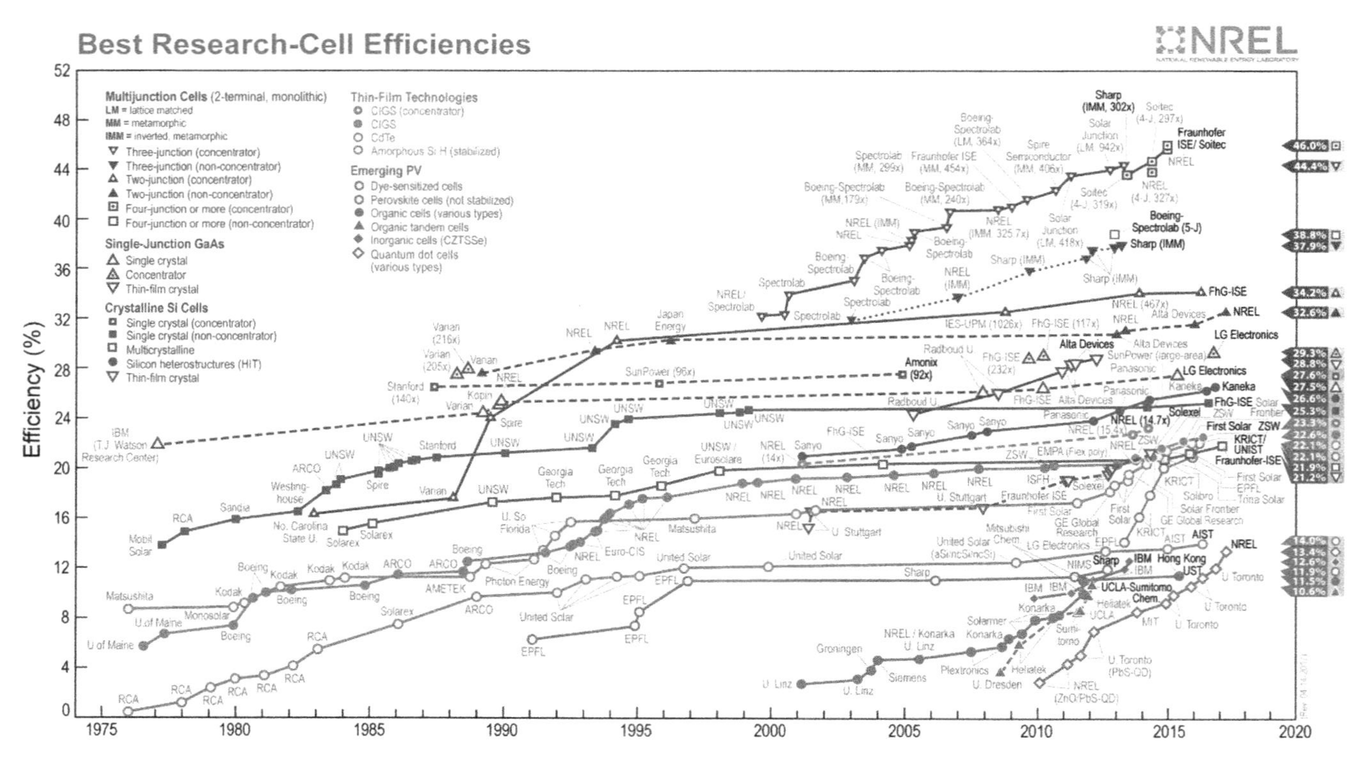

Fig. A.1 Conversion efficiencies (under standard condition) of best research solar cells worldwide for various photovoltaic technologies since 1976 [https://www.nrel.gov/pv/cell-efficiency.html (accessed 13-Dec-2019)]

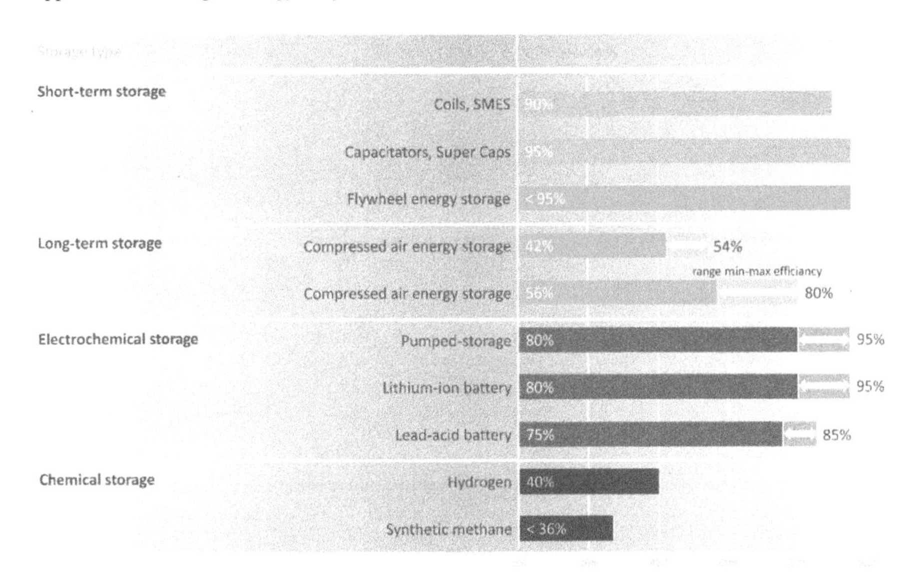

Fig. A.2 Efficiencies of different electricity storage systems (Mahnke et al. 2014)

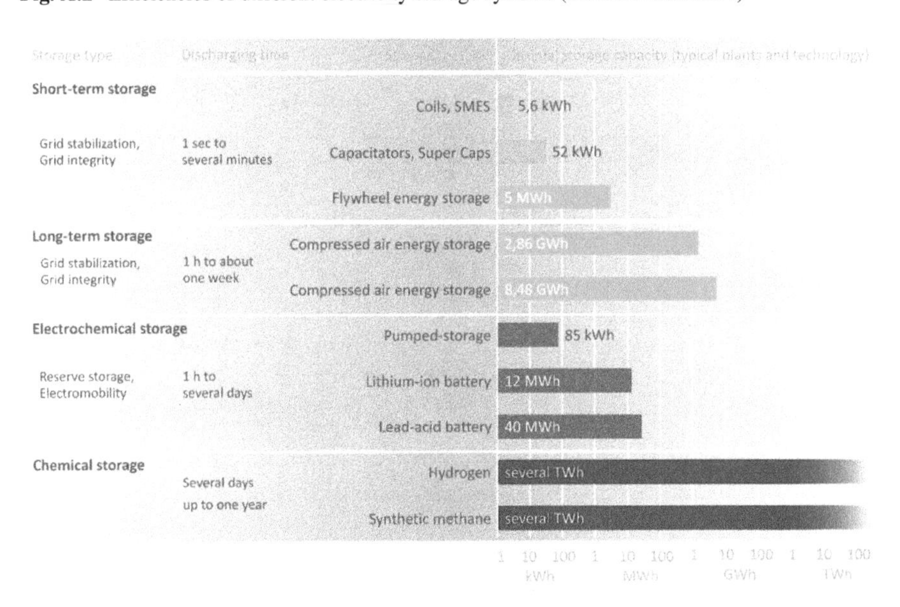

Fig. A.3 Capacities of different accumulators (Mahnke et al. 2014)

Appendix B
Network Actors

C. F. Gethmann et al., *Global Energy Supply and Emissions*,
Ethics of Science and Technology Assessment 47,
https://doi.org/10.1007/978-3-030-55355-5

Table B.1 Network actors China-EU energy dialogue

ABB	ABB
ACFIC	All-China Federation of Industry and Commerce
Alstom	Alstom
BP	British Petroleum
CASS	Chinese Academy of Social Sciences
CBC	Central Bank China
CGNPG	China Guangdong Nuclear Power Group
CN Miss EU	Mission of China to the European Union
CNOOC	China National Offshore Oil Corporation
CNPC	China National Petroleum Corp
CSECL	China Shenhua Energy Company Limited
CSPG	China State Grid Corporation
CWEA	Chinese Wind Energy Association
Dongfang	Dongfang Electric
EC2	EU-China Clean Energy Centre
EC DG Climate	European Commission, Directorate-General for Climate Action
EC DG DevCo	European Commission, Directorate-General for Development and Cooperation
EC DG Energy	European Commission, Directorate-General for Energy
EC DG Env	European Commission, Directorate-General for Environment
EC DG Research	European Commission, Directorate-General for Research and Innovation
EC DG Trade	European Commission, Directorate-General for Trade
EEAS	European External Action Service
Enercon	Enercon
EP	European Parliament
ERI	Energy Research Institute at NDRC
EUCCC	European Union Chamber of Commerce in China
EU Del CN	Delegation of the European Union to China
Goldwind	Goldwind
MFA	Ministry of Foreign Affairs
MOF	Ministry of Finance
MOFCOM	Ministry of Commerce

(continued)

Table B.1 (continued)

MoHURD	Ministry of Housing and Urban-Rural Development
MOST	Ministry of Science and Technology
NDRC	National Development and Reform Commission
NEA	National Energy Administration
NEC	National Energy Commission
Schneider	Schneider Electric
SERC	State Electricity Regulatory Commission
SGCC	State Grid Corporation China
Shell	Shell
Siemens	Siemens
SINOPEC	SINOPEC
Sinovel	Sinovel
State Council	State Council
Suntech	Suntech Power Co. Ltd
Total	Total
Tsinghua	Tsinghua University
Vestas	Vestas
Yangtze	China Yangtze Power Co.

Table B.2 Network actors India-EU energy dialogue

Alstom	Alstom
Areva	Areva
BP	British Petroleum
CII	Confederation of Indian Industry
CIL	Coal India Limited
CMPDIL	Central Mine Planning and Design Institute Limited
DAE	Department of Atomic Energy
EC DG Climate	European Commission, Directorate-General for Climate Action
EC DG DevCo	European Commission, Directorate-General for Development and Cooperation
EC DG Energy	European Commission, Directorate-General for Energy
EC DG Env	European Commission, Directorate-General for Environment
EC DG Trade	European Commission, Directorate-General for Trade
EEAS	European External Action Service
EIB	European Investment Bank
Essar	Essar
EU Chamber	European Union Chamber of Commerce in India
EU Del IND	Delegation of the European Union to India
FICCI	Federation of Indian Chambers of Commerce and Industry
MC	Ministry of Coal
MCI	Ministry of Commerce and Industry
MEA	Ministry of External Affairs
MEF	Ministry of Environment and Forests
MNRE	Ministry of New and Renewable Energy
MP	Ministry of Power
MPNG	Ministry of Petroleum and Natural Gas
NTPC	National Thermal Power Corporation
ONGC	Oil and Natural Gas Cooperation
President EC	President of the European Commission
RIL	Reliance Industries Limited
Schlumberger	Schlumberger
Shell	Shell
Suzlon	Suzlon
Vattenfall	Vattenfall

Table B.3 Network actors Brazil-EU energy dialogue

ABB	ABB
ABIOVE	Brazilian Vegetable Oil Industry Association
ANEEL	Brazilian Agency for Electric Energy
ANP	Brazilian Agency for Oil, Gas and Biofuels
APROBIO	Brazilian Biodiesel Association
Arcelor-Mittal	Arcelor-Mittal
Banco do Brasil	Banco do Brasil
BNDES	Brazilian National Development Bank
BRA	Mission Mission of Brazil to the European Union
Business Europe	Business Europe
CGEE	Centre for Strategic Studies
CNI	National Confederation of Industries
EC DG Climate	European Commission, Directorate-General for Climate Action
EC DG DevCo	European Commission, Directorate-General for Development and Cooperation
EC DG	Energy European Commission, Directorate-General for Energy
EC DG Env	European Commission, Directorate-General for Environment
EC DG	Research European Commission, Directorate-General for Research and Innovation
EC DG Trade	European Commission, Directorate-General for Trade
EC President	President of the European Commission
EEAS	European External Action Service
Embraer	Embraer
EP	European Parliament
EPE	Energy Research Company
E.ON	E.ON
EUBrasil	EUBrasil
EU Del BRA	Delegation of the European Union to Brazil
Eurochambres	Eurochambres
Greenpeace	Greenpeace
MDIC	Ministry of Development, Industry and Foreign Trade

(continued)

Table B.3 (continued)

MMA	Ministry of Environment
MME	Ministry of Mines and Energy
MRE	Ministry of External Relations
PETROBRAS	Petróleo Brasileiro S.A.
Planalto	Brazilian Presidency
Shell	Shell
Siemens	Siemens
UNICA	Sugarcane Producers Union
UNICAMP	University of Campinas
UNIFEI	University of Itajuba
USP	University of São Paulo
WWF	World Wildlife Fund

References

acatech (Deutsche Akademie der Technikwissenschaften e. V.) (2020) Zentrale und dezentrale Elemente im Energiesystem. Der richtige Mix für eine stabile und nachhaltige Versorgung. München (Germany)

Acemoglu D, Robinson J (2012) Why nations fail: the origins of power, prosperity, and poverty, New York, NY

AEE (Agentur für Erneuerbare Energien e.V. (2014) Potenziale der Bioenergie. Forschungsradar Energiewende. Agetur für Erneuerbare Energien (Germany)

AEE (Agentur für Erneuerbare Energien e.V.) (2009) Der volle Durchblick in Sachen Bioenergie. Agentur für Erneuerbare Energien e.V., Berlin, Germany

AEE (Agentur für Erneuerbare Energien e.V.) (2010) Erneuerbare Energien 2020. Potenzialatlas Deutschland, Berlin, Germany

Agora Energiewende (2014) Stromspeicher in der Energiewende. Untersuchung zum Bedarf an neuen Stromspeichern in Deutschland für den Erzeugungsausgleich, Systemdienstleistungen und im Verteilnetz. Agora Energiewende Berlin, Germany

Ahmad K (2002) Pollution cloud over south Asia is increasing ill health. Lancet 360:549

Akademien der Wissenschaften Schweiz (2012) Zukunft Stromversorgung Schweiz. Akademien der Wissenschaften, Bern, Switzerland

Allen MR, Stott P, Mitchell J, Schnur R, Delworth T (2000) Quantifying the uncertainty in forecasts of anthropogenic climate change. Nature 407:617–620

Andonova L, Chelminski K (2016) Emergence of a regime complex for clean energy: the critical role of legitimacy. Working paper, Graduate Institute Geneva, Switzerland

ANEEL (Agência Nacional de Energia Elétrica) (2016) ANEEL, 03.05.2016. Available at http://www.aneel.gov.br/. Accessed 13 Dec 2019

ANP (The Brazilian National Agency of Petroleum, Natural Gas and Biofuels) (2016) Competências da ANP. Available at http://www.anp.gov.br/. 3 May 2016

Ariely D (2008) Predictably irrational. Reviewed and expanded edition, New York, NY

Aristotle (E. Nic.) (1984) Nicomachean ethics. Transl. Ross WD, rev. Urmson JO. In: The complete works of aristotle. The Revised Oxford Translation, vol. 2, Barnes J (ed.), Princeton, NJ

Arnstein SR (1969) A ladder of citizen participation. J Am Inst Planners 35(4):216–224

Asafu-Adjaye J, Blomqvist L, Brand S, Brook B, Defries R, Ellis E, Keith D, Foreman C, Lewis M, Lynas M, Nordhaus T, Pielke R Jr, Sagoff M, Pritzker R, Roy J, Teague P, Stone R, Shellenberger M, An ECOMODERNIST Manifesto. Oakland Breakthrough Institute, CA, USA

Atkinson RW, Anderson HR, Sunyer J, Ayres J, Baccini M, Vonk JM, Boumghar A, Forastiere F, Forsberg B, Touloumi G, Schwartz J, Katsouyanni K (2001) Acute effects of particulate air pollution on respiratory admissions: results from APHEA 2 project. Air pollution and health: a European approach. Am J Respir Crit Care Med. 164:1860–1866

C. F. Gethmann et al., *Global Energy Supply and Emissions*,
Ethics of Science and Technology Assessment 47,
https://doi.org/10.1007/978-3-030-55355-5

Baccini L, Lenzi V, Thurner PW (2011) Global energy governance: bilateral trade and the diffusion of international organizations. Int Interact 3922:192–216

Barkhordarian A, von Storch H, Zorita E, Gómez-Navarro J (2016) An attempt to deconstruct recent climate change in the Baltic Sea Basin. J Geophys Res Atmos 121(3), 207–217. https://doi.org/10.1002/2015JD024648

Barteková E (2014) An introduction to the economics of rare earths. United Nations University Merit working papers series 2014-043. Maastricht Economic and social Research institute on Innovation and Technology, Masstricht, NL

Barteková E, Ziesemer THW (2019) Impact of electricity prices on foreign direct investment. Evidence from the European Union. Appl Econ 51(11):1183–1198

Bava US (2007) New powers for global change? India's role in the emerging world order. FES briefing paper 4. Friedrich Ebert Foundation, New Delhi, India

BBC-News (2002) Historic smog death toll rises (5 Dec)

Bennauer M, Egener EG, Schlehuber R, Werthes H, Zimmer G (2009) Art. Automation and control of electric poer generation and distribution sysetms. Steam turbines. In: Encyclopedia of control systems, robotics, and automation (=Encyclopedia of Life Support Systems EOLSS, vol XVIII). Ch. 201. UNESCO-EOLSS Joint Committee, Paris, France

Betz J, Hanif M (2010) The formation of preferences in two-level games: an analysis of India's domestic and foreign energy policy. GIGA working paper No. 142

Bilgili F, Tülüce NSH, Doğan İ (2012) The determinants of FDI in Turkey: a Markov regime-switching approach. Econ Modell 29(4):1161–1169

Blühdorn I (2013) Simulative Demokratie. Frankfurt, Germany

BMBF [Bundesministerium für Bildung und Forschung] (2014) Die neue Hightech-Strategie – Innovationen für Deutschland. https://www.bmbf.de/pub_hts/HTS_Broschure_Web.pdf. Accessed 13 Dec 2019

BMUD (Bundesministerium für Umwelt, Naturschutz, Bau und Reaktorsicherheit) (2015) Brazilian German joint statement on climate change. http://www.bmub.bund.de/fileadmin/Daten_BMU/Download_PDF/Klimaschutz/klimakonsultation_deutschland_brasilien_en_bf.pdf. Accessed 13 Dec 2019

Boezen HM, van der Zee SC, Postma DS, Vonk JM, Gerritsen J, Hoek G, Brunekreef B, Rijcken B, Schouten JP (1999) Effects of ambient air pollution on upper an lower respiratory symptoms and peak expiratory flow in children. Lancet 353:874–878

BP (British Petroleum) (2015) BP statistical review of world energy, 10 June 2015. https://www.bp.com/content/dam/bp/business-sites/en/global/corporate/pdfs/news-and-insights/speeches/speech-archive/statistical-review-of-world-energy-2015-bob-dudley-speech.pdf. Accessed 13 Dec 2019

BP (British Petroleum) (2017) BP statistical review of world energy, June 2017. https://www.connaissancedesenergies.org/sites/default/files/pdf-actualites/bp-statistical-review-of-world-energy-2017-full-report.pdf. Accessed 13 Dec 2019

BP (British Petroleum) (2019) BP statistical review of world energy, June 2019. https://www.bp.com/content/dam/bp/business-sites/en/global/corporate/pdfs/news-and-insights/press-releases/bp-statistical-review-of-world-energy-2019.pdf. Accessed 13 Dec 2019

Bray D (2010) The scientific consensus of climate change revisited. Environ Sci Policy 13(5):340–350

Bray D, von Storch H (1999) Climate science. An empirical example of postnormal science. Bull Am Meteorol Soc 80:439–456

Bray D, von Storch H (2009) 'Prediction' or 'Projection'? The nomenclature of climate science. Sci Commun 30:534–543. https://doi.org/10.1177/1075547009333698

Brillinger D, Curry J, Jacobsen R, Muller E, Muller R, Perlmutter S, Rohde R, Rosenfeld A, Wickham C, Wurtele J. Berkeley earth surface temperature analysis. http://static.berkeleyearth.org/pdf/berkeley-earth-summary.pdf. Accessed 13 Dec 2019

Brundtland G (1987) Report of the world commission on environment and development: our common future, Geneva, Switzerland

Brunekreef B, Holgate ST (2002) Air pollution and health (review). Lancet 360(10):1231–1242
Buchanan AE (1982) Marx and justice: the radical critique of liberalism. Law Philos 3(1):147–153
Buecheler S (2012) Photovoltaik-Technologien im Überblick.Tec21. 45(138):33–36
Cai Y, Judd KL, Lontzek TS (2015) The social cost of carbon with economic and climate risks. J Polit Econ. https://doi.org/10.1086/701890
Carpenter JR, Merckelbach L, Callies U, Clark S, Gaslikova L, Baschek B (2016) Potential impacts of offshore wind farms on North Sea stratification. PLoS ONE 11(8):e0160830. https://doi.org/10.1371/journal.pone.0160830
CEA (Central Electricity Authority) (2018) Ministry of power, Government of India. Installed capacity. http://www.cea.nic.in/installed_capacity.html. Accessed 13 Dec 2019
Chatterjee DK (2011) (ed) Encyclopedia of global justice, Dordrecht, NL
Chatterjee E (2017) Electricity and India's weak-strong state, paper presented at the Workshop on 'Energy and state capacities in BRIC countries'. Technische Universität Darmstadt, Germany, 21–22 Sept 2017
Clancy L, Goodmann P, Sinclair H, Dockery D (2002) Effect on air-pollution control on death rates in Dublin, Ireland: an intervention study. Lancet 360(10):1210–1214
Council of the European Union (2009) Third European Union-Brazil Summit. Joint Statement. [press release] 6 Oct 2009. Available at http://www.consilium.europa.eu/uedocs/cms_Data/docs/pressdata/en/er/110440.pdf. Accessed 13 Dec 2019
de Wild-Scholten MJ (2013) CPV data: "Environmental sustainability of concentrator PV systems: preliminary LCA results of the Apollon Project". In: 5th world conference on PV energy conversion, Valencia, Spain, 6–10 Sept 2010. Graph: PSE 2014
Deaton A (2013) The great escape: health, wealth, and the origins of inequality, Princeton, NJ
Department of Energy and Climate Change (2015) UK-India joint statement on energy and climate change. https://www.gov.uk/government/publications/uk-india-joint-statement-on-energy-and-climate-change. Accessed 13 Dec 2019
Deutsch M, Ess F, Hobohm J, Rits V, Schlesinger M, Strassburg S, Böllhoff C (2011) Analysis and comparison of relevant mid-and long-term energy scenarios for EU and their key underlying assumptions. PROGNOS, Basel, Switzerland
Diederen P, Kemp R, Verberne P, Ziesemer T, van Zon A (1995) Energy technologies, environmental policy and competitiveness. Final report for the JOULE II program of the European Commission, DG XII, MERIT and Department of Economics, Rijksuniversiteit Limburg, Maastricht, NL
Dolan TJ (2012) Nuclear fusion. In: Meyers RA (ed) Encyclopedia of sustainability science and technology, New York, NY, USA
Downing T, Anthoff D, Butterfield R, Ceronsky M, Grubb M, Guo J, Hepburn C, Hope C, Hunt A, Li A, Markandya A, Moss S, Nyong A, Tol R, Watkiss P (2005) Social cost of carbon: a closer look at uncertainty. Final project report. Stockholm Environment Institute, Oxford, UK
Downs ES (2008) China's new energy administration: China's National Energy Administration will struggle to manage the energy sector effectively. China Bus Rev. 42–45 (Nov–Dec)
Droste-Franke B, Paal BP, Rehtanz C, Sauer DU, Schneider JP, Schreurs M (2012). Balancing renewable electricity. Energy storage, demand side management and network extension from an interdisciplinary perspective, Heidelberg, Germany, New York, NY
Dubash NK (2009) Toward a progressive Indian and global climate politics. Working Paper 2009/1, Centre for Policy Research, Climate Initiative, Delhi, India, Sept
Dubash NK (2011) From norm taker to norm maker? Indian energy governance in global context. Glob Policy 2(Special Issue):66–79
Edenhofer O, Pichs-Madruga R, Sokona Y, Seyboth K, Eickemeier P, Matschoss P, Hansen G, Kadner S, Schlömer S, Zwickel T, von Stechow C (2011) Direct solar energy [= Ch. 3 in IPCC 2011].
EEAS (European External Action Service) (2012) Joint declaration for enhanced cooperation on energy between the European Union and the Government of India, Brussels and Delhi.
EEAS (European External Action Service) (2013) EU-China 2020 Strategic Agenda for Cooperation.

EFDA (European Fusion Development Agreement) (2001) Safety and environmental impact of fusion. EFDA-S-RE-1. EUR 8.5
Elster J (1989) The cement of society: a survey of social order, New York, NY
Enerdata (2016) Global energy statistical yearbook. Available at https://yearbook.enerdata.net. Accessed 11 Aug 2016
Engel JR (1995) Sustainable development. In: Warren TR (ed) Encyclopedia of bioethics, New York, NY, 2456 ff.
Engels F (1847) Preface to the first German edition. In: Marx (ed) The poverty of philosophy. Marx/Engels Internet Archive 1999
EPA (US Environmental Protection Agency) (2002) Health and environmental effects of particulate matter, Fact Sheet, 17.7.97. https://www.epa.gov/pm-pollution/health-and-environmental-effects-particulate-matter-pm. Accessed 13 Dec 2019
Eriksen EO, Fossum JE (ed) (2000) Democracy in the European Union, London, UK
Esty D (1999) Greening of the GATT (trade, environment and the future), Washington DC, USA
European Commission (2006) A European Strategy for sustainable, competitive and secure energy. Green paper. COM(2006) 105 final, Brussels.
European Commission (2007) An energy policy for Europe. Communication from the Commission to the European Council and the European Parliament, COM(2007) 1 final, Brussels, Belgium 10.01.2007
European Commission (2011) Energy 2020. A strategy for competitive, sustainable and secure energy. Publications Office of the European Union, Luxembourg
European Commission (2012) Energy roadmap 2050. Publications Office of the European Union, Luxembourg
European Commission (2013) A 2030 framework for climate and energy policy. Green paper, COM(2013), 169 final, Brussels, Belgium, 27 Mar 2013
European Commission (2014a) A policy framework for climate and energy in the period from 2020 to 2030', communication from the Commission to the European Parliament, the Council, the European Economic and Social Committee and the Committee of the Regions, COM(2014) 15 final, Brussels, Belgium, 22 Jan 2014
European Commission (2014b) Roadmaps for international cooperation. Accompanying the document: Report from the Commission to the European Parliament, the Council, the European Economic and social Committee and the Committee of the Regions. Report on the implementation of the strategy for international cooperation in research and innovation, Brussels, Belgium
European Commission (2014c) Energy from abroad. stakeholders for EU-China energy cooperation. Available at http://ec.europa.eu/energy/international/bilateral_cooperation/china/stakeholders_en.htm. Accessed 5 Nov 2014
European Commission (2015) EU-China summit. Press Release.
European Commission (2016a) Joint communication to the European Parliament and the council. Elements for a new EU strategy on China
European Commission (2016b) EU-China Roadmap on energy cooperation (2016–2020)
European Commission (2016c) Energy in Europe. https://ec.europa.eu/energy/energy_newsletter/newsletter-april-2016_en?redir=1. Accessed 13 Dec 2019
European Commission (2016d) Joint declaration between the EU and India on a clean energy and climate partnership. https://ec.europa.eu/clima/news/articles/news_2016033101_en. Accessed 13 Dec 2019
European Commission (2016e) Proposal for a regulation on the Governance of the Energy Union. COM(2016) 759 final. European Commission, Brussels, Belgium
European Commission (2016f) Memo. New Energy Union Governance to deliver common goals, 30 Nov 2016. https://ec.europa.eu/clima/sites/clima/files/%202016113001_governance_technical_en.pdf. Accessed 13 Dec 2019
European Commission (2018) The European green deal. Brussels, 11.12.2019, COM(2019) 640 final

European Council (2015a) Presidency conclusions', EUCO 11/15, CO EUR 1, CONCL 1, Brussels (Belgium), 20 Mar 2015
European Council (2015b) EU-China Summit joint statement, 29.06.15, Brussels, Belgium
European Council (2016a) EU-India agenda for action-2020. EU-India Summit, Brussels, Belgium, 30 Mar 2016
European Council (2016b) Joint statement 13th EU-India summit, Brussels, 30 Mar 2016. http://www.consilium.europa.eu/en/meetings/internationalsummit/2016/03/20160330-joint-statement-eu-india.pdf. Accessed 23 May 2016
European Council (2017) Joint statement 14th EU-India summit, New Delhi, 6 Oct 2017. http://www.consilium.europa.eu/media/23515/eu-india-joint-statement.pdf. Accessed 13 Dec 2019
EEAS (European External Action Service) (2016) Shared vision, common action: a stronger Europe. A Global Strategy for the European Union's Foreign And Security Policy
Federal Ministry for Economic Affairs and Energy (2017) Exchange of views on energy transition with China, Press release, 2 June 2017. https://www.bmwi.de/Redaktion/EN/Pressemitteilungen/2017/20170602-baake-trifft-ndrc-vize-hu-zucai.html. Accessed 13 Dec 2019
Fh-ISE (Fraunhofer Institute for Solar Energy Systems) (2015) Photovoltaics Report 2015. Fh-ISE, Freiburg, Germany
Fh-ISE (Fraunhofer Institute for Solar Energy Systems) (2015) Stromgestehungskosten Erneuerbare Energien. Fh-ISE, Freiburg, Germany
Figueres C, Schellnhuber HJ, Rockström J, Hobley A, Rahmstorf S (2017) Three years to safeguard our climate. Nature 546:593–595
Fischedick M, Borbonus S, Scheck H (2011) Towards global energy governance. Strategies for Equitable Access to Sustainable Energy, Policy paper, 34. Development and Peace Foundation, Bonn
Fischer S (2009) Energie- und Klimapolitik im Vertrag von Lissabon: Legitimationserweiterung für wachsende Herausforderungen. Integration, I/2009, Berlin, Germany
Fischer S (2011) Auf dem Weg zur gemeinsamen Energiepolitik. Strategien, Instrumente und Politikgestaltung in der Europäischen Union, Baden-Baden, Germany
Fischer S (2014) The EU's new energy and climate policy framework for 2030', SWP Comments, 55, Berlin, Germany
Fischer S, Geden O (2015a) Die Grenzen der Energieunion, SWP-Aktuell, 36, Berlin, Germany
Fischer S, Geden O (2015b) Europäische Energieunion. Für jeden ist was dabei, Neu Zürcher Zeitung, 2 Dec 2015
Florini A, Sovacool BK (2011) Bridging the gaps in global energy governance. Glob Governance 17(1):57–74
Fosu, AK, Getachew Y, Ziesemer T (2016) Optimal public investment, growth, and consumption: evidence from African countries. Macroecon Dyn 20(8):1957–1986. Also UNU-MERIT WP 2014-057. http://www.merit.unu.edu/publications/wppdf/2014/wp2014-057.pdf
Frankfurt School-UNEP Centre/BNEF (2016) Global trends in renewable energy investment 2016. Frankfurt a.M., Germany. http://www.fs-unep-centre.org
Funtowicz SO, Ravetz JR (1985) Three types of risk assessment: a methodological analysis. In: Whipple C, Covello VT (eds) Risk analysis in the private sector, New York, NY, USA, pp 217–231
Gadamer HG (1960) Wahrheit und Methode, Tübingen [Neuausgabe als Gesammelten Werke, Bd. 1, Frankfurt a. M. 2010] engl. Truth and Method, 2nd edn, Sheed and Ward, London, 1989
Gauthier D (1982) Justified inequality? Dialogue (Canada) 21:431–434
Gethmann CF (1995) Ethische Probleme der Verteilungsgerechtigkeit im Umweltstaat. In: Gethmann CF, Kloepfer M, Reinert S (eds) Verteilungsgerechtigkeit im Umweltstaat, Bonn, Germany, pp 1–22
Gethmann CF (2006) Das abendländische Vernunftprojekt und die Pluralität der Kulturen. In: Krois JM, Meuter N (eds) Kulturelle Existenz und Symbolische Form. Philosophische Essays zu Kultur und Medien. Festschrift für Oswald Schwemmer, Berlin, Germany, pp 17–39
Gethmann CF (2017) Theoretische und Praktische Probleme transdisziplinärer Forschung angesichts kollektiver Handlungsprobleme von globaler Reichweite. In: Decker M, Lindner

R, Lingner S, Scherz C, Sotoudeh M (eds) Grand challenges meistern – der Beitrag der Technikfolgenabschätzung. Baden-Baden, Germany, 2018

Gethmann CF, Kamp G (2020) Climate justice. In: Oxford research encyclopedia climate science. https://oxfordre.com/climatescience

Gethmann CF, Sander T (1999) Rechtfertigungsdiskurse. In: Grunwald A, Saupe S (eds) Ethik in der Technikgestaltung, Berlin, Germany, pp 117–151

Goldemberg J (2007) Ethanol for a sustainable energy future. Science 315(5813):808–810

Goldemberg J, Coelho ST, Guardabassi P (2008) The sustainability of ethanol production from sugarcane. Energy Policy 36(6):2086–2097

Gorke M (2003) The death of our plant's species. A Challenge to Ecology and Ethics, Washington, DC

Gosepath S (2015) The principles and the presumption of equality. In: Fourie C, Schuppert F, Wallimann-Helmer I (eds) Social equality. On what it means to be equals, Oxford, UK, pp 167–185

Gosseries A (2008) Theories of intergenerational justice: a synopsis. Surv Perspect Integrating Environ Soc 1:39–49

Government of India, Ministry of External Affairs (2017) India-Germany Joint Statement during the visit of Prime Minister to Germany. http://www.mea.gov.in/bilateral-documents.htm?dtl/28496/IndiaGermany+Joint+Statement+during+the+visit+of+Prime+Minister+to+Germany. Accessed 12 Jan 2018

Grant EJ (2017) Solid cancer incidence among the life span study of atomic bomb survivors: 1958–2009. Radiat Res 187:513–537

Gratius S (2012) Brazil and the European Union: between balancing and bandwagoning. ESPO working paper No. 02.07.12.

Greenpeace (2005) 2.000 Megawatt – sauber. Berlin (Germany)

Greenpeace (2011) Battle of the grids. How Europe can go 100% renewable and phase out dirty energy, Amsterdam, NL

Greenpeace/EPIA (European Photovoltaic Industry Association) (2006) Solar generation V-2008. Greenpeace/EPIA Amsterdam, Brussels

Grunwald A, Grünwald R, Oertel D, Paschen H (2002) Kernfusion Sachstandsbericht (= Arbeitsbericht Nr. 75 of Büro für Technikfolgen-Abschätzung beim Deutschen Bundestag). Berlin, Germany

Grünwald R (2008) CO_2-Abscheidung und -Lagerung bei Kraftwerken (= Sachstandsbericht zum Monitoring „Nachhaltige Energieversorgung". Büro für Technikfolgen-Abschätzung beim Deutschen Bundestag, Berlin, Germany

Grunwald Armin (2012) Ende einer Illusion. Warum ökologischer Konsum uns nicht retten kann, München

GWEC (Global Wind Energy Council) (2016) Global wind report 2015, Brussels, Belgium

Habermas J (1982) Diskursethik. Notizen zu einem Begründungsprogramm, Manuskript 1982. Veröffentlichungen in: ders., Moralbewußtsein und kommunikatives Handeln, Frankfurt, Germany 1983, 53–125; engl.: Discourse Ethics: Notes on a Program of Philosophical Justification. In: Moral Consciousness and Communicative Action. Cambridge, MS 1990, 43–115

Hall CAS, Klitgaard KA (2012) Energy and the wealth of nations: understanding the biophysical economy, New York, NY, Dordrecht, NL

Hansen J (2016) Europe's cloud—how coal-burning countries are making neighbours sick, prepared by working groups of the "Climate Action Network Europe" (CAN)

Hardin G (1968) The tragedy of the commons. Science 162(3859):1243–1248

Hare RM (1952) The language of morals, Oxford, UK

Hargreaves J (2010) Skill and uncertainty in climate models. Wileys Interdisc Rev Clim Change. https://doi.org/10.1002/wcc.58

Hartig GL (1795) Anweisungen zur Taxation der Forste zur Bestimmung des Holzertrags der Wälder. Berlin [quoted after Spektrum der Wissenschaft, Mar 2002, p 102]

Hasselmann K (1976) Stochastic climate models. Part I. Theory. Tellus 28:473–485

Hedley A, Wong C, Thach T, Ma S, Lam T, Anderson H (2002) Cadiorespiratory and all-cause mortality after restrictions on sulphur content of fuel in Hong Kong: an intervention study. Lancet 360:1646–1652
Heidegger M (1927) Sein und Zeit. Tübingen. 19. Auflage. Tübingen 2006; engl. Being and time. trans: Macquarrie J, Robinson E. Oxford, UK, Cambridge, MS
Heinelt H, Getimis P, Kafkalas G, Smith R, Swyngedouw E (ed.) (2002) Participatory governance in multi-level context. Wiesbaden, Germany
Herbert GMJ, Iniyan S, Amutha D (2014) A review of technical issues on the development of wind farms. Renew Sustain Energy Rev 32:619–641
Hinsch W (2002) Gerechtfertigte Ungleichheiten. Grundsätze sozialer Gerechtigkeit. Berlin, Germany, New York, NY
Hira A, de Oliveira LG (2008) No substitute for oil? How Brazil developed its ethanol industry. Energy Policy 37(6):2450–2456
Hirschberg S, Bauer C, Burgherr P et al (2016) Health effects of technologies for power generation: contributions from normal operation, severe accidents and terrorist threat. Reliab Eng Syst Saf 145:337–348
Hirschberg S, Burgherr P (2015) Sustainability assessment for energy technologies. Handbook clean energy systems. Hoboken, NJ, https://doi.org/10.1002/9781118991978.hces070
Hirschberg S, Burgherr P (2015) Sustainability assessment for energy technologies, Chap. XX. In: Kovacevic R, Pflug GC, Vespucci MT (eds) Handbook of risk management in energy production and trading, New York, NY, pp 1–22
Hirschberg S, Eckle P, Bauer C, Schenler W, Simons A, Körbel O, Dreier J, Prasser HM, Zimmermann M (2012) Bewertung aktueller und zukünftiger Kernenergietechnologien. Erweiterte Zusammenfassung des Berichts „Current and Future Nuclear Technologies“. Bundesamt für Energie BFE, Switzerland
HIS CERA (IHS Cambridge Energy Research Associates) (2011) Sound Energy policy for Europe. Pragmatic Pathways to a Low-Carbon Economy. HIS CEAR, Cambridge MS
Hobbes J (1651) Leviathan or the matter, forme and power of a common wealth ecclesiasticall and civil. Andrew Crooke, London
Hochstetter K (2017) Green industrial policy and the renewable energy transition: can it be good industrial policy. Paper presented at the Workshop on 'Energy and state capacities in BRIC countries' Technische Universität Darmstadt, Germany, 21–22 Sept 2017
Höffe O (1981) Sittlich-politische Diskurse. Frankfurt, Germany
Höffe O (2007) Democracy in an age of globalisation, Dordrecht, NL
Höhener E (2014) China entwickelt die Nukleartechnologien von morgen. Bull Nuklearforum Schweiz 6:14–17
Hu A, Guan Q (2017) China: tackle the challenge of global climate change, London, UK, New York, NY
Hulme M (2014) Can science fix climate change? A case against climate engineering, Hoboken, NJ, USA
Hume D (1751) An enquiry concerning the principles of morals, Oxford, UK
Hylton S (2015) What the railways did for us: the making of modern Britain, Gloucestershire, UK
IAEA (International Atomic Energy Agency) (2018) Nuclear power reactors in the world. (= Reference Data Series No. 2). IAEA, Vienna, Austria
IAEA (International Atomic Energy Agency) (2009) Passive safety systems and natural circulation in water cooler nuclear power plants. IAEA-TECDOC-1624. IAEA, Vienna, Austria
ICRP (1991) Recommendations of the International Commission on Radiological Protection. ICRP Publications 60, Oxford, UK, New York, NY, Frankfurt a. M., Germany
ICRP (2007) Recommendations of the International Commission on Radiological Protection. ICRP Publications 103. Oxford, UK, New York, NY, Frankfurt a. M., Germany
ICRP (2012) Recommendations of the International Commission on Radiological Protection. ICRP Publications 116. Oxford, UK, New York, NY, Frankfurt a. M., Germany

IEA (International Energy Agency) (2011) Technology roadmap biofuels for transport. ECD/IEA, Paris, France
IEA (International Energy Agency) (2012) Technology roadmap bioenergy for heat and power. OECD/IEA, Paris, France
IEA (International Energy Agency) (2013) World energy outlook 2013, Paris, France
IEA (International Energy Agency) (2014) Energy technology perspectives 2014 (ETP14). OECD/IEA, Paris, France
IEA (International Energy Agency) (2014) Technology roadmap: solar photovoltaic energy. OECD/IEA, Paris
IEA (International Energy Agency) (2015) Medium-term coal market report 2015. Market analysis and forecast to 2020. OECD/IEA, Paris, France
IEA (International Energy Agency) (2015) Energy technology perspectives (ETP15). OECD/IEA, Paris, France
IEA (International Energy Agency) (2015c) India energy outlook. World energy outlook special report. OECD/IEA, Paris, France. http://www.worldenergyoutlook.org/media/weowebsite/2015/IndiaEnergyOutlook_WEO2015.pdf. 23. May 2016
IEA (International Energy Agency) (2015) World energy outlook 2015. OECD/IEA, Paris, France
IEA (International Energy Agency) (2016a) World energy outlook 2016. OECD/IEA, Paris, France
IEA (International Energy Agency) (2016b) Tracking clean energy progress (TCEP 2016). OECD/IEA, Paris (France)
IEA (International Energy Agency) (2016c) India energy outlook, world energy outlook and special report. OECD/IEA, Paris, France
IEA (International Energy Agency) (2016d) Energy climate and change, world energy outlook special report. OECD/IEA, Paris, France
IEA (International Energy Agency) (2017) World energy outlook 2017. OECD/IEA, Paris, France
IEA (International Energy Agency) (2019) World energy outlook 2019. OECD/IEA, Paris, France
IEA-ETSAP (Energy Technology Systems Analysis Programme of IEA), IRENA (International Renewable Energy Agency) (2015) Biomass for heat and power. Technology Brief E05, Paris, France, Bonn, Germany
IHA (International Hydropower Asssociation) (2019) Hydropower status report. Sector trends and insights. IHA Central Office, London, UK
Intergovernmental Panel on Climate Change (IPCC) (2018) Global warming of 1.5 °C. Summary for Policymakers, Geneva, Switzerland
IPCC (Intergovernmental Panel on Climate Change) (2011) Renewable energy sources and climate change mitigation. In: Edenhofer O, Pichs-Madruga R, Sokona Y, Seyboth K, Matschoss P, Kadner S, Zwickel T, Eickemeier P, Hansen G, Schlömer S, von Stechow C (eds) Special report of the Intergovernmental Panel on Climate Change. Cambridge University Press, Cambridge, UK, New York, NY, USA
IPCC (Intergovernmental Panel on Climate Change) (2013) Summary for policymakers. In: Stocker TF, Qin D, Plattner GK, Tignor M, Allen SK, Boschung J, Nauels A, Xia Y, Bex V, Midgley PM (eds) Climate change 2013: the physical science basis. Contribution of working group I to the fifth assessment report of the intergovernmental panel on climate change, Cambridge, UK, New York, NY, USA
IPCC (Intergovernmental Panel on Climate Change) (2014a) Climate change 2014. Synthesis report. Summary for policymakers. https://www.ipcc.ch/pdf/assessment-report/ar5/syr/AR5_SYR_FINAL_SPM.pdf. Accessed 13 Dec 2019
IPCC (Intergovernmental Panel on Climate Change) (2014b) Mitigation of climate change. In: Edenhofer O, Pichs-Madruga R, Sokona Y, Farahani E, Kadner S, Seyboth K, Adler A, Baum I, Brunner S, Eickemeier P, Kriemann B, Savolainen J, Schlömer S, von Stechow C, Zwickel T, Minx JC (eds) Contribution of working group III to the fifth assessment report of the intergovernmental panel on climate change. Cambridge University Press, Cambridge, United Kingdom and New York, NY, USA

IPCC (Intergovernmental Panel on Climate Change) (2014c) Climate change 2014. Synthesis report. https://archive.ipcc.ch/news_and_events/docs/ar5/ar5_syr_headlines_en.pdf. Accessed 13 Dec 2019
IPCC Fourth Assessment Report 2007
IRENA (International Renewable Energy Agency) (2014) Tidal energy, technology brief. IRENA, Bonn, Germany
Irvine PJ, Kravitz B, Lawrence MG, Gerten D, Caminade C, Gosling SN, Hendy EJ, Kassie BT, Kissling WD, Muri H, Oschlies A, Smith SJ (2016a) Towards a comprehensive climate impacts assessment of solar geoengineering. Earth's Future 5. https://doi.org/10.1002/2016EF000389
Irvine PJ, Kravitz B, Lawrence MG, Muri H (2016) An overview of the Earth system science of solar geoengineering. WIREs Clim Change 7:816–833
Jaeger MD, Michaelowa K (2015) Energy poverty and policy coherence in India: norms as means in strategic two-level discourse. In: Knodt M, Piefer N, Müller F (eds) Challenges of European external energy governance with emerging powers, Fanham, UK, pp 235–247
Jaffrelot C (2006) Indien und die EU: Die Scharade einer strategischen Partnerschaft. GIGA Focus 5. German Institute of Global and Area Studies, Hamburg
Jänecke M, Kunig P, Stitzel M (2000) Environmental policy, Bonn, Germany
Janich P (1985) Protophysics of time. Constructive foundation of history of time measurement, Dordrecht, NL
Jänicke M (2013) Accelerators of global energy transition: horizontal and vertical reinforcement in multi-level climate governance. IASS working paper, Potsdam, Germany
Jordan DC, Kurtz SR (2013) Photovoltaic degradation rates—an analytical review. Prog Photovoltaics Res Appl 21(1):12–29. NREL, Fh-ISE
Joshi M, Ganeshan S (2015) India-EU energy relations: towards closer cooperation? In: Knodt M, Piefer N, Müller F (eds) Challenges of European external energy governance with emerging powers, Farnham, UK, pp 149–171
Joshi M, Khosla R (2016) India: meeting energy needs for development while addressing climate change. In: Sustainable energy in the G20: prospects for a global energy transition (IASS Study; Dec 2016, pp 57–63). Institute for Advanced Sustainability Studies (IASS), Potsdam, UK
Kahneman D (2011) Thinking. Fast and slow, London, UK
Kahneman D, Tversky A (2000) (eds) Choices, values, and frames, Cambridge, UK
Kaltschmitt M, Thrän D (2009) Art. Bioenergy. In: Bullinger HJ (ed) Technology guide: principles—applicatoin—trends, Berlin, Germany, pp 346–351
Kamp G (2015a) Ethik der langfristigen Planung. In: Kamp G (ed) Langfristiges Planen, Berlin, Germany, New York, NY, pp 165–214
Kamp G (2015b) Planungsrationalität und rationale Planer: Welches Akteurs- und Rationalitätsverständnis braucht die Planungstheorie? In: Kamp G (ed) Langfristiges Planen, Berlin, Germany, New York, NY, pp 9–38
Kant I (1785) Die Metaphysik der Sitten. Engl. The metaphysics of morals (trans: Gregor MJ). Cambridge University Press, Cambridge
Keith DW, Irvine PJ (2016) Solar geoengineering could substantially reduce climate risks—a research hypothesis for the next decade. Earth's Future 4:549–559. https://doi.org/10.1002/2016EF000465
Kelly JE (2014) Generation IV international forum: a decade of progress through international cooperation. Prog Nucl Energy 77:240–246
Kenigsberg Y (2004) Communication to the UNSCEAR Secretariat, New York, NY
Knodt M (2018) Energy Policy. In: Münch S (ed) Heinelt H. Handbook of EU policies, Cheltenham, UK, pp 224–240
Knodt M, Chaban N, Nielsen L (2017) Bilateral energy relations between the EU and emerging powers: mutual perceptions of the EU and Brazil, China, India and South Africa, Baden-Baden, Germany

Knodt M, Müller F, Piefer N (2010) Challenges of European external energy governance with emerging powers: meeting Tiger, Dragon, Lion and Jaguar, Panel on "A new mode of European External Energy Governance". In: 7th SGIR (Study Group on International Relations) Pan-European international relations conference, 9–11 Sept

Knodt M, Müller F, Piefer N (2012) Rising powers in international energy governance. Paper presented at the DVPW Konferenz 'Politik und Ökonomie in globaler Perspektive: Der (Wieder) Aufstieg des Globalen Südens', in Frankfurt, 5–7 Mar 2012

Knodt M, Müller F, Piefer N (2015a) Explaining European Union external energy governance with emerging powers. In: Knodt M, Piefer N, Müller F (eds) Challenges of European external energy governance with emerging powers, Farnham, UK, pp 57–74

Knodt M, Müller F, Piefer N (2015b) Understanding EU-emerging powers energy governance: form competition towards cooperation? In: Knodt M, Piefer N, Müller F (eds) Challenges of European external energy governance with emerging powers, Farnham, UK, pp 327–343

Knodt M, Ringel M (2017) Governance der Energieunion: Weiche Steuerung mit harten Zügen? In: Integration, Feb 2017, pp 125–140

Knodt M, Ringel M (2020) European Union energy policy—a discourse perspective. In: Knodt M, Kemmerzell J (eds) Handbook of energy governance in europe. Springer, Cham (Published online first)

Knodt M, Schaeffer R, Joshi M, Suetyi L, Maupin A (2019) EU energy cooperation with emerging powers: Brazil, India, China, and South Africa. In: Knodt M, Kemmerzell J (eds) Handbook of energy governance in Europe. Springer, Cham (Published online first)

Kong B (2011) Governing China's energy in the context of global governance. Glob Policy 2(1):51–65

Krestinina LY, Davis F, Ostroumova EV, Epifanova SB, Degteva MO, Preston DL (2007) Solid cancer incidence and low-dose-rate radiation exposures in the Techa River cohort: 1956–2002. Int J Epidemiol 36:1038–1046

Künzli J, Kaiser R, Medina S, Studnicka M, Chanel O, Fillinger P, Herry M, Horak F, Quenel P, Schneider J, Seethaler R. Vergnaud JC, Somme H (2000) Public health impact of outdoor and traffic related air pollution: a European assessment. Lancet 359:795–901

Lai SY, Shi Z (2016) How China views the EU in global energy governance: norm exporter, partner or outsider? In: Chaban N, Knodt M, Verdun A (eds) External images of the EU—energy power across the globe. Special Issue. Comparative European Politics (Published online first). https://doi.org/10.1057/cep.2016.14

LeBlanc D (2010) Molten salt reactors: a new beginning for an old idea. Nucl Eng Des 240:1644–1656

Lee B, Mabey N, Preston F, Froggat A, Bradley S (2015) Enhancing engagement between China and the EU on resource government and low carbon development. Research paper, London, UK

Lee K, Brewer E, Christiano C, Meyo F, Miguel E, Podolsky M, Rosa J, Wolfram C (2014) Barriers to electrification for 'Under grid' Households in rural Kenya. The National Bureau of Economic Research (NBER) working paper 20327 Cambridge, MA. https://doi.org/10.3386/w20327

Leung GCK (2011) China's energy security: perception and reality. Energy Policy 39(3):1330–1337

Lichtenstein S, Slovic P (eds) (2006) The construction of preference, Cambridge, UK

Lucas H, Rabia F, Ferroukhi, Hawila D (2013) Renewable energy auctions in developing countries. IRENA (International Renewable Energy Agency). www.irena.org/Publications

Mahnke E, Mühlenhoff J, Lieblang L (2014) Strom Speichern. (= Renews Spezial 75/2014). Agentur für erneuerbare Energien e.V., Berlin, Germany

Maisonnier D (2008) European DEMO design and maintenance strategy. Fusion Eng Des 83(7-9):858–864

Mao J, Ribes A, Yan B, Shi X, Thornton PE, Séférian R, Ciais P, Myneni RB, Douville H, Piao S, Zhu Z, Dickinson RE, Dai Y, Ricciuto DM, Jin M, Hoffman FM, Wang B, Huang M, Xu L (2016) Human-induced greening of the northern extratropical land surface. Nat Clim Change 6:959–963. https://doi.org/10.1038/nclimate3056

Marquard O (1986) Apologie des Zufälligen. Philosophische Überlegungen zum Menschen. In: Ders.: Apologie des Zufälligen. Philosophische Schriften. Stuttgart (Germany) 117–139
Marx K (1875) Critique of the Gotha Programme. In: Marx/Engels selected works, vol 3, Moscow 1970, pp 13–30
Mayer T, Zignago S (2011) Notes on CEPII's distances measures : the GeoDist Database. Centre d'Etudes Prospectives et d'Informations Internationales(CEPII) Working Paper 2011-25. Paris (France)
Mazouz N (2012) Was ist gerecht? Was ist gut? Eine deliberative Theorie des Gerechten, Weilerswist, Germany
McDowell J (1981) Non-cognitivism and rule-following. In: Holtzman S, Leich C (eds) Wittgenstein—to follow a rule, London, UK
McGuffie K, Henderson-Sellers A (1997) A climate modelling primer, 2nd edn, Chichester, UK
McGuire MC (1982) Regulation, factor rewards, and international trade. J Public Econ 17(3):335–354
Mearns E (2014) Energy and mankind. Part 3: energy matters. http://euanmearns.com/energy-and-mankind-part-3/. Date of access 13 Dec 2019
Meidan M, Andrews-Speed P, Xin M (2009) Shaping China's energy policy: actors and processes. J Contemp China 18(61):591–616
Meiners HG, Denneborg M, Müller F, Bergmann A, Weber FA, Dopp E, Hansen C, Schüth C (2012) Umweltauswirkungen von Fracking bei der Aufsuchung und Gewinnung von Erdgas aus unkonventionellen Lagerstätten – Risikobewertung, Handlungsempfehlungen und Evaluierung bestehender rechtlicher Regelungen und Verwaltungsstrukturen. Gutachten Umweltbundesamt, Berlin, Germany
Meinshausen MN, Meinshausen W, Hare SC, Raper B, Frieler K, Knutti R, Frame DJ, Allen MR (2009) Greenhouse-gas emission targets for limiting global warming to 2 °C. Nature 458:1158–1162
Merrifield JD (1988) The impact of selected abatement strategies on transnational pollution. The terms of trade and factor rewards: a general equilibrium approach. J Environ Econ Manage 29:259–284
Meyer F (2007) Druckluftspeicher-Kraftwerke (= BINE Projektinfo 05/07). BINE Informationsdienst FIZ Karlsruhe, Bonn, Germany, pp 1–4
MME/SPE (Ministry of Mines and Energy/Secretariat of Energy Planning and Development (2015) Energy Expansion in Brazil – 2024 Investment Opportunities; Office of Strategic Energy Studies. http://www.mme.gov.br/documents/10584/3642013/03+-+Energy+Expansion+in+Brazil+Investiment+Opportunities+(PDF)/97e49acb-ee22-4c98-ad80-c70056288e89;jsessionid=EA174D8159C21B7C94B25A59F05C52EC.srv155?version. Accessed 24 Aug 2016
Moellendorf D, Widdows H. (2015) The Routledge handbook of global ethics, New York, NY
Mokyr J (2017) A culture of growth. The origin of the modern economy, Princeton, NJ
Montada L (2009) Gerechtigkeitsforschung. Themen, Erkenntnisse und ihre Relevanz. In: Krampen G (ed) Psychologie – Experten als Zeitzeugen. Göttingen, Germany, pp 275–288
MoSPI (Ministry of Statistics and Programme Implementation) (2018) Government of India. Energy Statistics, New Delhi. http://mospi.nic.in/sites/default/files/publication_reports/Energy_Statistics_2018.pdf. Accessed 13 Dec 2019
Moss RH, Edmonds JA, Hibbard HA, Manning MR, Rose SK, van Vuuren DP, Detlef P, Carter T, Emori S, Kainuma M, Kram T, Meehl GA, Mitchell JFB, Nakicenovic N, Riahi K, Shmith SJ, Stuffer RJ, Thomson AM, Weyant JP, Wilbanks TJ (2010) The next generation of scenarios for climate change research and assessment. Nature 463:747–756
Müller F (2015) IRENA's Renewable energy governance: institutional change, cooperation opportunities, and governance innovations. In: Knodt M, Piefer N, Müller F (eds) Challenges of European external energy governance with emerging powers, Farnham, UK, pp 307–325
Müller F, Knodt M, Piefer N (2015) Conceptionalizing emerging powers and EU energy governance: towards a research agenda. In: Knodt M, Piefer N, Müller F (eds) Challenges of European external energy governance with emerging powers, Farnham, UK, pp 17–32

Müller P, von Storch H (2004) Computer modelling in atmospheric and oceanic sciences—building knowledge, Berlin, Germany
Mutén L (1985) Limits to taxation in developing countries. In: Gebauer W (ed) Öffentliche Finanzen und monetäre Ökonomie, Frankfurt a.M., Germany
Nordhaus WD (1996) Do real-output and real-wage measures capture reality? In: Bresnahan TF, Gordon RJ (eds) The economics of new goods, Chicago, IL
Noronha L, Sharma D (eds) (2009) Energy, climate and security: the interlinkages. In: Proceedings of the 2nd conference jointly organized by TERI and KAF. Konrad-Adenauer-Foundation, New Delhi, India
Nuklearforum Schweiz (2019) Kernkraftwerke der Welt 2019, Olten, Switzerland
Nussbaum M (2011) Creating capabilities: the human development approach, Cambridge, MS
Nutzinger HG (2001) Sustainability and innovation: two conceptual levels and a double restriction analysis. In: Steger U (ed) Sustainable development and innovation in the energy sector, Graue Reihe No. 28, Bad-Neuenahr-Ahrweiler, Germany, pp 32–42
Olson M (1965) the logic of collective action: public goods and the theory of groups, Cambridge, MS
Oreskes N, Shrader-Frechette K, Beltz K (1994) Verification, validation, and confirmation of numerical models in earth sciences. Science 263:641–646
Organization for Economic Co-Operation and Development (OECD) (2019) Environment at a glance: climate change. http://www.oecd.org/environment/environment-at-a-glance/Climate-Change-Archive-December-2019.pdf. Accessed 13 Dec 2019
Ostrom E (1990) Governing the commons: the evolution of institutions for collective action, Cambridge, UK
Ott K (2000) Technikfolgenabschätzung und Ethik, Zürich, Switzerland
Pethig R (1976) Pollution, welfare, and environmental policy in the theory of comparative advantage. J Environ Econ Manage 2:160–169
Piefer N, Knodt M, Müller F (2015a) EU and emerging powers in energy governance: exploring the empirical puzzle. In: Knodt M, Piefer N, Müller F (eds) Challenges of European external energy governance with emerging powers, Farnham, UK, pp 33–53
Piefer N, Knodt M, Lai SY (2015b) Perceptions and challenges of China-EU energy cooperation. Paper presented at the UACES conference 2015, Bilbao, Spain
Piefer N, Knodt M, Müller F (2015c) Policy recommendations for enhanced EU-emerging powers energy cooperation. In: Knodt M, Piefer N, Müller F (eds) Challenges of European external energy governance with emerging powers, Farnham, UK, pp 345–354
Pioro IL (ed) (2016) Handbook of generation IV nuclear reactors, Duxford, UK, Cambridge, MA
Planning Commission (2006) Integrated energy policy—report of the expert committee. Government of India, New Delhi, India
Pope CA, Burnett RT, Thun MJ, Calle EE, Krewski D, Ito K, Thurston GD (2002) Lung cancer, cardiopulmonary mortality, and long-term exposure to fine particulate air pollution. J Am Med Assoc (JAMA) 287(9):1132–1141
Power A (1996) Ecological ethics II: justice, economy, politics. In: Nida-Rümelin J (ed) Applied ethics, Stuttgart, Germany, 434p
Prasser HM (2014) Innovation bei Kernreaktoren: Kugelhaufen Salzschmelze, Thorium. Bull Nuklearforum Schweiz 6:9–12
Prasser HM (2014b) Reaktoren der dritten Generation. Vertiefungskurs Nuklearforum Schweiz, Olten, Switzerland
Preston DL, Kusumi S, Tomonaga M, Izumi S, Ron E, Kuramoto A et al (1994) Cancer incidence in atomic bomb survivors. Part III. Leukemia, lymphoma and multiple myeloma, 1950–1987. Radiat Res 137:68–97
PSI (Paul Scherrer Institut) (2013) Energie-Spiegel. Facts for the energy decisions of tomorrow. no. 22, Oct 2013
Quaschning V (2019) Statistiken. Weltweit installierte Photovoltaikleistung. https://www.volker-quaschning.de/datserv/pv-welt/index.php. Accessed 13 Dec 2019

Ratter BMW, Philipp KHI, von Storch H (2012) Between hype and decline—recent trends in public perception of climate change. Environ Sci Policy 18:3–8
Rawls J (1971) A theory of justice, Cambridge, MA
Rawls J (1999) The law of peoples, Cambridge, MA
REN21 (Renewable Energy Policy Network for the 21st Century) (2015) Renewables 2015. Global status report 4.3 REN21 Secretariat, Paris, France
REN21 (Renewable Energy Policy Network for the 21st Century) (2016) Renewables 2016. Global status report. REN21 Secretariat Paris, France
Renn O (2008) white paper on risk governance: toward an integrative framework. In: Renn O (ed) Global risk governance. Concept and practice using the IRGC framework, pp 3–73
Rezai A, van der Ploeg F (2014) Abandoning fossil fuel: how fast and how much? Centre for Economic Policy Research (CEPR). International macroeconomics and public policy discussion paper series No. 9921, London, UK
Ribas A, Schaeffer R (2015) Brazil-EU energy governance: fuelling the dialogue through alternative energy sources. In: Knodt M, Piefer N, Müller F (eds) Challenges of European external energy governance with emerging powers, Farnham, UK, pp 173–194
Ringel M, Knodt M (2018) The governance of the European Energy Union: efficiency, effectiveness and acceptance of the Winter Package 2016. Energy Policy 112:209–220 (Published online first). https://doi.org/10.1016/j.enpol.2017.09.047
Röhrkasten S (2015) Antrieb für die deutsch-brasilianische Energiepartnerschaft, SWP-Aktuell 66, Juni 2015. https://www.swp-berlin.org/publikation/deutsch-brasilianische-energiepartnerschaft/. Accessed 13 Dec 2019
Romero M, Steinfeld A (2012) Concentrating solar thermal power and thermochemical fuels. Energy Environ Sci 5(11):9234–9245
Rubini L (2014) 'The good, the bad, and the ugly.' Lessons on methodology in legal analysis from the recent World Trade Organization litigation on renewable energy subsidies. J World Trade 48(5):895–938
Russel RR, Wilkinson M (1979) Microeconomics, New York, NY
Sachs J, Warner A (2001) The curse of natural resources. Eur Econ Rev 45(4–6):827–838
Sawhney A (2008) India-EU trade and investment agreement: environmental services sector study. Indian Council for Research on International Economic Relations, New Delhi. http://wtocentre.iift.ac.in/EU%20BTIA/EU%20BTIA/Report%20on%20Environmental%20Services%20India-EU%20BTIA.pdf. Accessed 13 Dec 2019
Schmitt TM (2018) (Why) did Desertec fail? An interim analysis of a large-scale renewable energy infrastructure project from a Social Studies of Technology perspective. Local Environ 23:747–776
Schonfeld SJ (2013) Solid cancer mortality in the Techa River Cohort (1950–2007). Radiat Res 179:183–189
Schwartz P (1991) The art of the long view, Hoboken, NJ USA
Sen A (1979) Equality of what? In: McMurrin S (ed) Tanner lectures on human values, vol 1, Cambridge, MS
Siebert H (1977) Environmental quality and the gains from trade. Kyklos 30:657–673
Simon J (1996) The ultimate resource 2, 2nd edn, Princeton, NJ
Singer P (2009) The life you can save, New York, NY
Smil V (1981) Energy development in China: the need for a coherent policy. Energy Policy 9(2):113–126
Smil V (2017) Energy and civilization: a history, Cambridge, MS
Smith A (1779) An inquiry into the nature and causes of the wealth of nations, New York, NY
Stemmer P (2000) Handeln zugunsten anderer. Eine moralphilosophische Untersuchung, Berlin, Germany, New York, NY
Soete LLG, Ziesemer T (1997) Gains from trade and environmental policy under imperfect competition and pollution from transport. In: Feser HD, von Hauff M (eds) Neuere Entwicklungen in der Umweltökonomie und -politik. Volkswirtschaftliche Schriften Universität Kaiserslautern. Regensburg (Germany), 249–268. Reprinted in: Singer H, Hatti N, Tandon R (eds) New world

order series, vol 21: Trade and environment; North-South perspectives, New Delhi, India, 2003; also UNU-MERIT WP 1997-003.

Sornette D, Kröger W, Wheatley S (2019) New ways and needs for exploiting nuclear power, Cham, Switzerland

SRU (German Advisory council on the Environment) (2011) Pathways towards a 100% renewable electricity system, Berlin, Germany

Stefansson V (2005) World geothermal assessment. In: Proceedings world geothermal congress, pp 24–29

Steger U, Achterberg W, Blok K, Bode H, Frenz W, Gather C, Hanekamp G, Imboden D, Jahnke M, Kost M, Kurz R, Nutzinger HG, Ziesemer T (2005) Sustainable development and innovation in the energy sector, Berlin Heidelberg

Steger U, Achterberg W, Blok K, Bode H, Frenz W, Gather C, Hanekamp G, Imboden D, Jahnke M, Kost M, Kurz R, Nutzinger HG, Ziesemer T (2002) Nachhaltige Entwicklung und Innovation im Energiebereich, Berlin, Germany, New York, NY, p 277

Stehr N, von Storch H (2010) Climate and society. Climate as a resource, climate as a risk, Singapore

Sterner M (2009) Bioenergy and renewable power methane in integrated 100% renewable energy systems. Limiting global Warming by Transforming Energy Systems (=Erneuerbare Energien und Energieeffizienz 14). Fraunhofer IWES, Kassel, Germany

Sterner M, Stadler I (2018) Urban energy storage and sector coupling. In: Droege P (ed) Urban energy transition. Renewable strategies for cities and regions, 2nd edn. Amsterdam, NL, Kidlington, UK, Cambridge, MA, USA, pp 225–244

Streffer C (2009) Radiological protection: challenges and fascination of biological research. Strahlenschutzpraxis 2:35–45

Streffer C, Bücker J, Cansier A, Cansier D, GEthmann CF, Guderian R, Hanekamp G, Henschler D, Pöch G, Rehbinder E, Renn O, Selsina M, Wuttke K (2003) Environmental standards. Combined exposures and their effects on humans and their environment, Berlin, Germany

Streffer C, Gethmann CF, Heinloth K, Rumpff K, Witt A (2005) Ethical problems of a long-term global energy supply, Berlin, Germany

Streffer C, Gethmann CF, Kamp G, Kröger W, Rehbinder E, Renn O, Röhlig KJ (2011) Radioactive waste—technical and normative aspects of its disposal, Berlin, Germany, New York, NY

Streffer C, Gethmann CF, Kamp G, Kröger W, Rehbinder E, Renn O, Röhlig KJ (2011) Radioactive waste. Technical and normative aspects of its disposal, Berlin, Germany

Stückelberger C (2016) (ed) Global ethics applied, vols 1–4. http://www.globethics.net/gel/10214920; http://www.globethics.net/gel/10214922; http://www.globethics.net/gel/10214923; http://www.globethics.net/gel/10214924

Sumjer J, Atkinson R, Ballester F et al (2003) Respiratory effects of sulphur dioxide: a hierarchical multicity analysis in the APHEA 2 study. Occup Environ Med 60:8. https://doi.org/10.1136/oem.60.8.e2

Tahvonen O, von Storch H, von Storch J (1994) Economic efficiency of CO_2 reduction programs. Clim Res 4:127–141

Thakur A (2018) SCADA systems. Engineers garage. The EE world online resource. https://www.engineersgarage.com/egblog/scada-systems/. Accessed 13 Dec 2019

The Hindu (2013a) India EU FTA talks gain momentum. http://www.thehindu.com/business/Economy/indiaeu-fta-talks-gain%20momentum/article4623470.ece. Accessed 13 Dec 2019

The Hindu (2013b) India-EU FTA talks likely to hit roadblock. http://www.thehindu.com/business/Economy/indiaeu-fta-talks-likely-to-hit-roadblock/article4695909.ece. Accessed 13 Dec 2019

Trapp RW (1998) Klugheitsdilemmata und die Umweltproblematik, Paderborn, Germany

UNEP (UN Environment Programme) (2002) The asian brown cloud: Climate and other environmental impacts, UNEP 2002, Nairobi, Kenya

UNEP (UN Environment Programme) (2008) 2008 ECOREA; environmental review 2008. United Nations (ed) (1987) Report of the World Commission on environment and development, A/42/427

United Nations Environment Program (UNEP) (2010) Information note: how close are we to the two degree limit? UNEP Governing Council Meeting & Global Ministerial Environment Forum 24–26 Feb 2010, Bali, Indonesia
United Nations Framework Convention on Climate Change (2015) Paris climate change conference, Nov 2015. http://unfccc.int/meetings/paris_nov_2015/meeting/8926.php. Accessed 13 Dec 2019
UNSCEAR (United Nations Scientific Committee on the Effects of Atomic Radiation) (2000) Sources and effects of ionizing radiation. United Nations, New York, NY
UNSCEAR (United Nations Scientific Committee on the Effects of Atomic Radiation) (2013) United Nations General Assembly, official records, sixty eighth session, Supplement No 46. Report of the United Nations Scientific Committee on the Atomic Radiation, 27–31 May 2013, New York, NY
US DOE (Department of Energy) Nuclear Energy Research Advisory Committee (2002) A technology roadmap for generation IV nuclear energy systems. US DOE, Washington D.C., USA
van der Heijden J (2011) Institutional layering: a review of the use of the concept. Politics 31(1):9–18
Vaze P (2009) The economical environmentalist. My attempt to live a low-carbon life and what it cost. Earthscan, London, UK
VDI (Verein Deutscher Ingenieure e.V.) (2013) Statusreport Fossil befeuerte Großkraftwerke in Deutschland – Stand, Tendenzen, Schlussfolgerungen" VDI, Düsseldorf, Germany
Vermeend W, van der Vaart J (1997) Greening taxes: the Dutch model. Paper for the European Association of Environmental and Resource Economics (EAERE). Eight annual conference, Tilburg, The Netherlands, 26–28 June
von Storch H (2009) Climate research and policy advice: scientific and cultural constructions of knowledge. Environ Sci Policy 12:741–747. https://doi.org/10.1016/j.envsci.2009.04.008
von Storch H, Flöser G (eds) (2001) Models in environmental research. In: Proceedings of the second GKSS school on environmental research, Berlin, Germany
von Storch H, Güss S, Heimann M (1999) The climate system and its modelling. An introduction, Berlin, Germany
von Storch H, Krauss W (2013) Die Klimafalle. Die gefährliche Nähe von Politik und Klimaforschung, München, Germany
von Storch H, Tol RSJ, Flöser G (eds) (2007) Environmental crises. Science and policy, Berlin, Germany
WBG (World Bank Group) (2016a) Brazil. Overview. https://www.worldbank.org/en/country/brazil/overview. Accessed 13 Dec 2019
WBG (World Bank Group) (2016b) https://data.worldbank.org/indicator/NY.GDP.MKTP.KD.ZG?locations=IN. Accessed 13 Dec 2019
WBGU (Wissenschaftlicher Beirat der Bundesregierung Globale Umweltveränderungen) (2011) Welt im Wandel: Gesellschaftsvertrag für eine Große Transformation, Hauptgutachten, Berlin, Germany. https://www.wbgu.de/de/publikationen/publikation/welt-im-wandel-gesellschaftsvertrag-fuer-eine-grosse-transformation. Accessed 13 Dec 2019
Wenar L (2015) Blood oil: tyrants, violence, and the rules that run the world, Oxford UK
Westphale K (2015) Germany's energy cooperation with emerging powers: internationalizing the Energiewende? In: Knodt M, Piefer N, Müller F (eds) Challenges of European external energy governance with emerging powers, Farnham, UK, pp 87–100
WHO (World Health Organization) (2011) Exposure of children to air pollution (particulate matter) in outdoor air. WHO European Centre for Environment and Health, Fact Sheet 3.3, Bonn, Germany
WHO (World Health Organization) (2014) Burden of disease from ambient air pollution for 2012, Geneva, Switzerland)
WHO (World Health Organization) (2016) Ambient air pollution: a global assessment of exposure and burden of disease. World Health Organization, Geneva, Switzerland
Wissenschaftsrat (2015) Positionspapier "Zum wissenschaftspolitischen Diskurs über große gesellschaftliche Herausforderungen", Stuttgart, Germany

Wittgenstein L (1953) Philosophische Untersuchungen, Frankfurt, Germany. In: von Schulte J (Hrsg) Kritisch-genetische Edition. Frankfurt a. M. 2001; Engl. Hacker PMS, Schulte J (eds and trans) (2009) Philosophical investigation, 4th edn (2009), Oxford, UK
Yan X, Yang L, Zhang X, Zhan W (2017) Concept of an accelerator-driven advanced nuclear energy system. Energies 10:944. https://doi.org/10.3390/en10070944
Yepes T, Pierce J, Foster V (2008) Making sense of Africa's infrastructure endowment: a benchmarking approach. The World Bank (Africa Infrastructure Country Diagnostic (AICD)) working paper No. 1
Yergin D (2011) The quest: energy, security, and the remaking of the modern world, New York, NY
Yu X (2010) An overview of legislative and institutional approaches to China's energy development. Energy Policy 38:2161–2167
Zha D (2013) China's perspective: the search for energy security. In: Tellis AJ, Mirski S (eds) Crux of Asia: China, India, and the emerging global order. Carnegie Endowment for International Peace, Washington, D.C., pp 209–220
Zha D, Lai SY (2015) China-EU energy governance: what lessons to be drawn. In: Knodt M, Piefer N, Müller F (eds) Challenges of European external energy governance with emerging powers, Farnham, UK, pp 129–147
Zhong S (2016) Structural decompositions of energy consumption, energy intensity, emissions and emission intensity—a sectoral perspective: empirical evidence from WIOD over 1995 to 2009. United Nations University Merit working papers series 2016-015. Maastricht Economic and social Research institute on Innovation and Technology, Masstricht, NL
Ziesemer T (2000) Reconciling environmental policy with employment, international competitiveness and participation requirements. Konjunkturpolitik 46(3):241–273

CPSIA information can be obtained
at www.ICGtesting.com
Printed in the USA
LVHW082023210921
698347LV00001B/33

9 783030 553579